THE UNDECIDABLE

THE UNDECIDABLE

Basic Papers On Undecidable Propositions, Unsolvable Problems And Computable Functions

Edited by

MARTIN DAVIS

Professor of Mathematics
Yeshiva University

RAVEN PRESS
HEWLETT, NEW YORK

Printed in the United States of America

TO

E. L. POST

1897 - 1954

Preface

This book is an anthology of fundamental papers dealing with undecidability and unsolvability. It begins with Gödel's epoch-making paper of 1931 in which it was shown for the first time that systems of logic, no matter how powerful, could never admit proofs of all true assertions of arithmetic. In the anthology, there are the basic papers of Gödel, Church, Turing, and Post in which the class of recursive functions was singled out and seen to be just the class of functions that can be computed by finite algorithms. Also included is the work of Church, Turing, and Post in which problems from the theory of abstract computing machines, from mathematical logic, and finally from algebra are shown to be unsolvable in the sense that there is no finite algorithm for dealing with them. Finally, included in the anthology is work of Kleene and of Post initiating the classification theory of unsolvable problems.

Most of the papers collected here have been previously available— but only in the periodicals in which they originally appeared. However, it has fortunately also been possible to include previously unpublished lectures by Gödel and a hitherto unpublished account by Post of his early work in this area.

This collection should prove valuable, not only as a reference work for specialists, but also as a text or supplementary text: it could be used as a supplementary text in courses in logic or in philosophy or foundations of mathematics. An instructor who prefers to work with original sources could design a course in advanced logic or recursive function theory around this book in various ways. It should also be useful for self-study. With these possibilities in mind, I have prefaced most of the papers in this anthology with brief editorial remarks. In these, I aim to tell the student what I would have liked to have been told before reading the paper in question.

Thanks are due to the authors and publishers who have kindly permitted us to reprint these papers. I am especially grateful to Professors Gödel and Kleene who have furnished supplementary material to be included with their papers. Finally, it is a pleasure to thank Alan M. Edelson of Raven Press for his patience and cooperation.

Martin Davis

Ravello, Italy
June 1964

EDITORIAL NOTE

The symbol *a* on a reference to an article indicates that the article in question also appears (perhaps in translation) in the present anthology. Footnotes indicated by lower case Roman numerals give the page references in the present anthology corresponding to original page references in the text.

Contents

ON FORMALLY UNDECIDABLE PROPOSITIONS OF THE PRINCIPIA MATHEMATICA AND RELATED SYSTEMS. I.

This remarkable paper is not only an intellectual landmark, but is written with a clarity and vigor that makes it a pleasure to read.

The reader should be warned that what Gödel calls recursive functions are now called primitive recursive functions. (The revised terminology was introduced by Kleene, this anthology, pp. 237-253).

For the reader who wishes to complete the details of the proof of Theorem V (p. 22), it is suggested that for the purpose of making the induction easier, thc result be strengthened to demand that the formula in question provably represent a function.

For a remark on an application of this paper to the Entscheidungsproblem, cf. editorial remarks, p. 109.

Kurt Gödel

ON FORMALLY UNDECIDABLE PROPOSITIONS OF PRINCIPIA MATHEMATICA AND RELATED SYSTEMS I [1]* **

{1}

It is well known that the development of mathematics in the direction of greater precision has led to the formalization of extensive mathematical domains, in the sense that proofs can be carried out according to a few mechanical rules. The most extensive formal systems constructed up to the present time are the system of Principia Mathematica (PM)[2], on the one hand, and, on the other hand, the Zermelo-Fraenkel axiom system for set theory[3] (which has been developed further by J.v. Neumann). Both of these systems are so broad that all methods of proof used in mathematics today can be formalized in them, i.e. can be reduced to a few axioms and rules of in-

*This translation was prepared especially for this anthology by Professor Elliott Mendelson of Queens College, New York City, with the kind permission of the Monatshefte für Mathematik und Physik, and Springer-Verlag. The original German title is: "Uber formal unentscheidbare Sätze der Principia Mathematica und verwandter Systeme I."; the article appeared in vol. 38 (1931) pp. 173-198.

**Received November 17, 1930.

1. Cf. the summary of the results of this paper which appeared in the Anzeiger der Akad. d. Wiss. in Wien (math.-naturw. Kl.) 1930, Nr. 19.

2. A. Whitehead and B. Russell, Principia Mathematica. 2nd edition. Cambridge, 1925. Among the axioms of the system PM we also include, in particular, the axiom of infinity (in the form: there exist precisely denumerably many individuals), the axiom of reducibility and the axiom of choice (for all types).

3. Cf. A. Fraenkel, "Zehn Vorlesungen über die Grundlegung der Mengenlehre." Wissensch. u. Hyp., Vol. XXXI. J.v. Neumann, "Die Axiomatisierung der Mengenlehre." Math. Zeitschr. 27 (1928). Journ. f. reine u. angew. Math. 154 (1925), 160 (1929). We note that, in order to complete the formalization, one must add the axioms and rules of inference of the logical calculus to the set-theoretic axioms given in the literature just cited. The arguments that follow also hold for the formal systems constructed recently by D. Hilbert and his co-workers (so far as

ference. It is reasonable therefore to make the conjecture that these axioms and rules of inference are also sufficient to decide all mathematical questions which can be formally expressed in the given systems. In what follows it will be shown that this is not the case, but rather that, in both of the cited systems, there exist relatively simple problems of the theory of ordinary whole numbers which cannot be decided on the basis of the axioms.[4] This situation does not depend upon the special nature of the constructed systems, but rather holds for a very wide class of formal systems, among which are included, in particular, all those which arise from the given systems by addition of finitely many axioms[5], assuming that no false sentences of the kind given in footnote 4 become provable by means of the additional axioms.

Before we go into details, let us first sketch the main ideas of the proof, naturally without making any claim to rigor. The formulas of a formal system (we limit ourselves here to the system PM) are, considered from the outside, finite sequences of primitive symbols (variables, logical constants, and parentheses or dots) and one can easily make completely precise which sequences of primitive symbols are meaningful formulas and which are not[6]. Analogously, from the formal standpoint,

these have been published up to the present). Cf. D. Hilbert, Math. Ann. 88, Abh. aus d. math. Sem. der Univ. Hamburg I (1922), VI (1928); P. Bernays, Math. Ann. 90; J. v. Neumann, Math. Zeitschr. 26 (1927); W. Ackermann, Math. Ann. 93.

4. More precisely, there exist undecidable sentences in which, other than the logical constants: $\overline{\ }$ (not), $\vee$ (or), (x) (for all), $=$ (identical with), the only concepts occurring are $+$ (addition), $\cdot$ (multiplication) (of natural numbers), and where the prefix (x) refers only to natural numbers.

5. In PM only those axioms are considered distinct which do not arise from each other by a change of types.

6. By a "formula of PM", we always understood here and in the sequel a formula written without abbreviations (i. e. without use of definitions). Definitions serve only to make writing briefer and are therefore theoretically superfluous.

proofs are nothing but finite sequences of formulas (with certain specifiable properties). Naturally, for metamathematical considerations, it makes no difference which objects one takes as primitive symbols, and we decide to use natural numbers[7] for that purpose. Accordingly, a formula is a finite sequence of natural numbers [8] and a proof-figure is a finite sequence of finite sequences of natural numbers. Metamathematical concepts (assertions) thereby become concepts (assertions) about natural numbers or sequences of such,[9] and therefore (at least partially) expressible in the symbolism of the system PM itself. It can be shown, in particular, that the concepts "formula", "proof-figure", "provable formula" are definable within the system PM, i.e. one can produce,[10] for example, a formula $F(v)$ of PM with one free variable v (of the type of a sequence of numbers) such that $F(v)$, when intuitively interpreted, says: v is a provable formula. Now we obtain an undecidable proposition of the system PM, i.e. a proposition A for which neither A nor *non*-A is provable, as follows:

A formula of PM with exactly one free variable, which is of the type of the natural numbers (class of classes), will be called a class-expression. We think of the class-expressions ordered in a sequence in some manner[11], we denote the n-th by $R(n)$, and we note that the concept "class-expression" as well as the ordering relation R can be defined in the system

7. That is, we map the primitive symbols in one-to-one fashion onto the natural numbers. (Cf. page 13) to see how this is done.)

8. That is, a mapping of a segment of the natural number sequence into the natural numbers. (Numbers, of course, cannot be spatially ordered.)

9. In other words: the process described above provides an isomorphic image of the system PM in the domain of arithmetic and one can just as well carry out all metamathematical arguments in this isomorphic image. This occurs in the following sketch of the proof, i. e. by "formula", "sentence", "variable", etc., one is always to understand the corresponding objects of the isomorphic image

10. It would be very easy (though somewhat tedious) actually to write this formula down.

11. Say, according to increasing sum of the terms, and lexicographically for equal sums.

PM. Let α be an arbitrary class-expression; by $[\alpha;n]$ we denote that formula which arises from the class-expression α by substitution of the symbol for the natural number n for the free variable. The ternary relation $x=[y;z]$ also turns out to be definable within PM. We now define a class K of natural numbers in the following way:

$$n \in K \equiv \overline{\text{Bew}}\,[R(n);n]^{11a} \qquad (1)$$

(where Bew x means: x is a provable formula). Since the concepts occurring in the definiens are all definable in PM, so also is the concept K which is built up from them, i.e. there is a class-expression S[12] such that the formula $[S;n]$, intuitively interpreted, says that the natural number n belongs to K. As a class-expression, S is identical with some definite $R(q)$, i.e.

$$S=R(q)$$

holds for some definite natural number q. We now show that the proposition $[R(q);q]$[13] is undecidable in PM. For, if the proposition $[R(q);q]$ were assumed to be provable, then it would be true, i.e. according to what was said above, q would belong to K, i.e. according to (1), $\overline{\text{Bew}}\,[R(q);q]$ would hold, contradicting our assumption. On the other hand, if the negation of $[R(q);q]$ were provable, then $n \in K$ would hold, i.e. Bew $[R(q);q]$ would be true. Hence, $[R(q);q]$ together with its negation would be provable, which is again impossible.

11a. The bar above denotes negation.

12. Again there is not the slightest difficulty in actually writing down the formula S.

13. One should observe that "$[R(q);q]$" (or the synonymous "$[S;q]$") is merely a metamathematical description of the undecidable proposition. Nevertheless, as soon as one has obtained the formula S, one can, of course, also determine the number q, and thereby effectively write down the undecidable proposition itself.

The analogy of this result with Richard's antinomy is immediately evident; there is also a close relationship[14] with the Liar Paradox, for the undecidable proposition $[R(q);q]$ says that q belongs to K, i.e. according to (1), that $[R(q);q]$ is not provable. Thus we have a proposition before us which asserts its own unprovability[15]. The method of proof which has just been explained can obviously be applied to every formal system which, first, possesses sufficient means of expression when interpreted according to its meaning to define the concepts (especially the concept "provable formula") occurring in the above argument; and, secondly, in which every provable formula is true. In the precise execution of the above proof, which now follows, we shall have the task (among others) of replacing the second of the assumptions just mentioned by a purely formal and much weaker assumption.

From the remark that $[R(q);q]$ asserts its own unprovability it follows immediately that $[R(q);q]$ is true, since $[R(q);q]$ is indeed unprovable (because it is undecidable). The proposition undecidable in the system PM is thus decided by metamathematical arguments. The precise analysis of this remarkable circumstance leads to surprising results concerning consistency proofs of formal systems, which will be treated in more detail in Section 4 (Theorem XI).

{2}

We pass now to the rigorous execution of the proof sketched above, and we first give a precise description of the formal system P for which we wish to prove the existence of unde—

14. Every epistemological antinomy can be used for a similar proof of undecidability.

15. Contrary to appearances, such a proposition is not circular, for, to begin with, it asserts the unprovability of a quite definite formula (namely, the q-th in the lexicographical ordering, after a certain substitution) and only subsequently (accidentally, as it were) does it turn out that this formula itself is precisely the one whose unprovability is expressed.

cidable propositions. P is essentially the system which one obtains by building the logic of PM around Peano's axioms (numbers as individuals, successor relation as undefined primitive concept).[16]

The primitive symbols of the system P are the following:

I. Constants: "$\sim$" (not), "$\vee$" (or), "Π" (for all), "0" (zero), "f" (the successor of), "(", ")" (parentheses).

II. Variables of the first type (for individuals, i.e. natural numbers including 0): "x_1", "y_1", "z_1", ...

Variables of the second type (for classes of individuals): "x_2", "y_2", "z_2", ...

Variables of the third type (for classes of classes of individuals): "x_3", "y_3", "z_3", ...

– Etc., for every natural number as type.[17]

Remark: Variables for functions (relations) of two or more arguments are superfluous as primitive symbols, since one can define relations as classes of ordered pairs and ordered pairs, in turn, as classes of classes, e.g. define the ordered pair a, b by $((a),(a, b))$, where (x,y) denotes the class whose only elements are x and y, and (x) that whose only element is x.[18]

By a term of the first type we mean a combination of symbols of the form:

$$a, \quad fa, \quad ffa, \quad fffa, \ \ldots, \text{etc.},$$

where a is either 0 or a variable of the first type. In the first case we call such an expression a numeral. For $n>1$ we mean

16. The addition of Peano's axioms, as well as all other changes made in the system PM, serve only to simplify the proof and are theoretically dispensable.

17. It is assumed that, for each type, denumerably many variables are at our disposal.

18. Inhomogeneous relations can also be defined in this way, e.g. a relation between individuals and classes as a class of elements of the form $((x_2),((x_1), x_2))$. All theorems about relations provable in PM are, as is easily seen, also provable under this method of treatment.

by a term of the n-th type just a variable of the n-th type. Combinations of symbols of the form $a(b)$, where b is a term of the n-th type and a is a term of the $(n+1)$st type, will be called elementary formulas. We define the class of formulas as the smallest class[18a] to which all elementary formulas belong and to which $\sim(a)$, $(a) \vee (b)$, $x\,\Pi(a)$ (where x is an arbitrary variable)[19] also belong whenever a and b belong. We call $(a) \vee (b)$ the disjunction of a and b, $\sim(a)$ the negation of a, and $x\,\Pi(a)$ a generalization of a. A sentence is a formula in which no free variables occur (free variables being defined in the usual way). A formula with exactly n free individual variables (and otherwise no free variables) is called an n-ary predicate, for $n = 1$ also a class expression.

By Subst $a\binom{v}{b}$ (where a is a formula, v is a variable, and b is a term of the same type as v) we understand the formula which arises from a when we replace v, wherever it is free, by b[20]. We say that a formula a is a type elevation of another formula b when a arises from b by raising the type of all variables occurring in b by the same number.

The following formulas (I through V) are called axioms (they are written with the help of the abbreviations: $.$, $\supset$, $\equiv$, (Ex), $=$[21], which are defined in the well-known way, and with the use of the usual conventions on the omission of parentheses[22]):

18a. With respect to this definition (and similar ones later), cf. J. Lukasiewicz and A. Tarski, "Untersuchungen über den Aussagenkalkül." Comptes Rendus des seances de la Societe des Sciences et des Lettres de Varsovie XXIII (1930) Cl. III.

19. Thus, $x\,\Pi\,(a)$ is also a formula when x does not occur or does not occur free in a. Naturally, in this case, $x\,\Pi\,(a)$ has the same meaning as a.

20. In case v does not occur as a free variable in a, then Subst $a\binom{v}{b} = a$. One should note that "Subst" is a metamathematical symbol.

21. As in PM I, *13, $x_1 = y_1$ is to be thought of as defined by $x_2\,\Pi(x_2(x_1) \supset x_2(y_1))$ (similarly for higher types).

22. In order to obtain the axioms from the schemata as written, one must there-

I. 1. $\sim(fx_1 = 0)$
2. $fx_1 = fy_1 \supset x_1 = y_1$
3. $x_2(0) . x_1 \Pi (x_2(x_1) \supset x_2(fx_1)) \supset x_1 \Pi(x_2(x_1))$.

II. Every formula which arises from the following schemata by substitution of arbitrary formulas for p, q, r.

1. $p \vee p \supset p$
2. $p \supset p \vee q$
3. $p \vee q \supset q \vee p$
4. $(p \supset q) \supset (r \vee p \supset r \vee q)$.

III. Every formula which results from one of the two schemata

1. $v \Pi (a) \supset \text{Subst}\ a \binom{v}{c}$
2. $v \Pi (b \vee a) \supset b \vee v \Pi (a)$

by making one of the following substitutions for a, v, b, c (and carrying out in 1. the operation indicated by "Subst"):

For a an arbitrary formula; for v an arbitrary variable; for b a formula in which v does not occur free; and for c a term of the same type as v, assuming that c contains no variable which is bound at a place in a at which v is free[23].

IV. Every formula which results from the schema

1. $(Eu)(v \Pi (u(v) \equiv a))$

by substituting for v (for u) an arbitrary variable of the type

fore (after performing the permitted substitutions in II, III, IV)

1. eliminate abbreviations,
2. add omitted parentheses.

One should observe that the resulting expressions must be "formulas" in the above sense. (Cf. also the precise definitions of the metamathematical concepts on page 17 ff.)

23. c is therefore either a variable or 0 or a term of the form $f \ldots fu$, where u is either 0 or a variable of the first type. With respect to the concept "free (bound) at a place of a", cf. I A5 of the paper cited in footnote 24.

n (of type $n+1$), and for a any formula which does not contain u free. This axiom represents the axiom of reducibility (comprehension axiom of set theory).

V. Every formula which arises from the following by type elevation (and this formula itself):

1. $x_1 \Pi (x_2(x_1) \equiv y_2(x_1)) \supset x_2 = y_2$.

This axiom asserts that a class is completely determined by its elements.

A formula c is called an immediate consequence of a and b (of a) if a is the formula $(\sim(b)) \vee (c)$ (if c is the formula $v \, \Pi(a)$, where v denotes an arbitrary variable). The class of provable formulas is defined as the smallest class of formulas which contains the axioms and is closed with respect to the relation "immediate consequence" [24].

We now set up a one-to-one correspondence of natural numbers to the primitive symbols of the system P in the following manner:

"0" . . . 1	"$\vee$" . . . 7	" (" . . . 11
"f" . . . 3	"Π" . . . 9	") " . . . 13
"$\sim$" . . . 5		

and furthermore, to the variables of n-th type we assign the numbers of the form p^n (where p is a prime number >13). Thus, to every finite sequence of primitive symbols (hence also to every formula), there corresponds in a one-to-one fashion a finite sequence of positive integers. We map (again in a one-to-one fashion) the finite sequences of positive integers into the natural numbers by letting the number $2^{n_1} \cdot 3^{n_2} \ldots p_k^{n_k}$ correspond to the sequence $n_1, n_2, \ldots, n_k$, where p_k de-

24. The rule of substitution has been rendered superfluous by our having carried out all possible substitutions in the axioms themselves (as in J.v. Neumann, "Zur Hilbertschen Beweistheorie", Math. Zeitschr. 26, 1927).

notes the k-th prime number (according to magnitude). Hence, a natural number is correlated in one-to-one fashion not only to every primitive symbol but also to every finite sequence of such symbols. The number corresponding to the primitive symbol (or sequence of primitive symbols) a will be written $\Phi(a)$. Assume given now any class or relation $R(a_1, a_2, \ldots, a_n)$ between primitive symbols or sequences of such symbols. We correlate to it that class (relation) $R'(x_1, x_2, \ldots, x_n)$ of natural numbers which holds for $x_1, x_2, \ldots, x_n$ when and only when there exist $a_1, a_2, \ldots, a_n$ such that $x_i = \Phi(a_i)$ $(i=1, 2, \ldots, n)$ and $R(a_1, a_2, \ldots, a_n)$ is true. Those classes and relations of natural numbers which correspond in this manner to the previously defined metamathematical concepts, e.g. "variable", "formula", "sentence", "axiom", "provable formula", etc., are denoted by the same words in small capital letters. [In the original italics were used.] For example, the proposition that there exist undecidable problems in the system P becomes: There exist SENTENCES a such that neither a nor the NEGATION of a is a PROVABLE FORMULA.

We now introduce a digression which, for the moment, has nothing to do with the system P, and, first, we present the following definition: A number-theoretic function[25] $\phi(x_1, x_2, \ldots x_n)$ is said to be recursively defined from the number-theoretic functions $\psi(x_1, x_2, \ldots, x_{n-1})$ and $\mu(x_1, x_2, \ldots, x_{n+1})$ if the following holds for all $x_2, \ldots, x_n, k$[26]:

$$\begin{aligned} \phi(0, x_2, \ldots, x_n) &= \psi(x_2, \ldots, x_n) \\ \phi(k+1, x_2, \ldots, x_n) &= \mu(k, \phi(k, x_2, \ldots, x_n), x_2, \ldots, x_n). \end{aligned} \tag{2}$$

A number-theoretic function ϕ is said to be recursive if there exists a finite sequence of number-theoretic functions

25. That is, its domain of definition is the class of non-negative integers (of n-tuples of such integers) and its values are non-negative integers.

26. Small Roman letters (possibly with subscripts) are, in what follows, always variables for non-negative integers (in case nothing is expressly said to the contrary).

$\phi_1, \phi_2, \ldots, \phi_n$ which ends with ϕ and has the property that each function ϕ_k of the sequence either is defined recursively from two of the preceding functions, or results[27] from one of the preceding functions by substitution, or, finally, is a constant or the successor function $x+1$. The length of the shortest sequence of ϕ_i's belonging to a recursive function ϕ is called its rank. A relation among natural numbers $R(x_1, \ldots, x_n)$ is called recursive[28] if there exists a recursive function $\phi(x_1, \ldots, x_n)$ such that, for all $x_1, x_2, \ldots, x_n$

$$R(x_1, \ldots, x_n) \sim [\phi(x_1, \ldots, x_n) = 0]$$ [29].

The following theorems hold:

I. Every function (relation) resulting from recursive functions (relations) by substitution of recursive functions for variables is recursive; likewise, every function which arises from recursive functions by recursive definition according to schema (2) is recursive.

II. If R and S are recursive relations, then so are $\bar{R}$ and $R \vee S$ (hence also $R \& S$).

III. If the functions $\phi(\mathfrak{x})$, $\psi(\mathfrak{y})$ are recursive, then so is the relation: $\phi(\mathfrak{x}) = \psi(\mathfrak{y})$[30].

IV. If the function $\phi(\mathfrak{x})$ and the relation $R(x, \mathfrak{y})$ are recursive, then so are the relations S, T

27. More precisely: by substitution of some of the preceding functions for the arguments of one of the preceding functions, e.g. $\phi_k(x_1, x_2) = \phi_p[\phi_q(x_1, x_2), \phi_r(x_2)]$ $(p, q, r < k)$. Not all variables of the left side have to occur on the right (likewise in the recursion schema (2)).

28. We consider classes as relations (one-place relations). Naturally, recursive relations R have the property that, for every particular n-tuple of numbers, one can decide whether or not $R(x_1, \ldots, x_n)$ holds.

29. In all informal (in particular, metamathematical) considerations Hilbert's symbolism is employed. Cf. Hilbert-Ackermann, Grundzüge der theoretischen Logik, Berlin 1928.

30. We use German letters $\mathfrak{x}$, $\mathfrak{y}$ as abbreviations for arbitrary n-tuples of variables, e.g. $x_1, x_2, \ldots, x_n$.

$$S(\mathfrak{x}, \mathfrak{y}) \sim (Ex)(x \leqslant \phi(\mathfrak{x}) \,\&\, R(x, \mathfrak{y}))$$
$$T(\mathfrak{x}, \mathfrak{y}) \sim (x)(x \leqslant \phi(\mathfrak{x}) \rightarrow R(x, \mathfrak{y}))$$

as well as the function

$$\psi(\mathfrak{x}, \mathfrak{y}) = \epsilon x (x \leqslant \phi(\mathfrak{x}) \,\&\, R(x, \mathfrak{y})),$$

where $\epsilon x F(x)$ denotes: the smallest number x for which $F(x)$ holds, and 0 if there is no such number.

Theorem I follows directly from the definition of "recursive". Theorems II and III depend upon the fact that the number-theoretic functions

$$\alpha(x),\ \beta(x,y),\ \gamma(x,y)$$

corresponding to the logical concepts $\overline{}$, $\vee$, $=$, namely:

$$\alpha(0)=1;\ \alpha(x)=0 \text{ for } x \neq 0$$
$$\beta(0,x)=\beta(x,0)=0;\ \beta(x,y)=1 \text{ if } x \text{ and } y \text{ are both} \neq 0$$
$$\gamma(x,y)=0 \text{ if } x=y;\ \gamma(x,y)=1 \text{ if } x \neq y$$

are recursive, as one can easily confirm. The proof of Theorem IV is briefly the following: By hypothesis, there exists a recursive $\rho(x, \mathfrak{y})$ such that

$$R(x, \mathfrak{y}) \sim [\rho(x, \mathfrak{y}) = 0].$$

We now define a function $\chi(x, \mathfrak{y})$, according to recursion schema (2), as follows:

$$\chi(0, \mathfrak{y}) = 0$$
$$\chi(n+1, \mathfrak{y}) = (n+1).a + \chi(n, \mathfrak{y}).\alpha(a)$$ [31]

where $a = \alpha(\alpha(\rho(0, \mathfrak{y}))).\alpha[\rho(n+1, \mathfrak{y})].\alpha[\chi(n, \mathfrak{y})]$.

$\chi(n+1, \mathfrak{y})$ is therefore either $= n+1$ (if $a=1$) or $= \chi(n, \mathfrak{y})$ (if $a = 0$).[32] Obviously the first case occurs when and only when all factors of a are 1, i.e., when

31. We assume known that the functions $x + y$ (addition) and $x \cdot y$ (multiplication) are recursive.

32. As is apparent from the definition of α, a cannot assume values other than 0 and 1.

$$\overline{R}(0, \mathfrak{y}) \,\&\, R(n+1, \mathfrak{y}) \,\&\, [\chi(n, \mathfrak{y}) = 0].$$

holds.

From this it follows that the function $\chi(n, \mathfrak{y})$ (considered as a function of n) remains 0 until the least value of n for which $R(n, \mathfrak{y})$ holds, and, from there on, is equal to this value (if $R(0, \mathfrak{y})$ already holds, then the corresponding $\chi(n, \mathfrak{y})$ is constant and $=0$). Hence we have:

$$\psi(\mathfrak{x}, \mathfrak{y}) = \chi(\phi(\mathfrak{x}), \mathfrak{y})$$
$$S(\mathfrak{x}, \mathfrak{y}) \sim R[\psi(\mathfrak{x}, \mathfrak{y}), \mathfrak{y}].$$

The relation T can, by negation, be reduced to a case analogous to that of S, thus proving Theorem IV.

The functions $x+y$, $x.y$, x^y and the relations $x<y$, $x=y$ are, as one can easily check, recursive, and we now define, starting from these concepts, a sequence of functions (relations) 1-45, of which each is defined from the preceding ones by the methods indicated in Theorems I-IV. In so doing, several of the definitional steps allowed by Theorems I-IV are often combined into one step. Each of the functions (relations) 1-45, among which occur, for example, the concepts "FORMULA", "AXIOM", "DIRECT CONSEQUENCE", is therefore recursive.

1. $x/y \equiv (Ez)[z \leqslant x \,\&\, x = y.z]$[33]
x is divisible by y[34].
2. $\mathrm{Prim}(x) \equiv \overline{(Ez)}[z \leqslant x \,\&\, z \neq 1 \,\&\, z \neq x \,\&\, x/z] \,\&\, x > 1$
x is a prime number.

33. The symbol $\equiv$ will be used in the sense of "definitional equality", and therefore in definitions it represents either $=$ or $\sim$ (otherwise the symbolism is Hilbert's.).

34. Everywhere in the following definitions where one of the expressions (x), (Ex), ϵx occurs it is followed by a bound for x. This bound serves only to assure the recursive nature of the defined concept (cf. Theorem IV). On the other hand the extension of the defined concept would, in most cases, not be changed by omission of this bound.

3. $0\ Pr\ x \equiv 0$

$(n+1)\ Pr\ x \equiv \epsilon y[y \leqslant x \ \&\ \text{Prim}(y) \ \&\ x/y \ \&\ y > n\ Pr\ x]$

$n\ Pr\ x$ is the n-th prime factor of x (according to magnitude).[34a]

4. $0! \equiv 1$

$(n+1)! \equiv (n+1).n!$

5. $Pr(0) \equiv 0$

$Pr(n+1) \equiv \epsilon y[y \leqslant \{Pr(n)\}! + 1 \ \&\ \text{Prim}(y) \ \&\ y > Pr(n)]$

$Pr(n)$ is the n-th prime number (according to magnitude).

6. $n\ Gl\ x \equiv \epsilon y[y > x \ \&\ x/(n\ Pr\ x)^y \ \&\ \overline{x/(n\ Pr\ x)^{y+1}}]$

$n\ Gl\ x$ is the n-th term of the sequence of numbers corresponding to the number x (for $n > 0$ and n not greater than the length of this sequence).

7. $l(x) \equiv \epsilon y[y \leqslant x \ \&\ y\ Pr\ x > 0 \ \&\ (y+1)\ Pr\ x = 0]$

$l(x)$ is the length of the sequence of numbers correlated with x

8. $x * y \equiv \epsilon z\{z \leqslant [Pr(l(x)+l(y))]^{x+y} \ \&$

$\quad (n)[n \leqslant l(x) \rightarrow n\ Gl\ z = n\ Gl\ x] \ \&$

$\quad (n)[0 < n \leqslant l(y) \rightarrow (n+l(x))\ Gl\ z = n\ Gl\ y]\}$

$x * y$ corresponds to the operation of juxtaposing two finite sequences of numbers.

9. $R(x) \equiv 2^x$

$R(x)$ corresponds to the sequence of numbers consisting of only the number x (for $x > 0$).

10. $E(x) \equiv R(11) * x * R(13)$

$E(x)$ corresponds to the operation of placing in parentheses (11 and 13 are correlated with the primitive symbols "(" and ")").

11. $n\ \text{Var}\ x \equiv (Ez)[13 < z < x \ \&\ \text{Prim}(z) \ \&\ x = z^n] \ \&\ n \neq 0$

x is a VARIABLE OF THE n-TH TYPE.

12. $\text{Var}(x) \equiv (En)[n \leqslant x \ \&\ n\text{Var}\ x]$

x is a VARIABLE.

13. $\text{Neg}(x) \equiv R(5) * E(x)$

$\text{Neg}(x)$ is the NEGATION of x.

34a. For $0 < n \leqslant z$, where z is the number of distinct prime numbers dividing x. Observe that, for $n = z+1$, $n\ Pr\ x = 0$.

14. $x \text{ Dis } y \equiv E(x) * R(7) * E(y)$
x Dis y is the DISJUNCTION of x and y.
15. $x \text{ Gen } y \equiv R(x) * R(9) * E(y)$
x Gen y is the GENERALIZATION of y by means of the VARIABLE x (assuming that x is a VARIABLE).
16. $0 \, N x \equiv x$
$(n+1) N x \equiv R(3) * n N x$
$n \, Nx$ corresponds to the n-fold prefixing of the symbol "f" in front of x.
17. $Z(n) \equiv n N [R(1)]$
$Z(n)$ is the NUMERAL for the number n.
18. $\text{Typ}_1'(x) \equiv (En, n)\{m, n \leqslant x \,\&\, [m=1 \vee 1 \text{ Var } m] \,\&\, x = nN[R(m)]\}$[34b]
x is a TERM OF THE FIRST TYPE.
19. $\text{Typ}_n(x) \equiv [n=1 \,\&\, \text{Typ}_1'(x)] \vee [n>1 \,\&\, (Ev)\{v \leqslant x \,\&\, n \text{Var } v \,\&\, x = R(v)\}]$
x is a TERM OF THE n-TH TYPE.
20. $Elf(x) \equiv (Ey, z, n)[y, z, n \leqslant x \,\&\, \text{Typ}_n(y) \,\&\, \text{Typ}_{n+1}(z) \,\&\, x = z * E(y)]$
x is an ELEMENTARY FORMULA.
21. $Op(x, y, z) \equiv x = \text{Neg}(y) \vee x = y \text{ Dis } z \vee$
$(Ev)[v \leqslant x \,\&\, \text{Var}(v) \,\&\, x = v \text{ Gen } y]$
22. $FR(x) \equiv (n)\{0 < n \leqslant l(x) \rightarrow Elf(n \, Gl \, x) \vee$
$(Ep, q)[0 < p, q < n \,\&\, Op(n \, Gl \, x, p \, Gl \, x, q \, Gl \, x)]\} \,\&\, l(x) > 0$
x is a sequence of FORMULAS each one of which is either an ELEMENTARY FORMULA or comes from preceding ones by the operations of NEGATION, DISJUNCTION, or GENERALIZATION.
23. $\text{Form}(x) \equiv (En)\{n \leqslant (Pr[l(x)]^2) \, x.[l(x)]^2$
$\&FR(n) \,\&\, x = [l(n)] \, Gl \, n\}$[35]

34b. $m, n \leqslant x$ stands for: $m \leqslant x \,\&\, n \leqslant x$ (and similarly for more than two variables).

35. One finds the bound $n \leqslant (Pr[l(x)^2]) x.[l(x)]^2$ as follows: the length of the shortest sequence of formulas belonging to x can be at most equal to the number of SUBFORMULAS of x. There are, however, at most $l(x)$ subformulas of length 1, at most $l(x)-1$ of length 2, etc., and, therefore, all together, at most $\frac{l(x)[l(x)+1]}{2} \leqslant [l(x)]^2$. The prime divisors of n can therefore all be taken smaller than $Pr\{[l(x)]^2\}$, their number $\leqslant l(x)^2$ and their exponents (which are SUBFORMULAS of x) $\leqslant x$.

x is a FORMULA (i.e. last term of a SEQUENCE OF FORMULAS n).

24. $v \text{ Geb } n, x \equiv \text{Var}(v) \,\&\, \text{Form}(x) \,\&$
$$(Ea, b, c)[a, b, c \leq x \,\&\, x = a * (v \text{ Gen } b) * c \,\&\, \text{Form}(b) \,\&\, l(a) + 1 \leq n \leq l(a) + l(v \text{ Gen } b)]$$
The VARIABLE v is BOUND at the n-th place in x.

25. $v \; Fr \; n, x \equiv \text{Var}(v) \,\&\, \text{Form}(x) \,\&\, v = n \; Gl \; x \,\&\, n \leq l(x) \,\&\, \overline{v \text{ Geb } n, x}$
The VARIABLE v is FREE at the n-th place in x.

26. $v \; Fr \; x \equiv (En)[n \leq l(x) \,\&\, v \; Fr \; n, x]$
v occurs in x as a FREE VARIABLE.

27. $Su \; x \binom{n}{y} \equiv \epsilon z \{ z \leq [Pr(l(x) + l(y))]^{x+y} \,\&\, [(Eu, v)(u, v \leq x \,\&\, x = u * R(n \; Gl \; x) * v \,\&\, z = u * y * v \,\&\, n = l(u) + 1] \}$
$Su \; x \binom{n}{y}$ arises from x by substituting y in place of the n-th term of x (assuming that $0 < n \leq l(x)$).

28. $0 \; St \; v, x \equiv \epsilon n \{ n \leq l(x) \,\&\, v \; Fr \; n, x \,\&\, \overline{(Ep)}[n < p \leq l(x) \,\&\, v \; Fr \; p, x] \}$
$(k+1) \; St \; v, x \equiv \epsilon n \{ n < k \; St \; v, x \,\&\, v \; Fr \; n, x \,\&\, \overline{(Ep)}[n < p < k \; St \; v, x \,\&\, v \; Fr \; p, x] \}$
$k \; St \; v, x$ is the $(k+1)$st place in x (counting from the end of the FORMULA x) at which v is FREE in x (and 0, in case there is no such place).

29. $A(v, x) \equiv \epsilon n \{ n \leq l(x) \,\&\, n \; St \; v, x = 0 \}$
$A(v, x)$ is the number of places at which v is FREE in x.

30. $Sb_0(x \, {}^{v}_{y}) \equiv x$
$Sb_{k+1}(x \, {}^{v}_{y}) \equiv Su[Sb_k(x \, {}^{v}_{y})] \binom{k \; St \; v, x}{y}$

31. $Sb(x \, {}^{v}_{y}) \equiv Sb_{A(v,x)}(x \, {}^{v}_{y})$[36]
$Sb(x \, {}^{v}_{y})$ is the concept $Subst \; a \binom{v}{b}$ defined above.[37]

32. $x \text{ Imp } y \equiv [\text{Neg}(x)] \text{ Dis } y$
$x \text{ Con } y \equiv \text{Neg}\{[\text{Neg}(x)] \text{ Dis}[\text{Neg}(y)]\}$
$x \text{ Aeq } y \equiv (x \text{ Imp } y) \text{ Con}(y \text{ Imp } x)$
$v \text{ Ex } y \equiv \text{Neg}\{v \text{ Gen}[\text{Neg}(y)]\}$

33. $n \; Th \; x \equiv \epsilon y \{ y \leq x^{(x^n)} \,\&\, (k)[k \leq l(x) \rightarrow (k \; Gl \; x \leq 13 \,\&\, k \; Gl \; y = k \; Gl \; x) \vee (k \; Gl \; x > 13 \,\&\, k \; Gl \; y = k \; Gl \; x . [1 \; Pr(k \; Gl \; x)]^n)] \}$

36. In case v is not a variable or x is not a formula, then $Sb(x \, {}^{v}_{y}) = x$

37. Instead of $Sb[Sb(x \, {}^{v}_{y}) \, {}^{w}_{z}]$ we write $Sb(x \, {}^{v}_{y} \, {}^{w}_{z})$ (and similarly for more than two VARIABLES).

$n\ Th\ x$ is the n-th TYPE ELEVATION of x (in case x and $n\ Th\ x$ are FORMULAS).

To the Axioms I, 1-3 correspond three definite numbers, which we denote by z_1, z_2, z_3, and we define:

34. $Z\text{-}Ax(x) \equiv (x = z_1 \vee x = z_2 \vee x = z_3)$
35. $A_1\text{-}Ax\ (x) \equiv (Ey)[y \leqslant x\ \&\ \mathrm{Form}(y)\ \&\ x = (y\ \mathrm{Dis}\ y)\ \mathrm{Imp}\ y]$
x is a FORMULA arising from a substitution in Axiom schema II, 1. Similarly, $A_2\text{-}Ax$, $A_3\text{-}Ax$, and $A_4\text{-}Ax$, corresponding to axioms II, 2-4, are defined.
36. $A\text{-}Ax(x) \equiv A_1\text{-}Ax(x) \vee A_2\text{-}Ax(x) \vee A_3\text{-}Ax(x) \vee A_4\text{-}Ax(x)$
x is a FORMULA resulting from substitution in a sentential axiom.
37. $Q(z, y, v) \equiv \overline{(En, m, w)[n \leqslant l(y)\ \&\ m \leqslant l(z)\ \&\ w \leqslant z\ \&\ w = m\ Gl\ z\ \&\ w\ \mathrm{Geb}\ n, y\ \&\ v\ Fr\ n, y]}$
z contains no VARIABLE which is BOUND at a place in y at which v is FREE
38. $L_1\text{-}Ax(x) \equiv (Ev, y, z, n)\{v, y, z, n \leqslant x\ \&\ n\ \mathrm{Var}\ v\ \&\ \mathrm{Typ}_n(z)\ \&\ \mathrm{Form}(y)\ \&\ Q(z, y, v)\ \&\ x = (v\ \mathrm{Gen}\ y)\ \mathrm{Imp}\ [Sb(y\,{}^{v}_{z})]\}$
x is a FORMULA arising from axiom schema III, 1 by substitution.
39. $L_2\text{-}Ax(x) \equiv (E v, q, p)\{v, q, p \leqslant x\ \&\ \mathrm{Var}(v)\ \&\ \mathrm{Form}(p)\ \&\ \overline{v\ Fr\ p}\ \&\ \mathrm{Form}(q)\ \&\ x = [v\ \mathrm{Gen}\ (p\ \mathrm{Dis}\ q)]\ \mathrm{Imp}[p\ \mathrm{Dis}(v\ \mathrm{Gen}\ q)]\}$
x is a FORMULA arising from axiom schema III, 2 by substitution.
40. $R\text{-}Ax(x) \equiv (Eu, v, y, n)[u, v, y, n \leqslant x\ \&\ n\ \mathrm{Var}\ v\ \&\ (n+1)\ \mathrm{Var}\ u\ \&\ \overline{u\ Fr\ y}\ \&\ \mathrm{Form}(y)\ \&\ x = u\,\mathrm{Ex}\{v\,\mathrm{Gen}[[R(u) * E(R(v))]\ \mathrm{Aeq}\ y]\}]$
x is a FORMULA arising from axiom schema IV, 1 by substitution.

To axiom V, 1 corresponds a definite number z_4, and we define:

41. $M\text{-}Ax(x) \equiv (En)[n \leqslant x\ \&\ x = n\ Th\ z_4]$
42. $Ax(x) \equiv Z\text{-}Ax(x) \vee A\text{-}Ax(x) \vee L_1\text{-}Ax(x) \vee L_2\text{-}Ax(x) \vee R\text{-}Ax(x) \vee M\text{-}Ax(x)$
x is an AXIOM.

43. $Fl(x, y, z) \equiv y = z \text{ Imp } x \vee (Ev)[v \leqslant x \,\&\, \text{Var}(v) \,\&\, x = v \text{ Gen } y]$
x is an IMMEDIATE CONSEQUENCE of y and z.

44. $Bw(x) \equiv (n)\{0 < n \leqslant l(x) \rightarrow A\,x\,(n\,Gl\,x) \vee$
$(Ep, q)[0 < p, q < n \,\&\, Fl(n\,Gl\,x, p\,Gl\,x, q\,Gl\,x)]\} \,\&\, l(x) > 0$
x is a PROOF FIGURE (a finite sequence of FORMULAS each of which is either an axiom or an IMMEDIATE CONSEQUENCE of two preceding ones).

45. $x\,B\,y \equiv Bw(x) \,\&\, [l(x)]\,Gl\,x = y$
x is a PROOF of the FORMULA y.

46. $\text{Bew}(x) \equiv (Ey)\,y\,B\,x$
x is a PROVABLE FORMULA. [$\text{Bew}(x)$ is the only one of the concepts 1-46 which cannot be asserted to be recursive.]

The fact which can be vaguely formulated as the assertion that every recursive relation is definable within the system P (under its intuitive interpretation), is rigorously expressed by the following theorem, without reference to the intuitive meaning of the formulas of P:

Theorem V: For every recursive relation $R(x_1, \ldots, x_n)$, there is an n-ary PREDICATE r (with the FREE VARIABLES[38] $u_1, u_2, \ldots, u_n$) such that, for all n-tuples of numbers $(x_1, \ldots, x_n)$, we have:

$$R(x_1, \ldots, x_n) \longrightarrow \text{Bew}\left[Sb\left(r\,{}^{u_1}_{Z(x_1)} \cdots {}^{u_n}_{Z(x_n)}\right)\right] \tag{3}$$

$$\overline{R}(x_1, \ldots, x_n) \longrightarrow \text{Bew}\left[\text{Neg}\; Sb\left(r\,{}^{u_1}_{Z(x_1)} \cdots {}^{u_n}_{Z(x_n)}\right)\right] \tag{4}$$

38. The VARIABLES $u_1, \ldots, u_n$ can be arbitrarily prescribed. There always exists, e.g. some r with the FREE VARIABLES 17, 19, 23, etc., for which (3) and (4) hold.

We shall be content here to indicate the outline of the proof of this theorem, since it offers no theoretical difficulties and is fairly tedious.[39] We shall prove the theorem for all relations $R(x_1, x_2, \ldots, x_n)$ of the form $x_1 = \phi(x_2, \ldots, x_n)$[40] (where ϕ is a recursive function) and we shall use complete induction on the rank of ϕ. For functions of rank one (i.e. constants and the function $x+1$) the theorem is trivial. Therefore let ϕ have rank m. It results from functions of lower rank $\phi_1, \phi_2, \ldots, \phi_k$ by the operations of substitution or recursive definition. Since, by inductive hypothesis, everything is already proved for $\phi_1, \ldots, \phi_k$, there exist corresponding PREDICATES $r_1, \ldots, r_k$ for which (3) and (4) hold. The definitional procedures by which ϕ arises from $\phi_1, \ldots, \phi_k$ (substitution and recursive definition) can both be formally imitated in the system P. If one does this, then one obtains from $r_1, \ldots, r_k$ a new PREDICATE r[41] for which one can prove without difficulty the validity of (3) and (4) by using the inductive hypothesis. A PREDICATE r which corresponds in this way to a recursive relation[42] shall be called recursive.

We now come to the goal of our work. Let κ be an arbitrary class of formulas. We denote by Flg(κ) (consequence set of κ) the smallest set of formulas which contains all FORMULAS of κ and all AXIOMS and is closed with respect to the relation of "IMMEDIATE CONSEQUENCE". We say that κ is ω-consistent if there is no CLASS EXPRESSION a such that

39. Theorem V depends of course upon the fact that, for a recursive relation R, it is decidable on the basis of the axioms of the system P whether or not R holds for any given n-tuple of numbers.

40. From this, its validity follows immediately for every recursive relation, since such a relation is equivalent to $0 = \phi(x_1, \ldots, x_n)$, where ϕ is recursive.

41. When this proof is rigorously carried out, r will naturally not be defined by this shortcut through the intuitive interpretation, but rather by its purely formal structure.

42. Which, therefore, expresses intuitively that this relation holds.

$$(n)[\,Sb(a\ {}^{v}_{Z(n)}) \in \mathrm{Flg}(\varkappa)]\ \&\ [\mathrm{Neg}(v\ \mathrm{Gen}\ a)] \in \mathrm{Flg}(\varkappa)$$

where v is the FREE VARIABLE of the CLASS EXPRESSION a.

Every ω-consistent system is obviously also consistent. However, as will be shown later, the converse does not hold.

The general result on the existence of undecidable propositions reads:

Theorem VI: For every ω-consistent recursive class $\varkappa$ of FORMULAS, there exists a recursive CLASS EXPRESSION r such that neither v Gen r nor Neg(v Gen r) belongs to Flg($\varkappa$) (where v is the FREE VARIABLE of r).

Proof. Let $\varkappa$ be an arbitrary recursive ω-consistent class of FORMULAS. We define:

$$Bw_{\varkappa}(x) \equiv (n)[n \leq l(x) \rightarrow A\,x(n\,Gl\,x) \vee (n\,Gl\,x) \in \varkappa \vee \tag{5}$$
$$(Ep,q)\{0 < p, q < n \,\&\, Fl(n\,Gl\,x, p\,Gl\,x, q\,Gl\,x)\}]\,\&\, l(x) > 0$$

(cf. the similar concept 44)

$$x\,B_{\varkappa}\,y \equiv Bw_{\varkappa}(x)\ \&\ [l(x)]\,Gl\,x = y \tag{6}$$
$$Bew_{\varkappa}(x) \equiv (Ey)(y\,B_{\varkappa}\,x) \tag{6.1}$$

(cf. the similar concepts 45, 46).

Obviously, we have:

$$(x)[\mathrm{Bew}_{\varkappa}(x) \sim x \in \mathrm{Flg}(\varkappa)] \tag{7}$$
$$(x)[\mathrm{Bew}(x) \rightarrow \mathrm{Bew}_{\varkappa}(x)] \tag{8}$$

Now we define the relation:

$$Q(x,y) \equiv \overline{x\,B_{\varkappa}\,[Sb(y\ {}^{19}_{Z(y)})]}. \tag{8.1}$$

Since $x\,B_{\varkappa}\,y$ (according to (6), (5)) and $Sb(y\ {}^{19}_{Z(y)})$ (according to Definitions 17, 31) are recursive, so also is $Q(x,y)$. According to Theorem V and (8), there exists therefore a PRED-

ICATE q (with the FREE VARIABLES 17 and 19) such that the following hold:

$$\overline{x B_\kappa [Sb(y\, {}^{19}_{Z(y)})]} \rightarrow Bew_\kappa [Sb(q\, {}^{17}_{Z(x)}\, {}^{19}_{Z(y)})] \tag{9}$$

$$x B_\kappa [Sb(y\, {}^{19}_{Z(y)})] \rightarrow Bew_\kappa [\text{Neg}\ Sb(q\, {}^{17}_{Z(x)}\, {}^{19}_{Z(y)})] \tag{10}$$

We set:

$$p = 17 \text{ Gen } q \tag{11}$$

(p is a CLASS EXPRESSION with the FREE VARIABLE 19) and

$$r = Sb(q\, {}^{19}_{Z(p)}) \tag{12}$$

(r is a recursive CLASS EXPRESSION with the FREE VARIABLE 17).[43] Then we have:

$$Sb(p\, {}^{19}_{Z(p)}) = Sb([17 \text{ Gen } q]\, {}^{19}_{Z(p)}) = 17 \text{ Gen } Sb(q\, {}^{19}_{Z(p)}) = 17 \text{ Gen } r^{44} \tag{13}$$

(by virtue of (11) and (12)); and, furthermore:

$$Sb(q\, {}^{17}_{Z(x)}\, {}^{19}_{Z(p)}) = Sb(r\, {}^{17}_{Z(x)}) \tag{14}$$

(from (12)). If one now substitutes p for y in (9) and (10), then, taking into account (13) and (14), we have the result:

$$\overline{x B_\kappa (17 \text{ Gen } r)} \rightarrow \text{Bew}_\kappa [Sb(r\, {}^{17}_{Z(x)})] \tag{15}$$

$$x B_\kappa (17 \text{ Gen } r) \rightarrow \text{Bew}_\kappa [\text{Neg}\ Sb(r\, {}^{17}_{Z(x)})] \tag{16}$$

From this follows:

1. 17 Gen r is not κ-PROVABLE.[45] For, were this the case, then (according to 6.1) there would exist an n such that $n B_\kappa$(17

43. r arises from the recursive predicate q by replacing one VARIABLE by a definite numeral (p).

44. The operations Gen and Sb naturally always commute with each other, in case they refer to different variables.

45. x is κ-provable shall mean: $x \in \text{Flg}(\kappa)$, which, according to (7), has the same meaning as $\text{Bew}_\kappa(x)$.

Gen r). Hence, according to (16), $\text{Bew}_\kappa\,[\text{Neg}\ Sb(r^{17}_{Z(n)})]$ would hold, while, on the other hand, from the κ-PROVABILITY of 17 Gen r that of $Sb(r^{17}_{Z(n)})$ would also follow. Therefore, κ would be inconsistent (a fortiori, ω-inconsistent).

2. Neg(17 Gen r) is not κ-PROVABLE. Proof: As was just proved, 17 Gen r is not κ-PROVABLE,, i.e. (according to 6.1), $(n)\ n B_\kappa$(17 Gen r) holds. From this, we deduce, according to (15), $(n)\ \text{Bew}_\kappa\,[Sb(r^{17}_{Z(n)})]$, which, together with Bew_κ [Neg (17 Gen r)], would contradict the ω-consistency of κ.

Hence, 17 Gen r is undecidable from κ, which proves Theorem VI.

One can easily convince oneself that the proof we have just given is constructive;[45a] i.e. the following has been proved in an intuitionistically unobjectionable way: Assume given an arbitrary recursively defined class κ of FORMULAS. Then, if we are presented with a formal decision (from κ) of the (effectively presentable) SENTENCE 17 Gen r, we can effectively give:

1. A PROOF of Neg(17 Gen r).

2. For any arbitrary n, a PROOF of $Sb(r^{17}_{Z(n)})$, i.e. a formal decision for 17 Gen r would have as a consequence the effective exhibition of an ω-inconsistency.

We shall call a relation (class) among natural numbers $R(x_1, \ldots, x_n)$ decidable if there exists an n-place PREDICATE r such that (3) and (4) (cf. Theorem V) hold. In particular, according to Theorem V, every recursive relation is decidable. Similarly a PREDICATE will be called decidable when it corresponds in this way to a decidable relation. Now it suffices for the existence of undecidable sentences to assume that the class κ is ω-consistent and decidable. For the decidability carries over from κ

45a. For, all the existential assertions occurring in the proof rest upon Theorem V, which, as is easy to see, is intuitionistically unobjectionable.

to $x B_\kappa y$ (cf. (5), (6)) and to $Q(x, y)$ (cf. (8.1)), and only this was used in the proof above. The undecidable proposition has, in this case, the form v Gen r, where r is a decidable CLASS EXPRESSION (moreover, it even suffices that κ be decidable in the system extended by κ).

If one assumes merely the consistency of κ, instead of its ω-consistency, then, to be sure, the existence of an undecidable proposition does not follow; however, we do obtain the existence of a property (r) for which neither a counterexample can be given nor can it be proved that it holds for all numbers. For in the proof that 17 Gen r is not κ-PROVABLE, only the consistency of κ was used (cf. p. 25), and from $\overline{\mathrm{Bew}_\kappa}(17 \text{ Gen } r)$ it follows, according to (15), that $Sb(r\,{}^{17}_{Z(x)})$ holds for all numbers x; consequently, for no number x is $\mathrm{Neg}(Sb\; r\,{}^{17}_{Z(x)})$ κ-PROVABLE.

If one adjoins Neg(17 Gen r) to κ, then one obtains a consistent, but not an ω-consistent, class of FORMULAS κ'. κ' is consistent, for, otherwise, 17 Gen r would be κ-provable. κ' is however not ω-consistent, for, by virtue of $\overline{\mathrm{Bew}_\kappa}(17 \text{ Gen } r)$ and (15), we have $(x)\mathrm{Bew}_\kappa\, Sb(r\,{}^{17}_{Z(x)})$, and, a fortiori, $(x)\mathrm{Bew}_{\kappa'}\, Sb(r\,{}^{17}_{Z(x)})$. On the other hand, of course, $\mathrm{Bew}_{\kappa'}[\mathrm{Neg}(17 \text{ Gen } r)]$ holds.[46]

A special case of Theorem VI occurs when the class κ consists of finitely many FORMULAS (and possibly also those arising therefrom by TYPE ELEVATION). Of course every finite class κ is recursive. Let a be the largest number in κ. Then, in this case, we have for κ:

$$x \varepsilon \kappa \sim (Em, n)[m \leq x \,\&\, n \leq a \,\&\, n \varepsilon \kappa \,\&\, x = m \; Th \; n]$$

46. The existence of consistent and non-ω-consistent κ is, of course, only proved under the assumption that there exists any consistent κ at all (i.e. that P is consistent).

Hence $\varkappa$ is recursive. This allows us to deduce that even with the aid of the axiom of choice (for all types) or of the generalized continuum hypothesis not all sentences are decidable, assuming that these hypotheses are ω-consistent.

In the proof of Theorem VI no properties of the system P were used other than the following:

1. The class of axioms and the rules of inference (i.e. the relation "immediate consequence") are recursively definable (when the primitive symbols are replaced in some manner by natural numbers).

2. Every recursive relation is definable within the system P (in the sense of Theorem V).

Hence, in every formal system which satisfies assumptions 1, 2 and is ω-consistent, there exist undecidable propositions of the form $(x) F(x)$, where F is a recursively defined property of natural numbers, and likewise in every extension of such a system by a recursively definable ω-consistent class of axioms. To the systems which satisfy assumptions 1, 2 belong, as one can easily confirm, the Zermelo-Fraenkel and the v. Neumann axiom systems for set theory,[47] and, in addition, the axiom system for number theory which consists of Peano's axioms, recursive definitions (according to schema (2)) and the logical rules.[48] Assumption 1 is fulfilled in general by every system whose rules of inference are the usual ones and whose axioms (as in F) result from substitution in finitely many schemata.[48a]

47. The proof of assumption 1 turns out to be even simpler here than in the case of the system P, since there is only one kind of primitive variable (resp. two in J. v. Neumann's system).

48. Cf. Problem III in D. Hilbert's address: "Probleme der Grundlegung der Mathematik", Math. Ann. 102.

48a. The true reason for the incompleteness which attaches to all formal systems of mathematics lies, as will be shown in Part II of this paper, in the fact that the formation of higher and higher types can be continued into the transfinite

We shall now derive further consequences from Theorem VI, and, to this end, we give the following definition:

A relation (class) is called arithmetical if it can be defined by means of only the concepts $+$, $.$ (addition and multiplication of natural numbers[49]) and the logical constants $\vee, \overline{}, (x), =$, where (x) and $=$ are to refer to natural numbers.[50] The concept "arithmetical proposition" is defined in a corresponding manner. In particular, the relations "greater" and "congruent with respect to a modulus", for example, are arithmetical; for we have:

$$x > y \sim \overline{(Ez)(y = x + z)}$$

$$x \equiv y (\text{mod } n) \sim (Ez)(x = y + z.n \vee y = x + z.n).$$

The following proposition is true:
Theorem VII: Every recursive relation is arithmetical.

We prove the theorem in the form: Every relation of the form $x_0 = \phi(x_1, \ldots, x_n)$, where ϕ is recursive, is arithmetical and we apply complete induction on the rank of ϕ. Let ϕ have rank s $(s > 1)$. Then we have either:

(cf. D. Hilbert, "Uber das Unendliche", Math. Ann. 95, p. 184), while, in every formal system, only countably many are available. Namely, one can show that the undecidable sentences which have been constructed here always become decidable through adjunction of sufficiently high types (e.g. of the type ω to the system P). A similar result holds for the axiom systems of set theory.

49. Zero is, here and in the sequel, always counted among the natural numbers.

50. The definiens of such a concept must therefore be constructed only by means of the indicated symbols, variables for natural numbers x, y, ... and the symbols 0 and 1 (function variables and set variables must not occur). (Of course, any other number variable may occur in the prefixes instead of x.)

1. $\phi(x_1, \ldots, x_n) = \rho[\chi_1(x_1, \ldots, x_n), \chi_2(x_1, \ldots, x_n), \ldots \chi_m(x_1, \ldots, x_n)]$[51] (where ρ and all the χ_i have lower rank than s) or:

2. $\phi(0, x_2, \ldots, x_n) = \psi(x_2, \ldots, x_n)$
$\phi(k+1, x_2, \ldots, x_n) = \mu[k, \phi(k, x_2, \ldots, x_n), x_2, \ldots, x_n]$
(where ψ, μ have lower rank than s).

In the first case we have:

$$x_0 = \phi(x_1, \ldots, x_n) \sim (Ey_1, \ldots, y_m)[R(x_0, y_1, \ldots, y_m) \& S_1(y_1, x_1, \ldots, x_n) \& \ldots \& S_m(y_m, x_1, \ldots, x_n)],$$

where R, and the S_i are the arithmetical relations which, according to inductive hypothesis, are equivalent to $x_0 = \rho(y_1, \ldots, y_m)$, and $y = \chi_i(x_1, \ldots, x_n)$, respectively. Hence, in this case, $x_0 = \phi(x_1, \ldots, x_n)$ is arithmetical.

In the second case we apply the following procedure: one can express the relation $x_0 = \phi(x_1, \ldots, x_n)$ with the help of the concept "sequence of numbers" (f)[52] in the following manner:

$$x_0 = \phi(x_1, \ldots, x_n) \sim (Ef)\{f_0 = \psi(x_2, \ldots, x_n) \& (k)(k < x_1 \rightarrow f_{k+1} = \mu(k, f_k, x_2, \ldots, x_n) \& x_0 = f_{x_1}\}$$

If $S(y, x_2, \ldots, x_n)$, $T(z, x_1, \ldots, x_{n+1})$ are the arithmetical relations which, according to the inductive hypothesis, are equivalent to $y = \psi(x_2, \ldots, x_n)$, and $z = \mu(x_1, \ldots, x_{n+1})$ respectively, then we have:

$$x_0 = \phi(x_1, \ldots, x_n) \sim (Ef)\{S(f_0, x_2, \ldots, x_n) \& (k)[k < x_1 \rightarrow T(f_{k+1}, k, x_2, \ldots, x_n)] \& x_0 = f_{x_1}\} \qquad (17)$$

Now we replace the concept "sequence of numbers" by "pairs of numbers" by correlating with the number pair n, d the sequence of numbers $f^{(n,d)}$ $(f_k^{(n,d)} = [n]_{1+(k+1)d}$, where $[n]_p$ denotes the smallest non-negative remainder of n modulo p).

51. Naturally, not all the variables $x_1, \ldots, x_n$ need actually occur in the χ_i (cf. the example in footnote 27).

52. f denotes here a variable whose domain is the sequence of natural numbers. The $(k+1)$st term of a sequence f is designated f_k (and the first, f_0).

Then:

Lemma 1: If f is an arbitrary sequence of natural numbers and k is an arbitrary natural number, then there exists a pair of natural numbers n, d such that $f^{(n,d)}$ and f coincide in their first k terms.

Proof: Let l be the greatest of the numbers $k, f_0, f_1, \ldots, f_{k-1}$. Determine n so that

$$n \equiv f_i\,[\bmod\,(1+(i+1)\,l!)] \quad \text{for } i=0,1,\ldots,k-1,$$

which is possible, since any two of the numbers $1+(i+1)\,l!$ $(i=0,1,\ldots,k-1)$ are relatively prime. For, a prime dividing two of these numbers must also divide the difference $(i_1-i_2)\,l!$ and therefore, since $i_1-i_2<l$, must also divide $l!$, which is impossible. The number pair $n, l!$ fulfills our requirement.

Since the relation $x=[n]_p$ is defined by

$$x \equiv n(\bmod\, p)\,\&\, x<p$$

and is therefore arithmetical, then so also is the relation $P(x_0, x_1, \ldots, x_n)$ defined as follows:

$$P(x_0,\ldots,x_n) \equiv (En,d)\{S([n]_{d+1}, x_2,\ldots,x_n)\,\&\,(k)[k<x_1 \rightarrow T([n]_{1+d(k+2)}, k, [n]_{1+d(k+1)}, x_2,\ldots,x_n)]\,\&\,x_0=[n]_{1+d(x_1+1)}\}$$

which, according to (17) and Lemma 1, is equivalent to $x_0 = \phi(x_1,\ldots,x_n)$ (in the sequence f in (17) only its values up to the (x_1+1)th term matter). Thus, Theorem VII is proved.

According to Theorem VII, for every problem of the form $(x)F(x)$ (F recursive), there is an equivalent arithmetical problem, and since the whole proof of Theorem VII can be formulated (for each particular F) within the system P, this equivalence is provable in P. Therefore we have:

Theorem VIII: There exist undecidable arithmetical propositions in each of the formal systems[53] mentioned in Theorem VI.

53. They are those ω-consistent systems which result from P by addition of a recursively definable class of axioms.

The same holds also (according to the remark on page 28) for the axiom system of set theory and its extensions by ω-consistent recursive classes of axioms.

Finally, we derive the following result:

Theorem IX: In all of the formal systems[53] mentioned in Theorem VI there exist undecidable problems of the restricted functional calculus[54] (i.e. formulas of the restricted functional calculus for which neither the universal validity nor the existence of a counter-example is provable).[55]

This is based upon:

Theorem X: Every problem of the form $(x)F(x)$ (F recursive) can be reduced to the question of the satisfiability of a formula of the restricted functional calculus (i.e. for each recursive F one can produce a formula of the restricted functional calculus whose satisfiability is equivalent to the truth of $(x)F(x)$).

We consider as formulas of the restricted functional calculus (r.f.) those formulas which are built up from the primitive symbols: $-$, $\vee$, (x), $=$; x, y, ... (individual variables); $F(x)$, $G(x,y)$, $H(x,y,z)$, ... (variables for properties and relations), where (x) and $=$ refer only to individuals.[56] We add to these

54. Cf. Hilbert-Ackermann, Grundzüge der theoretischen Logik. In the system P by formulas of the restricted functional calculus we are to understand those which arise from formulas of the restricted functional calculus of PM by the substitution indicated on p. 10 of classes of higher type for relations.

55. In my paper: "Die Vollständigkeit der Axiome des logischen Functionenkalküls", Monatsch. f. Math. u. Phys. XXXVII, 2, I have shown that every formula of the restricted functional calculus either can be proved to be universally valid or has a counter-example; the existence of this counter-example is, however, according to Theorem IX, not always provable (in the given formal systems).

56. D. Hilbert and W. Ackermann, in the book cited above, do not consider the symbol $=$ as belonging to the restricted functional calculus. However, for every formula in which the symbol $=$ occurs, there exists a formula without this symbol which is satisfiable if and only if the original one is (cf. the paper cited in footnote 55).

symbols a third kind of variable $\phi(x)$, $\psi(x,y)$, $\chi(x,y,z)$, etc., which represent objective functions (i.e. $\phi(x)$, $\psi(x,y)$, etc. denote single valued functions whose arguments and values are individuals[57]). A formula which, in addition to the symbols of the r.f. initially mentioned above also contains variables of the third kind ($\phi(x)$, $\psi(x,y)$, ..., etc.), shall be called a formula in the wider sense (i.w.s.).[58] The concepts "satisfiable", "universally valid" carry over without any further ado to formulas i.w.s., and we have the theorem that, for every formula i.w.s. A, one can give an ordinary formula B of the r.f. such that the satisfiability of A is equivalent with that of B. One obtains B from A by replacing the variables of the third kind $\phi(x)$, $\psi(x,y)$, ... occurring in A by expressions of the form $(\iota z)F(z,x)$, $(\iota z)G(z,x,y)$, ..., then by eliminating the "descriptive" functions in the sense of PM. I*14, and by logically multiplying[59] the formula thus obtained by an expression which says that the F, G, ... replacing ϕ, ψ, ... are single valued with respect to the first argument

We shall now show that, for every problem of the form $(x)F(x)$ (F recursive), there is an equivalent problem concerning the satisfiability of a formula i.w.s., from which, according to the remark just made, Theorem X follows.

Since F is recursive, there is a recursive function $\Phi(x)$ such that $F(x) \sim [\Phi(x)=0]$, and for Φ there is a sequence of functions Φ_1, Φ_2, ..., Φ_n such that $\Phi_n = \Phi$, $\Phi_1(x) = x+1$ and, for every Φ_k $(1<k\leq n)$, either:

$$1.\ (x_2, \ldots, x_m)[\Phi_k(0, x_2, \ldots, x_m) = \Phi_p(x_2, \ldots, x_m)] \qquad (18)$$

$$(x, x_2, \ldots, x_m)\{\Phi_k[\Phi_1(x), x_2, \ldots, x_m] = \Phi_q[x, \Phi_k(x, x_2, \ldots, x_m), x_2, \ldots, x_m]\}$$

$$p, q < k$$

57. And, in addition, the domain of definition shall always be the entire domain of individuals.

58. Variables of the third kind are permitted to replace individual variables at all argument places, e.g.: $y=\phi(x)$, $F(x,\phi(y))$, $G(\psi(x,\phi(y)),x)$, etc.

59. I.e. forming the conjunction.

or:

$$2.\ (x_1, \ldots, x_m)[\Phi_k(x_1, \ldots, x_m) = \Phi_r(\Phi_{i_1}(\mathfrak{X}_1), \ldots, \Phi_{i_s}(\mathfrak{X}_s))]^{60}$$

$$r < k,\ i_v < k \text{ (for } v = 1, 2, \ldots, s) \qquad (19)$$

or

$$3.\ (x_1, \ldots, x_m)[\Phi_k(x_1, \ldots, x_m) = \Phi_1(\Phi_1 \ldots \Phi_1(0))] \qquad (20)$$

Furthermore, we form the sentences:

$$(x)\,\overline{\Phi_1(x) = 0}\ \&\ (x,y)[\Phi_1(x) = \Phi_1(y) \rightarrow x = y] \qquad (21)$$

$$(x)[\Phi_n(x) = 0] \qquad (22)$$

Now, in all the formulas (18), (19), (20) (for $k = 2, 3, \ldots, n$) and in (21), (22), we replace the functions Φ_i by function variables ϕ_i, the number 0 by an individual variable x_0 which does not occur elsewhere, and we form the conjunction C of all the formulas so obtained.

Then the formula $(Ex_0)C$ has the required property, i.e.

1. If $(x)[\Phi(x) = 0]$ holds, then $(Ex_0)C$ is satisfiable, for, when the functions $\Phi_1, \Phi_2, \ldots, \Phi_n$ are substituted for $\phi_1, \phi_2, \ldots, \phi_n$ in $(Ex_0)C$, then a true sentence obviously results.
2. If $(Ex_0)C$ is satisfiable, then $(x)[\Phi(x) = 0]$ holds.

Proof: Let $\Psi_1, \Psi_2, \ldots, \Psi_n$ be the functions which, according to our hypothesis, yield a true sentence when substituted for $\phi_1, \phi_2, \ldots, \phi_n$ in $(Ex_0)C$. Let their domain of individuals be $\mathfrak{J}$. Because $(Ex_0)C$ holds for the functions Ψ_i, there exists an individual a (in $\mathfrak{J}$) such that all the formulas (18)-(22) become true sentences (18′)-(22′) when the Φ_i are replaced by the Ψ_i and 0 by a. Now we form the smallest subclass of $\mathfrak{J}$ which contains a and is closed with respect to the operation $\Psi_1(x)$. This subclass ($\mathfrak{J}'$) has the property that each of the functions Ψ_i, when applied to elements of $\mathfrak{J}'$, yields an element of $\mathfrak{J}'$. For, this holds for Ψ_1 by definition of $\mathfrak{J}'$, and, by virtue of (18′), (19′),

60. $\mathfrak{X}_i$ $(i = 1, \ldots, s)$ represent any complexes made up of the variables $x_1, x_2, \ldots, x_m$; e.g. x_1, x_3, x_2.

(20′), this property is transmitted from those Ψ_i with lower subscript to those with higher subscript. The functions which arise from the Ψ_i by restriction to the domain of individuals $\mathfrak{J}'$ are called Ψ_i'. These functions also satisfy all the formulas (18)-(20) (after substitution of a for 0 and Ψ_i for Φ_i).

Because of the truth of (21) for Ψ_1' and a, one can map the individuals of $\mathfrak{J}'$ in a one-to-one manner onto the natural numbers, and, moreover, in such a way that a goes over into 0 and the function Ψ_1' into the successor function Φ_1. Under this mapping, however, all the functions Ψ_i' go over into the functions Φ_i, and by the truth of (22) for Ψ_n' and a, we have $(x)[\Phi_n(x) = 0]$ or $(x)[\Phi(x) = 0]$, which was to be proved.[61]

Since one can carry out within the system P the argument which led to Theorem X (for every special F), then the equivalence between a sentence of the form $(x)F(x)$ (F recursive) and the satisfiability of the corresponding formula of the r.f. is provable in P, and therefore the undecidability of the former implies that of the latter, which proves Theorem IX.[62]

4

From the results of Section 2 there follows a remarkable result concerning a consistency proof for the system P (and its extensions), which is expressed by the following theorem:

61. From Theorem X it follows, for example, that the Fermat and Goldbach problems would have been solvable, if one had solved the decision problem of the r. f.

62. Theorem IX holds naturally also for the axiom system of set theory and its extensions by recursively definable ω-consistent classes of axioms, since there also exist undecidable sentences of the form $(x)F(x)$ (F recursive) in these systems.

Theorem XI: Let $\varkappa$ be an arbitrary recursive consistent class[63] of FORMULAS. Then the SENTENCE which asserts that $\varkappa$ is consistent is not $\varkappa$-PROVABLE; in particular, the consistency of P is unprovable in P,[64] assuming that P is consistent (in the contrary case, of course, every statement is provable).

The proof is (in outline) the following: Let $\varkappa$ be an arbitrary recursive class of FORMULAS (in the simplest case, the empty class) which, for the following considerations, is chosen once and for all. In the proof of the fact that 17 Gen r is not $\varkappa$-PROVABLE,[65] only the consistency of $\varkappa$ is used, as can be seen from 1. on page 25; that is, we have:

$$\text{Wid}(\varkappa) \rightarrow \overline{\text{Bew}_\varkappa}(17 \text{ Gen } r)$$

i.e., by virtue of (6.1):

$$\text{Wid}(\varkappa) \rightarrow \overline{(x)\, x B_\varkappa (17 \text{ Gen } r)}$$

By (13), 17 Gen $r = Sb(p_{Z(p)}^{19})$ and therefore:

$$\text{Wid}(\varkappa) \rightarrow \overline{(x)\, x B_\varkappa\ Sb(p_{Z(p)}^{19})}$$

i.e., by (8.1):

$$\text{Wid}(\varkappa) \rightarrow (x)\, Q(x, p) \tag{24}$$

Now we establish the following: All the defined concepts (proved assertions) of Section 2[66] and Section 4 are expressible (provable) in P. For, we have used throughout only the ordinary methods of definition and proof of classical mathematics,

63. $\varkappa$ is consistent (abbreviated Wid($\varkappa$)) is defined as follows: $\text{Wid}(\varkappa) \equiv (Ex)\,[\text{Form}(x)\ \&\ \overline{\text{Bew}_\varkappa}(x)]$.

64. This follows when one substitutes for $\varkappa$ the empty class of FORMULAS.

65. Of course, r (as well as p) depends upon $\varkappa$.

66. From the definition of "recursive" on p. 14 until the proof of Theorem VI, inclusive.

as they are formalized in the system P. In particular, $\varkappa$ (like every recursive class) is definable in P. Let w be the SENTENCE by which Wid($\varkappa$) is expressed in P. The relation $Q(x,y)$ is, according to (8.1), (9), (10), expressed by the PREDICATE q, and, consequently, $Q(x,p)$ by r (since, by (12), $r = Sb(q^{19}_{Z(p)})$), and the sentence $(x)Q(x,p)$ by 17 Gen r.

Hence w Imp (17 Gen r) is, by virtue of (24), PROVABLE in P[67] (a fortiori, $\varkappa$-PROVABLE). Now, were w to be $\varkappa$-PROVABLE, then 17 Gen r would also be $\varkappa$-PROVABLE, whence, by (23), it would follow that $\varkappa$ is not consistent.

Notice that this proof is also constructive, i.e. it permits us to effectively derive a contradiction from $\varkappa$, if we are presented with a PROOF of w from $\varkappa$. The whole proof of Theorem XI can be carried over, word for word, to the axiom systems of set theory M and of classical mathematics[68] A, and yields here also the result: There exists no consistency proof for M which can be formalized within M assuming that M is consistent, and similarly for A. It should be expressly noted that Theorem XI (and the corresponding results about M and A) in no way contradicts Hilbert's formalistic standpoint. For the latter presupposes only the existence of a consistency proof carried out by finitary methods, and it is conceivable that there might be finitary proofs which cannot be represented in P (or in M or A).

Since w is not $\varkappa$-PROVABLE for any consistent class $\varkappa$, then there already will exist SENTENCES (namely, w) undecidable from $\varkappa$, if Neg(w) is not $\varkappa$-PROVABLE; in other words, in Theorem VI one can replace the assumption of ω-consistency by the

67. That the truth of w Imp (17 Gen r) can be deduced from (23) rests simply on the fact that the undecidable proposition 17 Gen r, as was remarked at the very beginning, asserts its own unprovability.

68. Cf. J. v. Neumann, "Zur Hilbertschen Beweistheorie", Math. Zeitschr. 26, 1927.

following: The proposition "$\varkappa$ is inconsistent" is not $\varkappa$-provable. (One should observe that there exist consistent $\varkappa$ for which this proposition is $\varkappa$-provable.)

We have limited ourselves in this paper essentially to the system *P* and have only indicated the applications to other systems. The results will be expressed and proved in full generality in a sequel to appear shortly. Also in that paper, the proof of Theorem XI, which has only been sketched here, will be presented in detail.

ON UNDECIDABLE PROPOSITIONS OF FORMAL MATHEMATICAL SYSTEMS

This article first appeared in the form of mimeographed notes on lectures given by Gödel at the Institute for Advanced Study during the spring of 1934. The notes were taken by S. C. Kleene and J. B. Rosser who are represented in this anthology by some of their own important papers. These lectures cover ground quite similar to that covered in Gödel's original 1931 paper on undecidability (this anthology, pp. 5-38). The reader should however note the following points of novelty:

(1) the specific formal system chosen as example (although again based on the theory of types) permits variables ranging over numerical-valued functions; this makes it possible to give a much simpler proof than before that recursive (i.e. primitive recursive) functions are representatable in the system.

(2) There is a discussion of the relation between the existence of undecidable propositions and the possibility of defining "true sentence" of a given language in the language itself. (This matter has been treated more fully by Tarski; cf. footnote 25).

(3) In addition to the specific formal system chosen for consideration, there is given an undecidability theorem for "formal mathematical systems" in general. Since Gödel's characterization of "formal mathematical system" uses the notion of a rule's being constructive, that is, of there existing a "finite procedure" for carrying out the rule, an exact characterization of what constitutes a finite procedure becomes a prime requisite for a completely adequate development. Church's thesis

(cf. this anthology, pp. 100–102) identifies the functions which can be computed by a finite procedure with the class of recursive functions (general recursive functions). In the present article Gödel shows how an idea of Herbrand's can be modified so as to give a general notion of recursive function (cf. §9 below). Gödel indicates (cf. footnote 3) that he believed that the class of functions obtainable by recursions of the most general kind were the same as those computable by a finite procedure. However, Dr. Gödel has stated in a letter that he was, at the time of these lectures, not at all convinced that his concept of recursion comprised all possible recursions; and that in fact the equivalence between his definition and Kleene's in Math. Ann. 112 [this anthology pp. 237-252] is not quite trivial. So, despite appearances to the contrary, footnote 3 of these lectures is not a statement of Church's thesis.

In their original mimeographed form, these lecture notes were supplemented by two pages of "notes and errata" which we have incorporated into the text proper either directly or as footnotes. Dr. Gödel has kindly prepared and made available a Postscriptum, as well as a substantial number of corrections and emendations for the present version. Except for this Postscriptum, for minor typographical corrections, and where the contrary is explicitly stated, we have enclosed these within curly braces {...} (see p. 74 for some additional items received too late for inclusion in the text proper). Phrases which we have added for editorial reasons have been enclosed within square brackets [...].

Kurt Gödel

ON UNDECIDABLE PROPOSITIONS OF FORMAL MATHEMATICAL SYSTEMS

1. Introduction

A ***formal mathematical system*** is a system of symbols together with rules for employing them. The individual symbols are called ***undefined terms***. ***Formulas*** are finite sequences of the undefined terms. There shall be defined a class of formulas called ***meaningful formulas***, and a class of meaningful formulas called ***axioms***. There may be a finite or infinite number of axioms. Further, there shall be specified a list of rules, called ***rules of inference***; if such a rule be called R, it defines the relation of ***immediate consequence by R*** between a set of meaningful formulas $M_1, \ldots, M_k$ called the ***premises***, and a meaningful formula N called the ***conclusion*** (ordinarily $k=1$ or 2). We require that the rules of inference, and the definitions of meaningful formulas and axioms, be constructive; that is, for each rule of inference there shall be a finite procedure for determining whether a given formula B is an immediate consequence (by that rule) of given formulas $A_1, \ldots, A_n$, and there shall be a finite procedure for determining whether a given formula A is a meaningful formula or an axiom.

A formula N shall be called an ***immediate consequence*** of $M_1, \ldots, M_n$ if N is an immediate consequence of $M_1, \ldots, M_n$ by any one of the rules of inference. A finite sequence of formulas shall be a ***proof*** (specifically, a proof of the last formula of the sequence) if each formula of the sequence is either an axiom, or an immediate consequence of one or more of the preceding formulas. A formula is ***provable*** if a proof of it exists. Let the symbol $\sim$ be one of the undefined terms, and suppose it to express negation. Then the formal system shall be said to be ***complete*** if for every meaningful formula A either A or $\sim A$ is provable. We shall prove later that (under conditions to be stated) a system in which all propositions of arithmetic can be expressed as meaningful formulas is not complete.

2. Recursive functions and relations [1]

Now we turn to some considerations which for the present have nothing to do with a formal system.

Small Roman letters $x, y, z, \ldots$ will denote arbitrary natural numbers (i.e. non-negative integers); and German letters will be used in abbreviation for finite sequences of the former, i.e. $\mathfrak{x}$ for $x_1, \ldots, x_n$; $\mathfrak{y}$ for $y_1, \ldots, y_m$. Greek letters $\phi, \psi, \chi, \ldots$ will represent functions of one or more natural numbers whose values are natural numbers. Roman capitals $R, S, T, \ldots$ will stand for classes of, or relations among, natural numbers. $R(x)$ shall stand for the proposition that x is in the class R, and $S(x_1, \ldots, x_n)$ for the proposition that $x_1, \ldots, x_n$ stand in the relation S. Classes may be considered as relations with only one term, and relations as classes of ordered n-tuples. There shall correspond to each class or relation R a ***representing function*** ϕ such that $\phi(x_1, \ldots, x_n) = 0$ if $R(x_1, \ldots, x_n)$ and $\phi(x_1, \ldots, x_n) = 1$ if $\sim R(x_1, \ldots, x_n)$.

We use the following notations as abbreviations (p, q are to be replaced by any propositions): $(x)[A(x)]$ (for every natural number x, $A(x)$), $(Ex)[A(x)]$ (there exists a natural number x such that $A(x)$), $\epsilon x[A(x)]$ (the least natural number x such that $A(x)$ if $(Ex)[A(x)]$; otherwise 0), $\sim p$ (not p), $p \vee q$ (p or q),

1.{What is called "recursive" in these lectures (except for § 9) is now called "primitive recursive".}

$p \& q$ (p and q), $p \rightarrow q$ (p implies q, i.e. $(\sim p) \vee q$), $p \equiv q$ (p is equivalent to q, i.e. $(p \rightarrow q) \& (q \rightarrow p)$).

The function $\phi(x_1, \ldots, x_n)$ shall be ***compound*** with respect to $\psi(x_1, \ldots, x_m)$ and $\chi_i(x_1, \ldots, x_n)$ $(i=1, \ldots, m)$ if, for all natural numbers $x_1, \ldots, x_n$,

(1) $\phi(x_1, \ldots, x_n) = \psi(\chi_1(x_1, \ldots, x_n), \ \ldots, \ \chi_m(x_1, \ldots, x_n))$.

$\phi(x_1, \ldots, x_n)$ shall be said to be ***recursive*** with respect to $\psi(x_1, \ldots, x_{n-1})$ and $\chi(x_1, \ldots, x_{n+1})$ if, for all natural numbers $k, x_2, \ldots, x_n$,

$$(2) \quad \begin{aligned} \phi(0, x_2, \ldots, x_n) &= \psi(x_2, \ldots, x_n) \\ \phi(k+1, x_2, \ldots, x_n) &= \chi(k, \phi(k, x_2, \ldots, x_n), x_2, \ldots, x_n). \end{aligned}$$

In both (1) and (2), we allow the omission of each of the variables in any (or all) of its occurrences on the right side (e.g. $\phi(x, y) = \psi(\chi_1(x), \chi_2(x, y))$ is permitted under (1)).[2] We define the class of ***recursive*** functions to be the totality of functions which can be generated by substitution, according to the scheme (1), and recursion, according to the scheme (2), from the successor function $x+1$, constant functions $f(x_1, \ldots, x_n) = c$, and identity functions $U_j^n(x_1, \ldots, x_n) = x_j$ $(1 \leqslant j \leqslant n)$. In other words, a function ϕ shall be recursive if there exists a finite sequence of functions $\phi_1, \ldots, \phi_n$ which terminates with ϕ such that each function of the sequence is either the successor function $x+1$ or a constant function $f(x_1, \ldots, x_n) = c$, or an identity function $U_j^n(x_1, \ldots, x_n) = x_j$, or is compound with respect to preceding functions, or is recursive with respect to preceding functions. A relation R shall be ***recursive*** if the representing function is recursive.

Recursive functions have the important property that, for each given set of values of the arguments, the value of the

2. [This sentence could have been] omitted, since the removal of any of the occurrences of variables on the right may be effected by means of the function U_j^n.

function can be computed by a finite procedure.[3] Similarly, recursive relations (classes) are decidable in the sense that, for each given n-tuple of natural numbers, it can be determined by a finite procedure whether the relation holds or does not hold (the number belongs to the class or not), since the representing function is computable.

The functions $x+y$, xy, x^y and $x!$ are clearly recursive. Hence $\phi(\mathfrak{x})+\psi(\mathfrak{y})$, $\phi(\mathfrak{x})\psi(\mathfrak{y})$, $\phi(\mathfrak{x})^{\psi(\mathfrak{y})}$, and $\phi(\mathfrak{x})!$ are recursive, if $\phi(\mathfrak{x})$ and $\psi(\mathfrak{y})$ are.

I. ***If the relations $R(\mathfrak{x})$ and $S(\mathfrak{y})$ are recursive, then $\sim R(\mathfrak{x})$, $R(\mathfrak{x}) \vee S(\mathfrak{y})$, $R(\mathfrak{x}) \,\&\, S(\mathfrak{y})$, $R(\mathfrak{x}) \rightarrow S(\mathfrak{y})$, $R(\mathfrak{x}) \equiv S(\mathfrak{y})$ are recursive.***

By hypothesis, the representing functions $\rho(\mathfrak{x})$ and $\sigma(\mathfrak{y})$ of R and S, respectively, are recursive. If

$$\alpha(0)=1, \quad \alpha(k+1)=0,$$

then $\alpha(x)$ and hence $\alpha(\rho(\mathfrak{x}))$ are recursive. But, since $\alpha(\rho(\mathfrak{x}))$ is 1 or 0 according as $\rho(\mathfrak{x})$ is 0 or 1, $\alpha(\rho(\mathfrak{x}))$ is the representing function of $\sim R(\mathfrak{x})$. Thus $\sim R(\mathfrak{x})$ is recursive. If $\beta(0,x)=0$, $\beta(k+1,x)=\alpha(\alpha(x))$, then

$$\beta(0,x)=\beta(x,0)=0 \text{ and } \beta(x,y)=1 \text{ when } x,y>0.$$

Hence $\beta(\rho(\mathfrak{x}),\sigma(\mathfrak{y}))$, which is recursive, represents $R(\mathfrak{x}) \vee S(\mathfrak{y})$; that is, $R(\mathfrak{x}) \vee S(\mathfrak{y})$ is recursive. Since $R(\mathfrak{x}) \,\&\, S(\mathfrak{y}) \equiv \sim(\sim R(\mathfrak{x}) \vee \sim S(\mathfrak{y}))$, it follows that $R(\mathfrak{x}) \,\&\, S(\mathfrak{y})$ is recursive. Similarly $R(\mathfrak{x}) \rightarrow S(\mathfrak{y})$, $R(\mathfrak{x}) \equiv S(\mathfrak{y})$, and all other relations definable from $R(\mathfrak{x})$ and $S(\mathfrak{y})$ by use of $\sim$ and $\vee$, are recursive.

II. ***If the functions $\phi(\mathfrak{x})$, $\psi(\mathfrak{y})$ are recursive, then the relations $\phi(\mathfrak{x})=\psi(\mathfrak{y})$, $\phi(\mathfrak{x})<\psi(\mathfrak{y})$, $\phi(\mathfrak{x}) \leq \psi(\mathfrak{y})$ are recursive.***

3. The converse seems to be true, if, besides recursions according to the scheme (2), recursions of other forms (e.g., with respect to two variables simultaneously) are admitted. This cannot be proved, since the notion of finite computation is not defined, but it serves as a heuristic principle.

Let

$$\delta(0)=0, \quad \delta(k+1)=k,$$

and $x \dot{-} 0 = x$, $x \dot{-} (k+1) = \delta(x \dot{-} k)$. Then

$$x \dot{-} y = x - y \text{ if } x \geqslant y \text{ and } x \dot{-} y = 0 \text{ if } x \leqslant y.$$

Hence $\alpha(y \dot{-} x)$ is a representing function for $x<y$, and $\alpha(\psi(\mathfrak{y}) \dot{-} \phi(\mathfrak{x}))$ for $\phi(\mathfrak{x})<\psi(\mathfrak{y})$. Thus $\phi(\mathfrak{x})<\psi(\mathfrak{y})$ is recursive. $\phi(\mathfrak{x})=\psi(\mathfrak{y})$ and $\phi(\mathfrak{x})\leqslant \psi(\mathfrak{y})$ are likewise recursive, as may be seen directly, or inferred from the theorem for $\phi(\mathfrak{x})<\psi(\mathfrak{y})$ by use of I.

III. *If the function* $\phi(\mathfrak{x})$ *and the relation* $R(x,\mathfrak{y})$ *are recursive, then the relations* S, T, *where*

$$S(\mathfrak{x},\mathfrak{y}) \equiv (Ex)[x \leqslant \phi(\mathfrak{x}) \,\&\, R(x,\mathfrak{y})],$$
$$T(\mathfrak{x},\mathfrak{y}) \equiv (x)[x \leqslant \phi(\mathfrak{x}) \rightarrow R(x,\mathfrak{y})],$$

and the function ψ, *where*

$$\psi(\mathfrak{x},\mathfrak{y}) = \epsilon x[x \leqslant \phi(\mathfrak{x}) \,\&\, R(x,\mathfrak{y})],$$

are recursive.

Let the representing function of $R(x,\mathfrak{y})$ be $\rho(x,\mathfrak{y})$. Let $\pi(0,\mathfrak{y}) = \rho(0,\mathfrak{y})$ and $\pi(k+1,\mathfrak{y}) = \pi(k,\mathfrak{y})\cdot\rho(k+1,\mathfrak{y})$. Then $\pi(x,\mathfrak{y}) = \rho(0,\mathfrak{y})\,\rho(1,\mathfrak{y}) \ldots \rho(x,\mathfrak{y})$. Hence $\pi(x,\mathfrak{y})$ is 0 or 1 according as some or none of $\rho(0,\mathfrak{y}), \ldots, \rho(x,\mathfrak{y})$ are 0; that is, according as there do or do not exist natural numbers $n \leqslant x$ for which $R(n,\mathfrak{y})$ holds. Hence $\pi(\phi(\mathfrak{x}),\mathfrak{y})$, which is recursive, represents $(Ex)[x \leqslant \phi(\mathfrak{x}) \,\&\, R(x,\mathfrak{y})]$. Thus $(Ex)[x \leqslant \phi(\mathfrak{x}) \,\&\, R(x,\mathfrak{y})]$ is a recursive relation. It follows from this result and I that $(x)[x \leqslant \phi(\mathfrak{x}) \rightarrow R(x,\mathfrak{y})]$ is recursive, since $(x)[x \leqslant \phi(\mathfrak{x}) \rightarrow R(x,\mathfrak{y})] \equiv \sim[(Ex)[x \leqslant \phi(\mathfrak{x}) \,\&\, \sim R(x,\mathfrak{y})]]$. Let $\mu(0,\mathfrak{y}) = 0$ and $\mu(k+1,\mathfrak{y}) = (k+1)[\pi(k,\mathfrak{y}) \dot{-} \pi(k+1,\mathfrak{y})] + \mu(k,\mathfrak{y})[\alpha(\pi(k,\mathfrak{y}) \dot{-} \pi(k+1,\mathfrak{y}))]$. Since $1 \geqslant \pi(k,\mathfrak{y}) \geqslant \pi(k+1,\mathfrak{y}) \geqslant 0$, $\mu(k+1,\mathfrak{y}) = k+1$ if $\pi(k,\mathfrak{y})=1$ and $\pi(k+1,\mathfrak{y})=0$, and otherwise $\mu(k+1,\mathfrak{y}) = \mu(k,\mathfrak{y})$. Both $\pi(k,\mathfrak{y})=1$ and $\pi(k+1,\mathfrak{y})=0$ hold only when $\sim R(1,\mathfrak{y}), \ldots, \sim R(k,\mathfrak{y})$ and $R(k+1,\mathfrak{y})$; that is, when $k+1$ is the least value x' of x such that $R(x,\mathfrak{y})$. Hence if such an x' exists and is >1, $\mu(0,\mathfrak{y}) = \ldots = \mu(x'-1,\mathfrak{y}) = 0$ and $\mu(x,\mathfrak{y}) = x'$ for all $x \geqslant x'$. If $x'=0$, or x' does not exist, all $\mu(x,\mathfrak{y})$ are

0. Hence $\mu(\phi(\mathfrak{x}),\mathfrak{y})$ is the least $x \leqslant \phi(\mathfrak{x})$ such that $R(x,\mathfrak{y})$, if such exists, otherwise 0; i.e. $\mu(\phi(\mathfrak{x}),\mathfrak{y}) = \epsilon x[x \leqslant \phi(\mathfrak{x}) \& R(x,\mathfrak{y})]$.

3. A formal system.

We now describe in some detail a formal system which will serve as an example for what follows. While a formal system consists only of symbols and mechanical rules relating to them, the meaning which we attach to the symbols is a leading principle in the setting up of the system.

We shall depend on the theory of types as our means for avoiding paradox. Accordingly, we exclude the use of variables running over all objects, and use different kinds of variables for different domains. Specifically, $p, q, r, \ldots$ shall be variables for propositions. Then there shall be variables of successive types as follows:

$x, y, z, \ldots$ for natural numbers,

$f, g, h, \ldots$ for functions (of one variable) whose domain and values are natural numbers,

$F, G, H, \ldots$ for functions (of one variable) whose domain and values are functions $f, g, h, \ldots$,

and so on.[4] Different formal systems are determined according to how many of these types of variables are used. We shall restrict ourselves to the first two types; that is, we shall use variables of the three sorts $p, q, r, \ldots;\ x, y, z, \ldots;\ f, g, h, \ldots$. We assume that a denumerably infinite number of each are included among the undefined terms (as may be secured, for example, by the use of letters with numerical subscripts).

The undefined terms, in addition to variables, shall be: 0 (the number 0), N ($N(x)$ denotes the next greater number than x, i.e. the successor of x), $\sim$, $\vee$, $\&$, $\rightarrow$, $\equiv$, Π ($\Pi x(F(x))$ means

4. Functions for several variables need not be provided separately, since n-tuples of objects of each of these types can be mapped one-to-one on single objects of the same type. {Moreover inhomogeneous functions can be represented by homogeneous ones.}

Variables for classes and relations are unnecessary, since we can use, instead of the classes and relations, their representing functions.

"$F(x)$ is true for all natural numbers x", and may be regarded as the logical product of $F(x)$ over all x), Σ ($\Sigma x\,(F(x))$ means "there is at least one natural number x such that $F(x)$ is true", and may be regarded as the logical sum of $F(x)$ over all x), ϵ, =(equals), (,) ($f(x)$ is the value of f for the argument x, (and) being then interpreted as symbols for the operation of ***application*** of a function to an argument. Parentheses are also used as signs of inclusion, as in $\Pi x(A)$, $(A)\rightarrow(B)$, etc.).[5]

Next, the class of meaningful formulas must be defined. To do this we describe two classes of formulas which have significance- formulas which denote numbers, and formulas which denote propositions. The first comprises ***numerical symbols*** or expressions representing numbers (as 0, $N(0)$, . . .) together with ***functional expressions*** or expressions which become numerical expressions when numerical expressions are substituted in a suitable manner for variables which occur in them (as $\epsilon x[y=N(x)]$).[6] The second comprises ***propositions*** (e.g. $\Pi x[\sim(0=N(x))]$), together with ***propositional functions*** or expressions which become propositions when numerical expressions are substituted in a suitable manner for variables which occur in them (e.g. $\Sigma x[y=N(x)]$).[6] The exact definitions we give by complete induction, thus:

1. 0 and x, y, z, . . . (variables for numbers) are expressions of the Ist kind, and p, q, r, . . . (variables for propositions) are expressions of the IInd kind.

2. If A and B are expressions of the Ist kind, then $A=B$ is an expression of the IInd kind.

3. If A exp. I, then $N(A)$ exp. I.

5. $\sim$, $\vee$, &, $\rightarrow$, $\equiv$, and ϵ have the significances assigned to them in §2. $\Pi x(A)$ and $\Sigma x(A)$, when A does not involve x, mean the same as A. The fact that the logical notions among our undefined terms are not independent does not matter for our purpose.

6. As indicated on page 48, the substitutions which are meant in the case of $\epsilon x[y=N(x)]$ and $\Sigma x[y=N(x)]$ are substitutions for y.

4. If A exp. I, and f is a variable for a function, then $f(A)$ exp. I.

5. If A and B exp. II, then $\sim(A)$, $(A)\vee(B)$, $(A)\&(B)$, $(A)\rightarrow(B)$ and $(A)\equiv(B)$ exp. II.

6. If A exp. II, and x is a variable for a number, then $\Pi x(A)$ and $\Sigma x(A)$ exp. II, and $\epsilon x(A)$ exp. I.

7. If A exp. II, and f is a variable for a function, then $\Pi f(A)$ and $\Sigma f(A)$ exp. II.

8. If A exp. II, and p is a variable for a proposition, then $\Pi p(A)$ and $\Sigma p(A)$ exp. II.

9. The class of expressions of the Ist (IInd) kind shall be the least class satisfying 1-8.

A formula shall be ***meaningful*** if it is either an expression of the Ist kind or an expression of the IInd kind.

The occurrences[7] of variables in a meaningful expression can be classified as free and bound in the following manner: There corresponds to each occurrence of Π in a meaningful expression A a unique part of A, beginning with the occurrence of Π, of the form $\Pi t(B)$, where t is a variable and B is meaningful. This part of A will be called the ***scope*** of the given occurrence of Π in A. Similarly we define the scope of an occurrence of Σ or ϵ in A. A given occurrence of the variable t in A shall be bound or free according as it is or is not in the scope of a Π, Σ or ϵ followed by t.

In the above definitions of functional expressions and propositional functions, the substitutions which are meant are substitutions for the free occurrences of variables. (y is free and x is bound in $\epsilon x[y=N(x)]$ and $\Sigma x[y=N(x)]$.)

We use Subst $\left[A\,{t \atop G}\right]$ to denote the expression obtained from A by substituting G for each occurrence of t in A as a free variable.[8]

7. By an <u>occurrence</u> of a symbol (or formula) K in an expression A, we shall mean a particular part of A of the form K.

8. <u>Subst</u> by itself is not a formula of our system. {"Subst", of course, is not a symbol of the formal system, but rather of metamathematics.}

We may use $F(t)$ to represent a meaningful formula in which t occurs as a free variable[9] and $F(A)$ to denote Subst $\left[F(t)\frac{t}{A}\right]$.

If A is a meaningful formula, and f a variable for a function, then the occurrences of f in A as a free symbol are as the first symbol of parts of A of the form $f(U)$. We may list these parts as $f(U_1), \ldots, f(U_n)$ in such an order that, if U_j contains $f(U_i)$, then $i<j$. Let $G(x)$ be a meaningful formula in which x occurs as a free variable. Let A', $f(U_2'), \ldots, f(U_n')$ be obtained from A, $f(U_2), \ldots, f(U_n)$ by substituting $G(U_1)$ for $f(U_1)$;[10] then let A'', $f(U_3''), \ldots, f(U_n'')$ be obtained from A', $f(U_3'), \ldots, f(U_n')$ by substituting $G(U_2')$ for $f(U_2')$, and so on. We shall denote by Subst $\left[A\frac{f}{G(x),x}\right]$ the expression $A^{(n)}$.[11]

An expression A in which the distinct free variables $t_1, \ldots, t_n$ occur shall mean the same as $\Pi t_1(\ldots(\Pi t_n(A))\ldots)$.

The axioms shall be the following formulas (A1-C2):[12]

9. Then F by itself does not represent a formula.

10. More explicitly, let A' be the expression obtained from A by substituting $G(U_1)$ for the part $f(U_1)$, and let $f(U_2'), \ldots, f(U_n')$ be the parts of A' into which the parts $f(U_2), \ldots, f(U_n)$ of A are transformed by the substitution.

11. Here we describe the proper method of substituting an expression $G(x)$ for a functional variable f in an expression A, and denote the result of the substitution by Subst '$\left[A\frac{f}{G(x),x}\right]$. If $G(x)$ does not contain f (which can always be made the case in the course of formal proofs by a change in the notation), Subt '$\left[A\frac{f}{G(x),x}\right]$ may also be defined as follows: Replace $f(U)$ by $G(U)$ for one of the free occurrences of f in A, do the same thing with the resulting expression, and so on, until an expression is obtained in which f no longer occurs as a free variable.

{x must be mentioned explicitly as the variable of $G(x)$ to be used in the process of substitution, since $G(x)$ may contain other free variables too.}

[In the original mimeographed version the operation was written Subst'$\left[A\frac{f}{g(x)}\right]$.]

12. In writing down these axioms and other meaningful formulas we employ the usual conventions concerning the omission of parentheses.

All abbreviation of formulas is to be regarded as extraneous to the formal system; and each statement about a formula of the system refers to its unabbreviated form.

A. Axioms concerning the notions of the calculus of propositions.

1. $(p \to q) \to ((q \to r) \to (p \to r))$.
2. $((\sim p) \to p) \to p$.
3. $p \to ((\sim p) \to q)$.
4. $p \,\&\, q . \equiv . \sim [(\sim p) \vee (\sim q)]$.
5. $p \vee q . \equiv . (\sim p) \to q$.
6. $p \equiv q . \equiv . (p \to q) \,\&\, (q \to p)$.
7. $p \equiv q . \to . p \to q$.
8. $p \equiv q . \to . q \to p$.

To give the theory of this group of axioms would require a study of the theory of the calculus of propositions.

B. Axioms concerning the notion of identity.

1. $x = x$.
2. $x = y . \to . f(x) = f(y)$.
3. $(x = y) \,\&\, (y = z) . \to . z = x$.

C. Axioms which correspond to certain of Peano's axioms for the natural numbers.

1. $\sim(0 = N(x))$
2. $N(x) = N(y) . \to . x = y$

To complete the definition of the formal system under consideration, it remains to list the rules of procedure. Each rule is to be interpreted as a statement of the conditions under which a formula N shall be an immediate consequence by that rule of the meaningful formula(s) M (M_1, M_2).[13]

1. If $(A) \to (B)$ and A, then B.

13. N also will be meaningful, whenever the conditions are realized. Rule 1, for example, can be written more explicitly thus: N shall be an immediate consequence by Rule 1 of M_1 and M_2, if and only if {M_1 and M_2 are meaningful formulas and} there exist formulas A and B such that M_1 is $(A) \to (B)$, M_2 is A, and N is B.

2. Suppose that A is meaningful,[14] that t is a variable, and that t does not occur in A.

a. If $(A) \rightarrow (B)$, then $(A) \rightarrow (\Pi\, t(B))$.

b. If $(A) \rightarrow (\Pi\, t(B))$, then $(A) \rightarrow (B)$.

3. Suppose that A is meaningful, that t is a variable, and t does not occur in B.

a. If $(A) \rightarrow (B)$, then $(\Sigma\, t(A)) \rightarrow (B)$.

b. If $(\Sigma\, t(A)) \rightarrow (B)$, then $(A) \rightarrow (B)$.

4a. Suppose that x is a variable for a number, that A contains x as a free variable, that G is an expression of the first kind, and that no free variable of G is bound in F.

If A, then Subst $\left[A^{x}_{G}\right]$.

4b. Suppose that f is a variable for a function, that A con tains f as a free variable, that x is a variable for a number, that $G(x)$ is an expression of the Ist kind in which x occurs as a free variable, and that no free variable of $G(x)$ {or A is bound in either $G(x)$ or A, and that, moreover, $G(x)$ and A have no common bound variables.}

If A, then Subst $\left[A^{f}_{G(x),x}\right]$.

4c. Suppose that p is a variable for a proposition, that A contains p as a free variable, that P is an expression of the IInd kind, and that no free variable of P is bound in A.

If A, then Subst $\left[A^{p}_{P}\right]$.

4d. Suppose that x is a variable for a number, and that $F(x)$ is meaningful and contains x as a free variable, {and no free variable of F occurs as a bound variable in F.}

If $(A) \rightarrow (F(x))$, then $(A) \rightarrow (F(\epsilon\, x[F(x)]))$.[15]

14. This condition ensures that, when $(A) \rightarrow (B)$ is a meaningful formula, the occurrence of $\rightarrow$ which separates (A) from (B) in $(A) \rightarrow (B)$ should be the last occurrence of $\rightarrow$ introduced in the construction of $(A) \rightarrow (B)$ according to the definition of meaningful formula. (We may say then that the main operation of $(A) \rightarrow (B)$ is an implication, whose first and second terms are A and B, respectively.) It excludes such possibilities as that A be $p \rightarrow q) \rightarrow ((q \rightarrow r$, when $(A) \rightarrow (B)$ is Axiom A1.

15. If $F(x)$ is an expression of the IInd kind containing the variable x for a num-

5. Suppose that x is a variable for a number, and that $F(x)$ is a meaningful formula in which x occurs as a free variable.

If $F(0)$ and $(F(x)) \to (F(N(x)))$, then $F(x)$.

6. Suppose that s and t are variables of the same kind, that s does not occur in A as a free variable, and that t does not occur in A. Let A' denote the result of substituting t for s throughout A. Suppose that A is meaningful, and let B' denote the expression obtained from B by the substitution of A' for a given occurrence of A in B.

If B, then B'.[16]

One process used in mathematical proof is not represented in this system, namely the definition and introduction of new symbols. However, this process is not essential, but merely a matter of abbreviation.

ber as a free variable, then, with the aid of this rule, $\Sigma x F(x) \to F(\epsilon x [F(x)])$ is provable in our formal system. For Rule 4c allows us to infer $(p \to [(\sim p) \to p]) \to (([(\sim p) \to p] \to p) \to (p \to p))$ from Axiom A1 (i.e., by substituting $(\sim p) \to p$ for q and p for r), and $p \to [(\sim p) \to p]$ from Axiom A3 (by substituting p for q). Then Rule 1 allows us from these two results to infer $([(\sim p) \to p] \to p) \to (p \to q)$, and then from the latter and Axiom A2 to infer $p \to p$. Thence we can successively infer $F(x) \to F(x)$ by Rule 4c (by substituting $F(x)$ for p), $F(x) \to F(\epsilon x [F(x)])$ by Rule 4d, $F(x) \to F(\epsilon y [F(y)])$ by one or more applications of Rule 6, $\Sigma x [F(x)] \to F(\epsilon y [F(y)])$ by Rule 3a, and $\Sigma x [F(x) \to F(\epsilon x [F(x)])$ by Rule 6. Thus the last formula, which expresses the essential property of ϵ, is proved in our formal system. If the system admitted the use of ϵ with variables for functions (i.e., if a rule of inference 4d', obtained from 4d by replacing "x" by "f" and "number" by "function", were added), then similarly, for any expression $G(x)$ containing the variable f for a function as a free variable, $\Sigma f [F(f) \to G(\epsilon f [G(f)])$ would be provable. The latter formula expresses the axiom of choice for classes of functions of integers.

Note that by our formal rule for ϵ we cannot prove that $\epsilon x F(x)$ is the <u>smallest</u> integer x for which $F(x)$, nor that $\epsilon x F(x) = 0$ if there is no such integer, but we can prove only that if there are integers x satisfying $F(x)$, then $\epsilon x F(x)$ is one of them (i.e. $\Sigma x [F(x) \to F(\epsilon x [F(x)])$). This however suffices for all applications.

16. Note that B may be A itself (then B' is A').

4. A representation of the system by a system of positive integers

For the considerations which follow, the meaning of the symbols is immaterial, and it is desirable that they be forgotten. Notions which relate to the system considered purely formally may be called ***metamathematical***.

The undefined terms (hence the formulas and proofs) are countable, and hence a representation of the system by a system of positive integers can be constructed, as we shall now do.

We order the numbers 1-13 to symbols thus:

0	N	=	∾	∨	&	→	≡	Π	Σ	ϵ	(	)
1	2	3	4	5	6	7	8	9	10	11	12	13,

the integers >13 and ≡0 (mod 3) to the variables for propositions, the integers >13, ≡1 (mod 3) to the variables for numbers, and the integers >13, ≡2 (mod 3) to the variables for functions. Thus a one-to-one correspondence is established between the undefined terms and the positive integers.

We order single integers to finite sequences of positive integers by means of the scheme

$$k_1, \ldots, k_n \text{ corresponds to } 2^{k_1}.\ 3^{k_2}.\ 5^{k_3} \ldots p_n^{k_n},$$

where p_i is the i-th prime number (in order of magnitude). A formula is a finite sequence of undefined terms, and a proof a finite sequence of formulas. To each formula we order the integer which corresponds to the sequence of the integers ordered to its symbols; and to each proof we order the integer which corresponds to the sequence of the integers which are then ordered to its member formulas. Then a one-to-one correspondence is determined between formulas (proofs) and a subset of the positive integers.

We may now define various metamathematical classes and relations of positive integers, including one corresponding to each class and relation of formulas. x shall be an f number ($\mathfrak{B}$ number), if there is a formula (proof) to which x corresponds.[17] The relation x,yUz between numbers shall mean that x,y and z are f numbers, and the formula which z represents is an immediate consequence of the formulas which x and y represent. $x\mathfrak{B}y$ shall mean that x is a $\mathfrak{B}$ number and y an f number, and the proof which x represents is a proof of the formula which y represents. Also there are metamathematical functions of integers such as the following: Neg(x)=the number representing $\sim(X)$, if x represents the formula X; and $=0$ if x is not an f number. $Sb\left[x^{y}_{z}\right]$= the number which represents the result Subst $\left[F^{t}_{G}\right]$ of substituting G for the free occurrences of t in F, if x, z represent the formulas F, G, respectively, and y the variable t; and=0 otherwise. These relations and functions, which we have defined indirectly by using the correspondence between formulas and numbers, are constructive. Hence it is not surprising to find that they are recursive. We shall show this for some of the more important of them, by defining them directly, from relations and functions previously known to be recursive, by methods shown in §2 to generate recursive relations and functions out of recursive relations and functions.[18]

1. $x \mid y \equiv (Ez)[z \leqslant x \,\&\, x = yz]$.

"$x|y$" means "x is divisible by y". (yz is recursive. Hence, by II of §2, $x=yz$ is recursive. It follows by III that $x|y$ is recursive. $z \leqslant x$ is inserted in the definition to make it clear that III applies and could be omitted without changing the meaning.)

17. $\mathfrak{B}$ for "Beweis". Below occur U for "unmittelbar Folge", Gl for "Glied", $\mathfrak{k}$ for "einklammern".

18. We use formal notations (including those explained in §2) in the following definitions for the purpose of abbreviating the discussion. These formal notations must not be confused with the formulas of the formal mathematical system under consideration.

2. $\mathrm{Prime}\ (x) \equiv x>1 \,\&\, \sim(Ez)[z \leqslant x \,\&\, \sim(z=1) \,\&\, \sim(z=x) \,\&\, x \mid z]$.

"x is a prime number".

3. $\mathrm{Pr}(0)=0$
$\mathrm{Pr}(n+1)=\epsilon y[y \leqslant \{\mathrm{Pr}(n)\}! + 1 \,\&\, \mathrm{Prime}(y) \,\&\, y>\mathrm{Pr}(n)]$.

$\mathrm{Pr}(n)$ is the n-th prime number (in order of magnitude).

4. $n\mathrm{Gl}x=\epsilon y[y \leqslant x \,\&\, x \mid \{\mathrm{Pr}(n)\}^y \,\&\, \sim(x \mid \{\mathrm{Pr}(n)\}^{y+1})]$.

$n\mathrm{Gl}x$ is the n-th member of the sequence of positive integers which x represents (i.e. $n\mathrm{Gl}x$ is k_n if $x=2^{k_1}.3^{k_2}.\ldots.p_n^{k_n}.\ldots.p_l^{k_l}$).

5. $L(x)=\epsilon y[y \leqslant x \,\&\, (y+1)\mathrm{Gl}x=0]$.

$L(x)$ is the number of members in the sequence represented by x (if x represents a sequence of positive integers).

6. $x*y=\epsilon z\{z \leqslant [\mathrm{Pr}(L(x)+L(y))]^{x+y} \,\&\, (n)[n \leqslant L(x) \rightarrow n\mathrm{Gl}z = n\mathrm{Gl}x] \,\&\, (n)[0<n \leqslant L(y) \rightarrow (n+L(x))\ \mathrm{Gl}z=n\mathrm{Gl}y]\}$.

$*$ represents the operation of joining one finite sequence to another (i.e., if $x=2^{k_1}\ldots p_r^{k_r}$ and $y=2^{l_1}\ldots p_s^{l_s}$, then $x*y=2^{k_1}\ldots p_r^{k_r} p_{r+1}^{l_1} \ldots p_{r+s}^{l_s}$).

Note that the number of the sequence consisting of the single number x is 2^x.

7. $\mathfrak{k}(x)=2^{12}*x*2^{13}$.

If x represents the formula A, $\mathfrak{k}(x)$ represents (A) (for then the sequence of the numbers ordered to the symbols of (A) is 12, $k_1, \ldots, k_n$, 13, where $k_1, \ldots, k_n$ is the sequence of the numbers ordered to the symbols of A).

8. $\mathrm{Neg}(x)=2^4*\mathfrak{k}(x)$.

If x represents the formula A, $\mathrm{Neg}(x)$ represents $\sim(A)$.

9. $\mathrm{Imp}(x,y)=\mathfrak{k}(x)*2^7*\mathfrak{k}(y)$.

If x, y represent the formulas A, B respectively, then Imp(x, y) represents $(A) \to (B)$.

10. uGen $x = 2^9 * 2^u * \mathfrak{k}(x)$

If u represents the variable t, and x the formula A, then u Gen x represents $\Pi\, t(A)$.

Similarly for $\Sigma t(A)$ and $\epsilon x(A)$.

11. $x \equiv y \pmod{n} . \equiv . (Ez)[z \leqslant x + y \,\&\, (x = y + zn \vee y = x + zn)]$.

$x \equiv y \pmod{n}$ has the usual significance.

$t > 13$ expresses "t represents a variable". Also, using 11, recursive classes $\mathrm{Var}_p(t)$, $\mathrm{Var}_x(t)$, $\mathrm{Var}_f(t)$ can be defined to express "t represents a variable for a proposition", "t represents a variable for a number", "t represents a variable for a function", respectively.

Recursive classes $M_I(x)$, $M_{II}(x)$, $M(x)$ expressing "x represents an exp. I", "x represents an exp. II", "x represents a meaningful formula", respectively, recursive relations corresponding to the relations "t occurs in A as a free (bound) variable", and recursive functions corresponding to the operations of substitution used in the rules of inference, can be defined.[19]

12. $x, y U_1 z \equiv x = \mathrm{Imp}(y, z) \,\&\, M(x) \,\&\, M(y)$.

"z represents a formula which is an immediate consequence by Rule 1 of the formulas represented by x, y".

13. $x U_2 z \equiv (Et, v, w)[t, v, w \leqslant z \,\&\, M(v) \,\&\, M(w) \,\&\, x = \mathrm{Imp}(v, w) \,\&\, z =$
$= \mathrm{Imp}(v, t \text{ Gen } w) \,\&\, t > 13 \,\&\, \sim(Ek)[k \leqslant L(v) \,\&\, k \,\mathrm{Gl}\, v = t]]$.[20]

19. For the details of the definition of classes, relations and functions of these sorts, relating to a formal system similar to the one under consideration, see K. Gödel, "Uber formal unentscheidbare Sätze der Principia Mathematica und verwandter Systeme I." Monatshefte für Mathematik und Physik, vol. 38 (1931) pp. 173-198 [English translation, this anthology, pp. 5-38]. Specifically, see the definitions 1-31, pp. 182-184 [this anthology, pp. 17-20].

20. $(Et, v, w)[t, v, w \leqslant z \,\&\, \ldots]$ stands for $(Et)[(Ev)[(Ew)[t \leqslant z \,\&\, v \leqslant z \,\&\, w \leqslant z \,\&\, \ldots]]]$. Similarly, $(x, y, z)[A]$ stands for $(x)[(y)[(z)[A]]]$.

"z represents a formula which is an immediate consequence by Rule 2 of the formula represented by x."

Similarly for each of the other rules of inference.

14. $x, yUz = x, yU_1z \vee xU_2z \vee \ldots \vee xU_6z$.

"z represents a formula which is an immediate consequence of the formula(s) represented by x (x and y)."

Each axiom is represented by a number. Let the numbers corresponding to the axioms be $\alpha_1, \ldots, \alpha_{13}$.

15. $\mathrm{Ax}(x) \equiv x=\alpha_1 \vee x=\alpha_2 \vee \ldots \vee x=\alpha_{13}$.

"x represents an axiom."

16. $\mathfrak{B}(x) \equiv (n)[0<n \leqslant L(x) \rightarrow \{\mathrm{Ax}(n \text{ Gl } x) \vee (Ep,q)[0<p,q<n \& p \text{ Gl} x, q \text{ Gl} x \; U \; n\text{Gl} x]\}] \& L(x)>0$.

"x represents a proof."

17. $x\mathfrak{B}y \equiv \mathfrak{B}(x) \& L(x) \text{ Gl } x = y$.

"x represents a proof and y a formula, and the proof which x represents is a proof of the formula which y represents."

The assertion that the system is free from contradiction can be written as a proposition of arithmetic thus:

$$(x, y, z)[\sim(x\mathfrak{B}z \& y\mathfrak{B}\mathrm{Neg}(z))]$$

(i.e. for all natural numbers x, y and z, x does not represent a proof of the formula A, and y of $\sim(A)$, where z represents A).

5. Representation of recursive functions by formulas of our formal system

We abbreviate certain formal expressions as follows: z_0 for 0, z_1 for $N(0)$, z_2 for $N(N(0))$, etc. The z's then ***represent*** the natural numbers in the formal logic. Again, if $\phi(x_1, x_2, \ldots)$ is a function of positive integers, we shall say that the formal functional expression $G(u_1, u_2, \ldots)$ ***represents*** $\phi(x_1, x_2, \ldots)$ if $G(z_m, z_n, \ldots)=z_{\phi(m,n\ldots)}$ is provable formally for each given set of natural numbers $m, n, \ldots$; in other words, if $G(z_m, z_n, \ldots)=z_k$ is provable formally whenever $\phi(m, n, \ldots)=k$ holds. If the value of $\phi(x_1, x_2, \ldots)$ is independent of some variable x_p, then $G(u_1, u_2, \ldots)$ need not contain the corresponding variable u_p. Similarly, if $R(x_1, x_2, \ldots)$ is a class or relation of natural numbers, we shall say that the formal functional expression $H(u_1, u_2, \ldots)$ ***represents*** $R(x_1, x_2, \ldots)$ if we can prove formally $H(z_m, z_n, \ldots)$ whenever $R(m, n, \ldots)$ holds, and $\sim H(z_m, z_n, \ldots)$ whenever $R(m, n, \ldots)$ does not hold.

We now sketch a proof that every recursive function, class, and relation is represented by some formula of our formal system.

The recursive function $x+1$ is represented by $N(w)$, because $N(z_n)=z_{n+1}$ can be proved formally for each natural number n. The proof is immediate, since under our abbreviations z_{n+1} *is* $N(z_n)$. The constant function $f(x_1, x_2, \ldots, x_n)=c$ is represented by z_c, and the identity function $U_j^n(x_1, \ldots, x_n)=x_j$ is represented by u_j.

If $\phi(x_1, \ldots, x_n)$ is compound with respect to $\psi(x_1, \ldots, x_m)$ and $\chi_i(x_1, \ldots, x_n)$ $(i=1, \ldots, m)$, and if $G(w_1, \ldots, w_m)$ represents $\psi(x_1, \ldots, x_m)$ and $H_i(w_1, \ldots, w_n)$ represents $\chi_i(x_1, \ldots, x_n)$, then $G(H_1(w_1, \ldots, w_n), \ldots, H_m(w_1, \ldots, w_n))$ represents $\phi(x_1, \ldots, x_n)$.

If $\phi(x_1, \ldots, x_n)$ is recursive with respect to $\psi(x_1, \ldots, x_{n-1})$ and $\chi(x_1, \ldots, x_{n+1})$, and if $G(w_1, \ldots, w_{n-1})$ represents $\psi(x_1, \ldots, x_{n-1})$ and $H(w_1, \ldots, w_{n+1})$ represents $\chi(x_1, \ldots, x_{n+1})$, then

$$\epsilon z[\Sigma f \{ f(0)=G(w_2, \ldots, w_n) \,\&\, \Pi u [f(N(u))=H(u, f(u), w_2, \ldots, w_n)] \\ \&\, f(w_1)=z\}]$$

represents $\phi(x_1, \ldots, x_n)$. This formula (call it $K(w_1, \ldots, w_n)$) intuitively has the desired significance. For each set of natural numbers $w_1, \ldots, w_n$, there is one and only one function f satisfying the conditions $f(0)=G(w_2, \ldots, w_n)$, $f(k+1)=H(k, f(k), \ldots, w_2, \ldots, w_n)$, and therefore $K(w_1, \ldots, w_n)$ means "The value, which the function f satisfying the above conditions takes on for the argument w_1". This value obviously is $\phi(w_1, \ldots, w_n)$. The proof that $K(w_1, \ldots, w_n)$ actually represents $\phi(x_1, \ldots, x_n)$, if G represents ψ and H represents χ, is too long to give here.[21]

If $R(x, y, \ldots)$ is a recursive class or relation, there is a recursive function $\phi(x, y, \ldots)$ such that $\phi(x, y, \ldots)=0$ if $R(x, y, \ldots)$ and $\phi(x, y, \ldots)=1$ if $\sim R(x, y, \ldots)$. Then there is a $G(u, v, \ldots)$ which represents $\phi(x, y, \ldots)$. $G(u, v, \ldots)=0$ represents $R(x, y, \ldots)$. For if $R(m, n, \ldots)$, then $G(z_m, z_n, \ldots)=z_0=0$ is provable formally; and if $\sim R(m, n, \ldots)$, then $G(z_m, z_n, \ldots)=z_1$, and therefore $\sim[G(z_m, z_n, \ldots)=0]$, is provable formally.

Because certain metamathematical relations and propositions about our formal system can be expressed by recursive relations and statements about them, these relations and propositions can be expressed in the formal system. Thus parts of the theory whose object is our formal system can be expressed in the same formal system. This leads to interesting results.

We have noted that $x\mathfrak{B}y$ is a recursive relation; and we can also prove that $\gamma(x,y)$ is recursive, where $\gamma(x,y)$ is the number of the formula which results when we replace all free occurrences of w by z_y in the formula whose number is x. (In fact, if a is the number of w and $\chi(n)$ the number of z_n, $\gamma(x,y)$ is $Sb\left[x\,{}^{a}_{\chi(y)}\right]$). Hence there is a formula $B(u, v)$ which represents $x\mathfrak{B}y$, and a formula $S(u, v)$ which represents $\gamma(x,y)$.

21. {Note that this proof is not necessary for the demonstration of the existence of undecidable arithmetic propositions in the system considered. For, if some recursive function were not "represented" by the corresponding formula constructed on pages 58-59, this would trivially imply the existence of undecidable propositions unless some wrong proposition on integers were demonstrable.}

Let $U(w)$ be the formula $\Pi v[\sim B(v, S(w,w))]$ and let p be the number of $U(w)$. Now $U(z_p)$ is the formula which results when we replace all free occurrences of w by z_p in the formula whose number is p, and hence has the number $\gamma(p,p)$. Hence if $U(z_p)$ is provable, there is a k such that $k \mathfrak{B} \gamma(p,p)$. But since $S(u,v)$ represents $\gamma(x,y)$ and $B(u,v)$ represents $x\mathfrak{B}y$, it follows that $B(z_k, S(z_p, z_p))$ is provable. Also, it is a property of our system that if $\Pi v F(v)$ is provable, then $F(z_l)$ is provable for all l; consequently, if $U(z_p)$ is provable, $\sim B(z_k, S(z_p, z_p))$, as well as $B(z_k, S(z_p, z_p))$, is provable, and the system contains a contradiction. Thus we conclude that $U(z_p)$ cannot be proved unless the system contains a contradiction.[22]

Next we raise the question of whether $\sim U(z_p)$ can be proved if the system is not contradictory. If the system is not contradictory, $U(z_p)$ cannot be proved, as just seen. But $U(z_p)$ is the formula with the number $\gamma(p,p)$, so that, for all k, $\sim k\mathfrak{B}\gamma(p,p)$. Therefore $\sim B(z_k, S(z_p, z_p))$ is provable for all k. If furthermore $\sim U(z_p)$, i.e. $\sim \Pi v[\sim B(v, S(z_p, z_p))]$, is provable, then we have that a formula is provable which asserts that $\sim B(z_k, S(z_p, z_p))$ is not true for all k, and this, together with the fact that $\sim B(z_k, S(z_p, z_p))$ is provable for all k, makes the system intuitively contradictory. In other words, if we consider the system to be contradictory not merely if there is an A such that both A and $\sim A$ are provable, but also if there is an F such that all of the formulas $\sim \Pi v F(v)$, $F(z_0)$, $F(z_1)$, ... are provable, then, if $\sim U(z_p)$ is provable, the system is contradictory in this weaker sense. Hence neither $U(z_p)$ nor $\sim U(z_p)$ is provable, unless the system is contradictory.

22{This version of the argument is along the lines of one used by Herbrand in an informal exposition in J. r. angew. Math. 166 (1932) p. 7. It is a little shorter than my original proof in Monh. Math. Phys. 38 (1931) this anthology, pp. 24–26). However, in order to make it completely precise, a few words would have to be added about the properties of the symbolism used to denote the formulas of the system. Note that "$\Pi v [\sim B(v, S(z_p\ z_p))]$" is not the undecidable sentence, but only denotes it.}

If our system is free from contradiction in the strong sense (i.e., if A and $\sim A$ are not both provable for any A), then $U(z_p)$ is not provable. But $(x, y, z)[\sim\{x\mathfrak{B}y \& z\mathfrak{B}\mathrm{Neg}\ y\}]$ is a statement that our system is free from contradiction in the strong sense. Hence we have shown that $(x, y, z)[\sim\{x\mathfrak{B}y \& z\mathfrak{B}\mathrm{Neg}\ y\}] \longrightarrow (x) \sim x\mathfrak{B}\gamma(p, p)$. The fairly simple arguments of this proof can be paralleled in the formal logic to give a formal proof of

$$\mathrm{Contrad} \longrightarrow \Pi v[\sim B(v, S(z_p, z_p))],$$

where Contrad is a formula of the system which expresses the proposition $(x, y, z)[\sim\{x\mathfrak{B}y \& z\mathfrak{B}\mathrm{Neg}\ y\}]$. Then if Contrad could also be proved formally, we could use Rule I to infer $\Pi v[\sim B(v, S(z_p, z_p))]$ or $U(z_p)$, in which case as we have seen, the system would contain a contradiction. Hence Contrad cannot be proved in the system itself, unless the system contains a contradiction.

6. Conditions that a formal system must satisfy in order that the foregoing arguments apply

Now consider any formal system (in the sense of §1) satisfying the following five conditions:

(1) Supposing the symbols and formulas to be numbered in a manner similar to that used for the particular system considered above, then the class of axioms and the relation of immediate consequence shall be recursive.

This is a precise {condition which in practice suffices as a substitute for the unprecise} requirement of §1 that the class of axioms and relation of immediate consequence be constructive.

(2) There shall be a certain sequence of meaningful formulas z_n (standing for the natural numbers n), such that the relation between n and the number representing z_n is recursive.

(3) There shall be a symbol $\sim$ (negation) and two symbols v and w (variables) such that to every recursive relation of two variables there corresponds a formula $R(v, w)$ of the system such that $R(z_p, z_q)$ is provable if the relation holds of p and q and $\sim R(z_p, z_q)$ is provable if the relation does not hold of p and q; or, instead of a single symbol $\sim$, there may be a formula $F(x)$ not containing v or w such that the foregoing holds when $\sim(A)$ stands for the formula $F(A)$.

The formulas $R(v,w)$ which represent recursive relations, and their negations $\sim R(v,w)$, shall be called ***recursive propositional functions of two variables***; and $R(v, z_n)$ and $\sim R(v, z_n)$ ***recursive propositional functions of one variable.***

(4) There shall be a symbol Π such that if $\Pi v F(v)$ is provable for a recursive propositional function $F(v)$ of one variable, then $F(z_k)$ shall be provable for all k; or, instead of a single symbol Π, there may be a formula $G(x)$ not containing w such that the foregoing holds when $\Pi v F(v)$ stands for $G(F(v))$.

(5) The system shall be free from contradiction in the two following senses:

(a) If $R(v, w)$ is a recursive propositional function of two variables, then $R(z_p, z_q)$ and $\sim R(z_p, z_q)$ shall not both be provable.

(b) If $F(v)$ is a recursive propositional function of one variable, then the formulas $\sim \Pi v F(v)$, $F(z_0)$, $F(z_1)$, $F(z_2)$, . . . shall not all be provable.

Now, using condition (1), $x \mathfrak{B} y$, $\gamma(x, y)$ (defined as before), and $k \mathfrak{B} \gamma(l, l)$ are recursive. Then, by (3), there is an $R(v, w)$ such that $R(z_k, z_l)$ is provable if $k \mathfrak{B} \gamma(l, l)$, and $\sim R(z_k, z_l)$ is provable if $\sim k \mathfrak{B} \gamma(l, l)$. Noting that $R(v, w)$ plays the same role as $B(v, S(w, w))$ in our special system, we can prove by reasoning similar to that of §5 that, if p is the number of $\Pi v \sim R(v, w)$, (5a) implies that $\Pi v \sim R(v, z_p)$ is not provable and (5b) in conjunction with (5a) implies that $\sim \Pi v \sim R(v, z_p)$ is not provable. Also, as before, $(x, y, z)[\sim\{x\mathfrak{B}y \,\&\, z\mathfrak{B}\mathrm{Neg}\, y\}] \rightarrow (x) \sim x\mathfrak{B}\gamma(p, p)$ can be established. We shall not list the further conditions under which it is possible to convert the intuitive proof of this into a formal proof of Contrad $\rightarrow \Pi v \sim R(v, z_p)$ (Contrad defined as before). However, they are conditions satisfied by all systems of the type under consideration which contain a certain amount of ordinary arithmetic, and these systems therefore cannot contain a proof of their own freedom from contradiction.

7. Relation of the foregoing arguments to the paradoxes

We have seen that in a formal system we can construct statements about the formal system, of which some can be proved and some cannot, according to what they say about the system. We shall compare this fact with the famous Epimenides paradox ("Der Lügner"). Suppose that on May 4, 1934, ***A*** makes the single statement, "Every statement which ***A*** makes on May 4, 1934, is false." This statement clearly cannot be true. Also it cannot be false, since the only way for it to be false is for ***A*** to have made a true statement in the time specified and in that time he made only the single statement.

The solution suggested by Whitehead and Russell, that a proposition cannot say something about itself, is too drastic. We saw that we can construct propositions which make statements about themselves, and, in fact, these are arithmetic propositions which involve only recursively defined functions, and therefore are undoubtedly meaningful statements. It is even possible, for any metamathematical property ***f*** which can be expressed in the system, to construct a proposition which says of itself that it has this property.[23] For suppose that $F(z_n)$

23. {This was first noted by R. Carnap in: Logische Syntax der Sprache, Wien, 1934, page 91.}

means that n is the number of a formula that has the property f. Then if $F(S(w,w))$ has the number p, $F(S(z_p, z_p))$ says that it itself has the property f.[24] This construction can only be carried out if the property f can be expressed in the system, and the solution of the Epimenides paradox lies in the fact that the latter is not possible for every metamathematical property. For consider the above statement made by A. A must specify a language B and say that every statement that he made in the given time was a false statement in B. But "false statement in B" cannot be expressed in B, and so his statement was in some other language, and the paradox disappears.

The paradox can be considered as a proof that "false statement in B" cannot be expressed in B.[25] We now will establish this fact in a more formal manner, and in doing so obtain a heuristic argument for the existence of undecidable propositions. Suppose that $T(z_n)$ means that the formula whose number is n is true. That is, if n is the number of N, $T(z_n)$ shall be equivalent to N. Then we could apply our procedure to $\sim T(S(w, w))$, obtaining $\sim T(S(z_p, z_p))$, which says that it is itself false, and this leads to a contradiction similar to the "Epimenides". But, on the other hand, $\Sigma\, v\, B(v, z_k)$ is a statement in the system of the fact that the formula with number k is provable. So we see that the class α of numbers of true formulas cannot be expressed by a propositional function of our system, whereas the class β of provable formulas can. Hence $\alpha \neq \beta$ and if we assume $\beta \subseteq \alpha$ (i.e. every provable formula is true) we have $\beta \subset \alpha$,

24. Of course we can find properties f such that $F(S(z_p, z_p))$ is provable, just as we found ones for which it was not provable.

25. {For a closer examination of this fact see A. Tarski's papers published in: Trav. Soc. Sci. Lettr. de Varsovie, Cl. III, No. 34, 1933 (Polish)(translated in: Logic, Semantics, Metamathematics, Papers from 1923 to 1938 by A. Tarski, see in particular p. 247 ff.) and in: Philosophy and Phenom. Res. 4 (1944), p. 341-376. In these two papers the concept of truth relating to sentences of a language is discussed systematically. See also: R. Carnap, Mon. Hefte f. Math. u. Phys. 4 (1934) p. 263.}

i.e. there is a proposition A which is true but not provable. $\sim A$ then is not true and therefore not provable either, i.e. A is undecidable.[26]

8. Diophantine equivalents of undecidable propositions

Suppose $F(x_1, \ldots, x_n)$ a polynomial with integral coefficients. By use of logical quantifiers (x) and (Ex),[28] we can make certain statements about the solutions in natural numbers of the Diophantine equation $F(x_1, \ldots, x_n)=0$. Thus $(Ex_1)(Ex_2)\ldots(Ex_n)(F(x_1, \ldots, x_n)=0)$ says that there is a solution; $(x_3)(Ex_1)(Ex_2)(Ex_4)\ldots(Ex_n)(F(x_1, \ldots, x_n)=0)$ says that for any assigned value of x_3, the resulting equation has a solution; etc. We wish to prove that there is a sequence of logical quantifiers, say (P), and a Diophantine equation, $F=0$, such that our undecidable proposition is equivalent to $(P)(F=0)$.

To prove this we find it convenient to make use of the intermediary concept of an arithmetical expression, that is an expression built up out of $\sim$, $\vee$, &, $\rightarrow$, $\equiv$, $+$, $\times$, $=$, natural numbers, variables running over natural numbers, and the quantifiers (x) and (Ex),[27] according to the following induction:

1. If f and g are built up out of variables, natural numbers, $+$, and $\times$,[28] then $f=g$ is an arithmetical expression.
2. If A and B are arithmetical expressions, then $\sim A$, $A \vee B$, $A \& B$, $A \rightarrow B$, and $A \equiv B$ are arithmetical expressions.
3. If A is an arithmetical expression which contains x as a free variable, then $(x)A$ and $(Ex)A$ are arithmetical expressions.

26. {Note that this argument can be carried through with full precision for any system whose formulas have a well-defined meaning, provided the axioms and rules of inference are correct for this meaning, and arithmetic is contained in the system. One thus obtains a proof for the existence of undecidable propositions in that system, but no individual instance of an undecidable proposition.}

27. Where x is any variable for a natural number.

28. That is, if f and g are polynomials with natural number coefficients.

We shall prove first that, if $\phi(x_1, \ldots, x_n)$ is recursive, then there is an arithmetical expression $A(x_1, \ldots, x_n, y)$ such that $A(x_1, \ldots, x_n, y) .=. \phi(x_1, \ldots, x_n) = y$; and second that if $B(x_1, \ldots, x_n)$ is an arithmetical expression, then there are polynomials $Q(x_1, \ldots, x_n, y_1, \ldots, y_m)$ and $R(x_1, \ldots, x_n, y_1, \ldots, y_m)$ with natural number coefficients and a sequence (P) of quantifiers such that $B(x_1, \ldots, x_n) .\equiv. (P)[Q(x_1, \ldots, x_n, y_1, \ldots, y_m) = R(x_1, \ldots, x_n, y_1, \ldots, y_m)]$, where the x's and y's range over the natural numbers. Since our undecidable proposition has the form $(x)F(x)$ where F is recursive, there is a recursive function $\phi(x)$ such that our undecidable proposition is equivalent to $(x)[\phi(x)=0]$. Then there is an arithmetical expression $A(x, y)$ such that $\phi(x) = y .\equiv. A(x, y)$, and there are polynomials $Q(x, y, z_1, \ldots, z_m)$ and $R(x, y, z_1, \ldots, z_m)$ with natural number coefficients, and a sequence of quantifiers (P) such that $A(x, y) .\equiv. (P)[Q(x, y, z_1, \ldots, z_m) = R(x, y, z_1, \ldots, z_m)]$. Then our undecidable proposition is equivalent to $(x)(P)[Q(x, 0, z_1, \ldots, z_m) = R(x, 0, z_1, \ldots, z_m)]$.

We prove first that recursive functions are expressible arithmetically.

If $f(x) = x + 1$, $f(x) = y .\equiv. x + 1 = y$.

If $f(x_1, \ldots, x_n) = c$, $f(x_1, \ldots, x_n) = w .\equiv. w = c$.

Similarly for the identity functions $U_j^n(x_1, \ldots, x_n)$.

If $\psi(x_1, \ldots, x_m) = y .\equiv. A(x_1, \ldots, x_m, y)$, $\chi_i(x_1, \ldots, x_n) = y .\equiv. B_i(x_1, \ldots, x_n, y)$, and $\phi(x_1, \ldots, x_n) = \psi(\chi_1(x_1, \ldots, x_n), \ldots, \chi_m(x_1, \ldots, x_n))$, then $\phi(x_1, \ldots, x_n) = y .\equiv. (Et_1) \ldots (Et_m)[B_1(x_1, \ldots, x_n, t_1) \& \ldots \& B_m(x_1, \ldots, x_n, t_m) \& A(t_1, \ldots, t_m, y)]$.

To handle the case where ϕ is recursive with respect to ψ and χ, we require an arithmetical expression for $\beta(c, d, i) = y$, where $\beta(c, d, i)$ is a certain function which has the property that if a function $f(i)$ of natural numbers and a natural number l are given, then natural numbers c and d such that $\beta(c, d, i) = f(i)$ $(i = 0, \ldots, l)$ can be found. We may define $x \equiv y \pmod{z}$ as

$(Et)[x=y+tz \vee y=x+tz]$, and $x \geqslant y$ as $(Et)[x=y+t]$.[29] Then we define $\beta(c,d,i)$ to be the least non-negative residue of c modulo $1+(i+1)d$, i.e. $\beta(c,d,i)=z.\equiv.z\equiv c \pmod{[1+(i+1)d]}\,\&\, z\leqslant(i+1)d$. To prove that $\beta(c,d,i)$ has the aforesaid property, suppose $f(i)$ and l given. Choose s greater than all of the numbers l, $f(0)$, $f(1)$, ..., $f(l)$. Then the numbers $1+s!$, $1+2s!$, ..., $1+(l+1)s!$ are relatively prime. For if a prime number divides two of them, it divides their difference $(i-j)s!$; but it cannot divide $s!$, since it divides $1+is!$; then also it cannot divide $i-j$, since $i-j \leqslant l < s$ and hence $i-j$ is a factor of $s!$. Then if we let $d=s!$, we can find a c such that $c\equiv f(i) \pmod{[1+(i+1)d]}$ $(i=0,\ldots,l)$ since $1+s!$, ..., $1+(l+1)s!$ are relatively prime. Since $s>f(i)$ and therefore $1+(i+1)s! > f(i)$, we have $f(i)=\beta(c,d,i)$ as was to be shown.

If $\psi(x_1,\ldots,x_{n-1})=y.\equiv.A(x_1,\ldots,x_{n-1},y)$, $\chi(x_1,\ldots,x_{n+1})=y$ $.\equiv.B(x_1,\ldots,x_{n+1},y)$, $\phi(0,x_2,\ldots,x_n)=\psi(x_2,\ldots,x_n)$, and $\phi(k+1,$ $x_2,\ldots,x_n)=\chi(k,\phi(k,x_2,\ldots,x_n),x_2,\ldots,x_n)$, then $\phi(x_1,\ldots,x_n)=$ $y.\equiv.(Ef)[A(x_2,\ldots,x_n,f(0))\,\&\,(t)\{t+1\leqslant x_1 \rightarrow B(t,f(t),x_2,\ldots,x_n,f(t+1))\}$ $\&\, f(x_1)=y]$. But if there is an f satisfying the condition in square brackets, then there is a c and a d such that $\beta(c,d,i)=$ $f(i)$ $(i=0,\ldots,x_1)$ and therefore $(Ec)(Ed)[A(x_2,\ldots,x_n,\beta(c,d,0))$ $\&(t)\{t+1\leqslant x_1\rightarrow B(t,\beta(c,d,t),x_2,\ldots,x_n,\beta(c,d,t+1))\}\&\,\beta(c,d,x_1)=y]$. Conversely, this obviously implies the original expression. The latter formula can be transformed into the arithmetical one $(Ec)(Ed)[(Ev)\{A(x_2,\ldots,x_n,v)\&\, v=\beta(c,d,0)\}\,\&\,(t)\{t+1\leqslant x_1\rightarrow$ $(Ev)(Ew)[B(t,v,x_2,\ldots,x_n,w)\&\, v=\beta(c,d,t)\&\, w=\beta(c,d,t+1)]\}\,\&\, y$ $=\beta(c,d,x_1)]$ by substituting $(Ev)[A(x_2,\ldots,x_n,v)\&\, v=\beta(c,d,0)]$ for $A(x_2,\ldots,x_n,\beta(c,d,0))$, and $(Ev)(Ew)[B(t,v,x_2,\ldots,x_n,w)\&\, v=$ $\beta(c,d,t)\&\, w=\beta(c,d,t+1)]$ for $B(t,\beta(c,d,t),x_2,\ldots,x_n,\beta(c,d,t+1))$. This completes the proof that all recursive functions are arithmetical.

29. If we were allowing the variables to run over the positive and negative integers instead of just the natural numbers we could define $x \geqslant y$ as $(Et_1)\ldots(Et_4)$ $[x=y+t_1^2+t_2^2+t_3^2+t_4^2]$, since every positive integer is the sum of four squares.

We next show that all arithmetical expressions can be given the equivalent normal form $(P)[Q=R]$ where Q and R are polynomials with natural number coefficients.

If $\sim$, $\vee$, &, $\rightarrow$, $\equiv$, and quantifiers do not occur in an arithmetical expression, then it has the required normal form by definition (page 65).

Suppose that $A \equiv (P)[Q=R]$ where x does not occur in the quantifiers denoted by (P). Then $(x)A \cdot \equiv \cdot (x)(P)[Q=R]$ and $(Ex)A \cdot \equiv \cdot (Ex)(P)[Q=R]$.

Suppose also that $B \equiv (P')[Q'=R']$ where the variables of (P') are distinct from those of (P). Then owing to the fact that $p \vee (Ex)F(x) \cdot \equiv \cdot (Ex)[p \vee F(x)]$ and $p \vee (x)F(x) \cdot \equiv \cdot (x)[p \vee F(x)]$,

$$\begin{aligned} A \vee B \cdot &\equiv \cdot (P)(P')[Q=R \vee Q'=R'] \\ &\equiv \cdot (P)(P')[Q-R=0 \vee Q'-R'=0] \\ &\equiv \cdot (P)(P')[(Q-R)(Q'-R')=0] \\ &\equiv \cdot (P)(P')[QQ'+RR'=Q'R+QR'] \end{aligned}$$

Moreover $\sim A \cdot \equiv \cdot \sim (P)[Q=R]$. Then, since $\sim (x)p \cdot \equiv \cdot (Ex) \sim p$ and $\sim (Ex)p \equiv \cdot (x) \sim p$, we can shift the negative sign through the prefix (P) and find a P'' such that

$$\begin{aligned} \sim A \cdot &\equiv \cdot (P'')[\sim Q \equiv R] \\ &\equiv \cdot (P'')[(Q-R)^2 > 0] \\ &\equiv \cdot (P'')[Q^2 + R^2 \geqslant 2QR + 1] \\ &\equiv \cdot (P'')(Et)[Q^2 + R^2 = 2QR + t + 1] \end{aligned}$$

&, $\rightarrow$, and $\equiv$ are expressible by means of $\sim$ and $\vee$.

If the argument is modified slightly, the variables can be allowed to run over the integers instead of just the natural numbers.

Thus there exists a statement about the solutions of a Diophantine equation which is not decidable in our formal system. It can be shown that it is decidable in the next higher type, but there is another such statement which is not decidable even in that type, but which is decidable by going into the next higher type; and so on {including transfinite iterations describable in set theory, such as occur, e.g., in the higher axioms of infinity}. In other words, {on the basis of the principles of proof used in mathematics today[29a]} there can be no complete theory of Diophantine analysis, {not even of the problems of the form $(P)[F=0]$.}[30]

Presburger has given a set of axioms for the relations built up out of $+$, $=$, and logical symbols, together with a method of deciding such relations.[31] Skolem has sketched a method of deciding relations constructed similarly using $\times$ instead of $+$.[32] {However, on the basis of the principles of proof used in mathematics today, no general method of deciding relations in which both $+$ and $\times$ occur can be established, since (as shown above) there can be, on this basis, no complete theory of the Diophantine problems of the form $(P)[F=0]$.}[30]

9. General recursive functions

If $\psi(y)$ and $\chi(x)$ are given recursive functions, then the function $\phi(x,y)$, defined inductively by the relations $\phi(0,y)=\psi(y)$, $\phi(x+1,0)=\chi(x)$, $\phi(x+1,y+1)=\phi(x,\phi(x+1,y))$, is not in general recursive in the limited sense of §2. This is an example of a definition by induction with respect to two variables simultaneously.[33]

To get arithmetical definitions of such functions, we have to generalize our β function. The consideration of various sorts of functions defined by inductions leads to the question what one would mean by "every recursive function".

29a, 30: See p. 74.

31. Presburger, "Uber die Vollständigkeit eines gewissen Systems der Arithmetik etc.", Comptes Rend. du I. Congres des Math. des Pays Slaves, Warszawa 1929.

32. Th. Skolem, "Uber einige Satzfunktionen in der Arithmetik", Vidensk. Akad. Skrifter, Mat.-Nat. Kl., 1930, No. 7.

33. {For a very similar function W. Ackermann in Math. Ann. 99 (1928) p. 118 proved that it cannot be defined by recursion with respect to one variable.}

One may attempt to define this notion as follows: If ϕ denotes an unknown function, and $\psi_1, \ldots, \psi_k$ are known functions, and if the ψ's and the ϕ are substituted in one another in the most general fashions and certain pairs of the resulting expressions are equated, then if the resulting set of functional equations has one and only one solution for ϕ, ϕ is a recursive function.[34]

Thus we might have

$$\begin{aligned}
&\phi(x, 0) = \psi_1(x),\\
&\phi(0, y+1) = \psi_2(y),\\
&\phi(1, y+1) = \psi_3(y),\\
&\phi(x+2, y+1) = \psi_4(\phi(x, y+2), \phi(x, \phi(x, y+2))).
\end{aligned}$$

We shall make two restrictions on Herbrand's definition. The first is that the left-hand side of each of the given functional equations defining ϕ shall be of the form

$$\phi(\psi_{i1}(x_1, \ldots, x_n), \psi_{i2}(x_1, \ldots, x_n), \ldots, \psi_{il}(x_1, \ldots, x_n)).$$

The second (as stated below) is equivalent to the condition that all possible sets of arguments $(n_1, \ldots, n_l)$ of ϕ can be so arranged that the computation of the value of ϕ for any given set of arguments $(n_1, \ldots, n_l)$ by means of the given equations requires a knowledge of the values of ϕ only for sets of arguments which precede $(n_1, \ldots, n_l)$.

From the given set of functional equations, we define by induction a set of derived equations, thus:

34. This was suggested by Herbrand in a private communication. {A slightly different definition was given by him in J. r. ang. Math. 166 (1932), p. 5, where he postulated "computability". However, also in this definition, he did not require computability by any definite formal rules (note the phrase "considérée intuitionnistiquement" and footnote 5). In intuitionistic mathematics the two Herbrand definitions are trivially equivalent. In classical mathematics the non-equivalence of general recursiveness with the first mentioned concept of Herbrand was proved by L. Kalmár in Zs. f. math. Log. u. Grundl. d. Math. 1 (1955) p. 93. Whether Herbrand's second concept is equivalent with general recursiveness is a largely epistemological question which has not yet been answered. See the postscript.}

(1a) Any expression obtained by replacing all the variables of one of the given equations by natural numbers shall be a derived equation.

(1b) $\psi_{ij}(k_1, \ldots, k_n) = m$ shall be a derived equation if $k_1, \ldots, k_n$, are natural numbers, and $\psi_{ij}(k_1, \ldots, k_n) = m$ is a true equality.

(2a) If $\psi_{ij}(k_1, \ldots, k_n) = m$ is a derived equation, the equality obtained by substituting m for an occurrence of $\psi_{ij}(k_1, \ldots, k_n)$ in a derived equation shall be a derived equation.

(2b) If $\phi(k_1, \ldots, k_l) = m$ is a derived equation where $k_1, \ldots, k_l, m$ are natural numbers, the expression obtained by substituting m for an occurrence of $\phi(k_1, \ldots, k_l)$ on the right-hand side of a derived equation shall be a derived equation.

Now our second restriction on Herbrand's definition of recursive function is that for each set of natural numbers $k_1, \ldots, k_l$ there shall be one and only one m such that $\phi(k_1, \ldots, k_l) = m$ is a derived equation.

Using this definition of the notion of a recursive function, we can prove that, if $\phi(x_1, \ldots, x_l)$ is recursive, there is an arithmetical expression $A(x_1, \ldots, x_l, y)$ such that $\phi(x_1, \ldots, x_l) = y . \equiv . A(x_1, \ldots, x_l, y)$.

-POSTSCRIPTUM-

In consequence of later advances, in particular of the fact that, due to A.M. Turing's work, a precise and unquestionably adequate definition of the general concept of formal system can now be given, the existence of undecidable arithmetical propositions and the non-demonstrability of the consistency of a system in the same system can now be proved rigorously for <u>every</u> consistent formal system containing a certain amount of finitary number theory.

Turing's work gives an analysis of the concept of "mechanical procedure" (alias "algorithm" or "computation procedure" or "finite combinatorial procedure"). This concept is shown to be equivalent with that of a "Turing machine". * A formal system can simply be defined to be any mechanical procedure for producing formulas, called provable formulas. For any formal system in this sense there exists one in the sense of page 41 above that has the same provable formulas (and likewise vice versa), provided the term "finite procedure" occurring on page 41 is understood to mean "mechanical procedure". This meaning, however, is required by the concept of formal system, whose essence it is that reasoning is completely replaced by mechanical operations on formulas. (Note that the question of whether there exist finite non-mechanical procedures** not equivalent with any algorithm, has nothing whatsoever to do with the adequacy of the definition of "formal system" and of "mechanical procedure".)

On the basis of the definitions just mentioned, condition (1) in §6 becomes superfluous, because for any formal system provability is a predicate of the form $(Ex)x\mathfrak{B}y$, where $\mathfrak{B}$ is primitive recursive. Moreover, the two incompleteness results mentioned in the end of §8 can now be proved in the definitive form: "There exists no formalized theory that answers all Diophantine questions of the form $(P)[F=0]$", and: "There is no algorithm for deciding relations in which both + and × occur." (For theories and procedures in the more general sense indicated in footnote ** the situation may be different. Note

* See A. Turing, Proc. London Math. Soc., vol. 42 (1937), p. 249 (this anthology, p. 135) and the almost simultaneous paper by E.L. Post in J. S. L. 1 (1936) p. 103 (this anthology, p. 289). As for previous equivalent definitions of computability, which, however, are much less suitable for our purpose, see A. Church, Am. J. Math., vol. 58 (1936), pp. 356-358 (this anthology, pp. 100-102). One of those definitions is given in §9 of these lectures.

** I.e., such as involve the use of abstract terms on the basis of their meaning. See my paper in Dial. 12 (1958), p. 280.

that the results mentioned in this postscript do not establish any bounds for the powers of human reason, but rather for the potentialities of pure formalism in mathematics.) Thirdly, if "finite procedure" is understood to mean "mechanical procedure", the question raised in footnote 3 can be answered affirmatively for recursiveness as defined in § 9, which is equivalent with general recursiveness as defined today (see S. C. Kleene, Math. Ann. 112 (1936), p. 730 [this anthology p. 240] and Introduction to Metamathematics, 1952, p. 220ff, p. 232ff).

As for the elimination of ω-consistency (first accomplished by J.B. Rosser [cf. this anthology, pp. 231-235]) see A. Tarski, Undecidable Theories, 1953, p. 49, Cor. 2. The proof of the unprovability in the same system of the consistency of a system was carried out for number theory in Hilbert-Bernays, Grundlagen der Mathematik, vol. 2 (1939), pp. 297-324. The proof carries over almost literally to any system containing, among its axioms and rules of inference, the axioms and rules of inference of number theory. As to the consequences for Hilbert's program see my paper in Dial. 12 (1958), p.280 and the material cited there. See also: G. Kreisel, Dial. 12 (1958), p. 346.

By slightly strengthening the methods used above in §8, it can easily be accomplished that the prefix of the undecidable proposition consists of only one block of universal quantifiers followed by one block of existential quantifiers, and that, moreover, the degree of the polynomial is 4. (unpublished result.)

A number of misprints and oversights in the original mimeographed lecture notes have been corrected in this volume. I am indebted to Professor Martin Davis for calling my attention to some of them.

Kurt Gödel

Princeton, N. J.
June 3, 1964

[EDITOR'S NOTE- In addition to the corrections and emendations supplied by Dr. Gödel for this anthology and incorporated into the body of the text, a number of additional items received too late for inclusion in the text proper are listed below.]

Page 43, footnote 2: replace "could have been" by "should be".

Page 44, footnote 3:{this statement is outdated; see the Postscript.}

Page 69, footnote 29a: "{Note that the axioms about all sets or about classes of sets that are assumed today do not carry any farther, because they are assumed to hold also for the sets of some definite type (or "rank", according to current terminology). See A. Levy, Fund. Math. 49 (1960), p. 1. The principles of proof of intuitionistic mathematics are not taken into account here, because, at any rate up to now, they have proved weaker than those of classical mathematics.}"

Page 69, footnote 30: substitute the following: "{By a complete theory of some class of problems on the basis of certain principles of proof we here mean a theorem, demonstrable on this basis, which states that (and how) the solution of any problem of the class can be obtained on this basis. For a different and more definitive version of this incompleteness result see the postscript.}"

Kurt Gödel

ON INTUITIONISTIC ARITHMETIC AND NUMBER THEORY *

If one lets correspond to the basic notions of Heyting's propositional calculus[1] the classical notions given by the same symbols and to "absurdity" ($\neg$), ordinary negation ($\sim$), then the intuitionistic propositional calculus A appears as a proper subsystem of the usual propositional calculus H. But, using a different correspondence (translation) of the concepts, the reverse occurs: the classical propositional calculus is a subsystem of the intuitionistic one. For, one has: Every formula constructed in terms of conjunction ($\wedge$) and negation ($\neg$) alone which is valid in A is also provable in H. For each such formula must be of the form: $\neg A_1 \wedge \neg A_2 \wedge \ldots \wedge \neg A_n$, and if it is valid in A, so must be each individual $\neg A_i$; but then by Glivenko[2] $\neg A_i$ is also provable in H and hence also the conjunction of the $\neg A_i$. From this, it follows that: if one translates the classical notions $\sim p$, $p \rightarrow q$, $p \vee q$, $p.q$ by the following intuitionistic notions: $\neg p$, $\neg (p \wedge \neg q)$, $\neg(\neg p \wedge \neg q)$, $p \wedge q$ then each classically valid formula is also valid in H.

The aim of the present investigation is to prove that something analogous holds for all of arithmetic and number theory, as given e.g. by the axioms of Herbrand.[3,4] Here also one can give an interpretation of the classical notions in terms of intuitionistic notions, so that all of the classical axioms become provable propositions for intuitionism as well.

*Translated by the editor from the original in Ergebnisse eines mathematischen Kolloquiums, Heft 4 (1933) pp. 34-38, with the kind permission of the publishers, Franz Deuticke, Vienna.

As basic symbols of Herbrand's system, we consider:

1. the operations of the propositional calculus: $\sim, \rightarrow, \vee, .$;
2. number variables: x, y, $z \ldots$;
3. the universal quantifier: $(x), (y), \ldots$;
4. $=$;
5. 0 and $+1$.
6. the countable set of function symbols f_i, introduced according to Axiom Group C, to each of which is associated an integer n_i (the number of arguments of f_i).

In order to make precise how formulas are to be constructed from these basic symbols, we first define the notion numerical expression by the following recursive prescription:

1. 0 and all of the variables x, $y \ldots$ are numerical expressions;
2. if Z is a numerical expression, then so is $Z+1$;
3. if $Z_1, Z_2, \ldots Z_{n_i}$ are numerical expressions, then so is $f_i(Z_1, Z_2, \ldots Z_{n_i})$.

By an elementary formula we understand an expression of the form $Z_1=Z_2$ where Z_1, Z_2 are numerical expressions. Meaningful formulas of number theory (called in what follows: Z-formulas) are either elementary formulas or expressions which are constructed from elementary formulas using the operations of the propositional calculus and the universal quantifier (x), $(y) \ldots$.

To Herbrand's Axiom Groups A-D, we adjoin the logical Axiom Groups E-G (which Herbrand did not explicitly introduce):

E. Every expression which results from the substitution of Z-formulas for variables in a valid formula [i.e. a tautology] of the propositional calculus, is an axiom.

F. All formulas of the form: $(x)F(x) \rightarrow F(Z)$, where $F(x)$ is an arbitrary Z-formula and Z is a numerical expression, are axioms (with the obvious restriction that the variables in Z may not be bound in $F(x)$).

G. All formulas of the form: $x = y. \rightarrow . F(x) \rightarrow F(y)$ (where $F(x)$ is an arbitrary Z-formula) are axioms.

As rules of inference, we take:

I. From A and $A \rightarrow B$, B follows.
II. From $A \rightarrow B$, $A \rightarrow (x)B$ follows if x has no free occurrences in A.

In F, G, and Rule II, x, y denote arbitrary variables.

Unlike Herbrand's system, Heyting's[5] has no provision for number variables, but rather has only variables x, y . . . for arbitrary objects. Because of this certain complications arise which we shall avoid by introducing variables x', y', z', . . . for natural numbers into Heyting's system. Then, the expression: $(x')F(x')$ shall be equivalent to $(x).\ x \epsilon N \supset F(x)$ and a formula $A(x'y' \ldots)$, which contains x', y', . . . as free variables, shall be equivalent to: $x, y, \ldots \epsilon N \supset A(x\, y \ldots)$.[5a] Thus each proposition containing variables of this new sort is equivalent to one containing only unprimed variables, Using the rules of translation just given, it can be shown, quite formally, that practically all of the theorems of H_2 §5, §6 remain valid. In particular this is the case for the following theorems which are used in what follows: H_2 5.4, 5.5, 5.8, 6.26, 6.3, 6.4, 6.78, where, however, in 5.4 it is necessary to replace $p = p$ by $p' \epsilon N$. In the theorems of H_2 §10, the hypotheses $p \epsilon N$, $q \epsilon N$ etc. may be eliminated on introducing number variables.

The definition of number-theoretic functions by recursion is intuitionistically unobjectionable (cf. H_2 10.03, 10.04). Then the functions f_i (Axiom Group C) occur also in intuitionistic mathematics, and we adjoin their defining formulas to Heyting's axioms. We also adjoin the formulas of Group D, which are obviously intuitionistically valid. Let H' be Heyting's system extended in this way. Now we associate to each Z-formula A, a formula A' of H'(its "translation") according to the following stipulations: The variables x, y, ... are to be translated as x', y', ...; each f_i by the equivalent symbol f_i of H'; $=$ by $=$; 0 by 1^6; $+1$ by seq ; the operations of the propositional calculus as indicated above. $(x)A$ is to be translated as $(x')A'$, where A' is the translation of A. A formula which is the translation of a Z-formula, will be called a Z'-formula.

Theorem I (still to be proved) states: If the formula A is provable in Herbrand's system, then its translation A' is provable in H'.

Lemma 1. <u>For each Z'-formula</u> A',

$$\neg\neg A' \supset A' \tag{1}$$

<u>is provable in</u> H'. Our proof is by complete induction:

α) (1) holds for A an elementary formula, because for numerical expressions Z, we have $Z' \epsilon N$, as can be shown analogously to H_2 10.4. Hence, by H_2 10.25, we have for elementary formulas $A' \vee \neg A'$ from which (by H_1 4.45) we obtain (1).

β) If (1) holds for two Z'-formulas A' and B', then it also holds for $A' \wedge B'$. For, by H_1 4.61, we have $\neg\neg(A' \wedge B') .\supset. \neg\neg A' \wedge \neg\neg B'$ and hence by the induction hypothesis and H_1 2.23:

$$\neg\neg(A' \wedge B') .\supset. A' \wedge B'$$

γ) If (1) holds for A', then it also holds for $\neg A'$. For, quite generally, we have $\neg\neg\neg A' \supset \neg A'$ (by H_1 4.32).

δ) If (1) holds for A', it holds also for $(x')A'$. Proof: By induction hypothesis $\neg\neg A' \supset A'$ holds. Hence, by H_2 5.8, $(x').\neg\neg A' \supset A'$ and by H_2 6.4, $(x')\neg\neg A' .\supset. (x')A'$. By H_2 6.78, we have furthermore $\neg\neg(x')A' \supset (x')\neg\neg A'$ and the two last formulas yield $\neg\neg(x')A' \supset (x')A'$, q. e. d.

Lemma 1 follows from α) - δ) since every Z'-formula is constructed from elementary formulas using the operations $\wedge$, $-$, (x').

Lemma 2. For arbitrary Z'-formulas, A', B',

$$A' \supset B'. \supset\subset . \neg(A' \wedge \neg B')$$

holds in the system H'.

Proof: By H_1 4.9, we have

$$A' \supset B'. \supset . \neg(A' \wedge \neg B'); \tag{2}$$

moreover $\neg(A' \wedge \neg B'). \supset . A' \supset \neg\neg B'$ by H_1 4.52, and, since B' is a Z'-formula, it follows from Lemma 1 that $\neg(A' \wedge \neg B'). \supset . A' \supset B'$, which together with (2) gives the desired result.

We now prove: The translation A' of each axiom A of Herbrand's system is provable in H'.

1) The axioms of Group A are translated into propositions which by Lemma 2 are equivalent to H_2 10.2, 10.22, 10.221, 10.25, 10.26.

2) The translation of an axiom of Group B is of the form:

$$\neg[\Phi(1) \wedge (x') \neg(\Phi(x') \wedge \neg\Phi(\text{seq}`x')) \wedge \neg(x')\Phi(x')] \tag{3}$$

and this follows by Lemma 2 from:

$$\Phi(1) \wedge (x'). \Phi(x') \supset \Phi(\text{seq}`x'): \supset (x')\Phi(x'), \tag{4}$$

because one is replacing $p \supset q$ everywhere by $\neg(p \wedge \neg q)$. But (4) is precisely H_2 10.14 in our notation.

3) The axioms of Groups C and D were directly adjoined to Heyting's system. Group E follows from what was proved above about Heyting's propositional calculus and F follows at once, using Lemma 2, from H_2 6.3 and 5.4; G follows in the same way from H_2 6.26 and 10.01.

It remains to be shown that the application of rules of inference I, II to formulas whose translations into H' are proveable, results in formulas for which the same holds. For Rule I this becomes: If A' and $(A \rightarrow B)'$ are provable in H', then so is B', i.e. from A' and $\neg(A' \wedge \neg B')$, B' follows. But this is obtained at once from Lemma 2 and H_1 1.3. The corresponding result for Rule II follows in the same way from H_2 5.5 and Lemma 2.

Theorem I, whose proof has now been completed, shows that intuitionistic arithmetic and number theory are only apparently narrower than the classical versions, and in fact contain them (using a somewhat deviant interpretation). The reason for this lies in the fact that the intuitionistic prohibition against negating universal propositions to form purely existential propositions is made ineffective by permitting the predicate of absurdity to be applied to universal propositions, which leads formally to exactly the same propositions as are asserted in classical mathematics. Intuitionism would seem to result in genuine restrictions only for analysis and set theory, and these restrictions are the result, not of the denial of tertium non datur, but rather of the prohibition of impredicative concepts. The above considerations, of course, yield a consistency proof for classical arithmetic and number theory. However, this proof is certainly not "finitary" in the sense given by Herbrand,[7] following Hilbert.

Footnotes

1. Heyting, "Die formalen Regeln der intuitionistischen Logik." Sitz.-Ber. preuss. Akad. Wiss. (phys.-math. Kl.), 1930, II. (Cited as H_1).

2. Glivenko, "Sur quelques points de la Logique de M. Brouwer." Akad. roy. de Belgique, Bull. de la Cl. des Sciences, Série 5, Tome 15, 1929.

3. Herbrand, "Sur la non-contradiction de l'Arithmétique." Crelles Journ., 166, 1931.
4. The result obtained by Glivenko l.c.[2] for the propositional calculus can not be extended to number theory.
5. Heyting, "Die formalen Regeln der intuitionistischen Mathematik." Sitz.-Ber. preuss. Akad. Wiss. (phys.-math. Kl.), 1930, II. (Cited as H_2)
5a. It is clear how (Ex') could be defined; but this will not be used in what follows.
6. In Heyting's system the number series begins with 1.
7. Herbrand, "Les bases de la logique Hilbertienne." Rev. de Mét. et de Mor., 37, 1930, p. 248 and "Thèses présentées à la Faculté des Sciences de Paris" (1930), p. 3ff.

Kurt Gödel

ON THE LENGTH OF PROOFS *

Let the length of a proof in a formal system S be understood to mean the number of formulas of which it consists. Furthermore, if every natural number is represented in S by a definite symbol, i.e. numeral, (e.g. by a symbol of the form $1 + 1 + 1 + \ldots + 1$) then a function $\phi(x)$ may be called computable in S if to each numeral m there corresponds a numeral n such that $\phi(m) = n$ is provable in S. In particular, e.g. all recursively defined functions are already computable in classical arithmetic (i.e. in the system S_1 of the sequence defined below).

Now, let S_i be the system of logic of the i-th order, where natural numbers are thought of as individuals. That is, more precisely: S_i shall contain variables and quantifiers for natural numbers, for classes of natural numbers, for classes of classes of natural numbers, etc. up to classes of the i-th type, but no variables of higher type. The corresponding logical axioms are also to be available in S_i. Then, as is known, there are propositions of S_i which are provable in S_{i+1}, but not in S_i. If we consider, on the other hand, those formulas f which are provable in both S_i and S_{i+1}, then the following result holds: To each function ϕ, computable in S_i, there correspond infinitely many formulas f with the property that if k is the length of a shortest proof of f in S_i and l is the length of a shortest proof of f in S_{i+1}, then $k > \phi(l)$. For example, if we set $\phi(n) = 10^6 n$, then we have: There are infinitely many formulas whose shortest proof in S_i is more than 10^6 times as long as in S_{i+1}. The transition to the logic of the next higher type not only results in certain previously unprovable prop-

*Translated by the editor from the original article in Ergebnisse eines mathematischen Kolloquiums, Heft 7 (1936) pp. 23-24, with the kind permission of the publishers, Franz Deuticke, Vienna.

ositions becoming provable, but also in it becoming possible to shorten extraordinarily infinitely many of the proofs already available.

The formulas f, for which the above inequality $k>\phi(l)$ holds are arithmetic sentences of the same character as the undecidable sentences of S_i constructed by me, i.e. they can be brought into the following normal form:

$$(\mathfrak{P})[\, Q\,(x_1,\; x_2,\; \ldots\; x_n\,)=0\,],$$

where x_1, $x_2 \ldots x_n$ are variables for integers, Q is a definite polynomial with integer coefficients in the n variables x_1, x_2, ... x_n, and $\mathfrak{P}$ denotes a prefix, i.e. a definite sequence of universal and existential quantifiers on the variables x_1, x_2, ... x_n. Such a sentence thus expresses a property of the Diophantine equation $Q=0$. E.g. the sentence

$$(x_1)(x_2)(E\,x_3)(E\,x_4)[\, Q\,(x_1,\; x_2,\; x_3,\; x_4\,)=0\,]$$

states that the Diophantine equation $Q\,(a,\; b,\; x,\; y\,)=0$ has integer solutions x, y for all [integral] values of the parameters a, b. If the prefix contains universal quantifiers on $x_{i_1}, x_{i_1}, \ldots,$ x_{i_k} and existential quantifiers on $x_{j_1}, x_{j_2}, \ldots, x_{j_{n-k}}$, then one should regard $x_{i_1}, \ldots, x_{i_k}$ as parameters and the above formula states that for arbitrary values of the parameters, solutions $x_{j_1}, x_{j_2}, \ldots, x_{j_{n-k}}$ exist, where the value of x_{j_r} depends only on the values of those parameters which precede x_{j_r} in the prefix.

(Remark added in proof [of the original German publication]):

It may also be shown that a function which is computable in one of the systems S_i or even in a system of transfinite type, is already computable in S_1. Thus, the concept "computable" is in a certain definite sense "absolute," while practically all other familiar metamathematical concepts (e.g. provable, definable, etc.) depend quite essentially on the system with respect to which they are defined.

Kurt Gödel

REMARKS BEFORE THE PRINCETON BICENTENNIAL CONFERENCE ON PROBLEMS IN MATHEMATICS
-1946-

Tarski has stressed in his lecture (and I think justly) the great importance of the concept of general recursiveness (or Turing's computability). It seems to me that this importance is largely due to the fact that with this concept one has for the first time succeeded in giving an absolute definition of an interesting epistemological notion, i.e., one not depending on the formalism chosen.* In all other cases treated previously, such as demonstrability or definability, one has been able to define them only relative to a given language, and for each individual language it is clear that the one thus obtained is not the one looked for. For the concept of computability however, although it is merely a special kind of demonstrability or decidability the situation is different. By a kind of miracle it is not necessary to distinguish orders, and the diagonal procedure does not lead outside the defined notion. This, I think, should encourage one to expect the same thing to be possible also in other cases (such as demonstrability or definability). It is true that for these other cases there exist certain negative results, such as the incompleteness of every formalism or the paradox of Richard, but closer examination shows that these results do not make a definition of the absolute notions concerned impossible under all circumstances, but only exclude certain ways of defining them, or, at least, that certain very closely related concepts may be definable in an absolute sense.

*To be more precise: a function of integers is computable in any formal system containing arithmetic, if and only if it is computable in arithmetic, where a function f is called computable in S if there is in S a computable term representing f. [Footnote added for this anthology.]

Let us consider, e.g., the concept of demonstrability. It is well known that, in whichever way you make it precise by means of a formalism, the contemplation of this very formalism gives rise to new axioms which are exactly as evident and justified as those with which you started, and that this process of extension can be iterated into the transfinite. So there cannot exist any formalism which would embrace all these steps; but this does not exclude that all these steps (or at least all of them which give something new for the domain of propositions in which you are interested) could be described and collected together in some non-constructive way. In set theory, e.g. the successive extensions can most conveniently be represented by stronger and stronger axioms of infinity. It is certainly impossible to give a combinational and decidable characterization of what an axiom of infinity is but there might exist, e.g. a characterization of the following sort: An axiom of infinity is a proposition which has a certain (decidable) formal structure and which in addition is true. Such a concept of demonstrability might have the required closure property, i.e. the following could be true: Any proof for a set-theoretic theorem in the next higher system above set theory (i.e. any proof involving the concept of truth which I just used) is replaceable by a proof from such an axiom of infinity. It is not impossible that for such a concept of demonstrability some completeness theorem would hold which would say that every proposition expressible in set theory is decidable from the present axioms plus some true assertion about the largeness of the universe of all sets.

Let me consider a second example where I can give somewhat more definite suggestions, namely the concept of definability (or to be more exact of mathematical definability). Here also you have corresponding to the transfinite hierarchy of formal systems, a transfinite hierarchy of concepts of definability. Again it is not possible to collect together all these languages in one, as long as you have a finitistic concept of language, i.e., as long as you require that a language must have a finite number of primitive terms. But if you drop this condition, it does become possible (at least as far as it is necessary for the purpose), namely by means of a language which has as many primitive terms as you wish to consider steps in this hierarchy of languages, i.e., as many as there are ordinal numbers. The simplest way of doing it is to take the ordinals themselves as primitive terms. So one is led to the concept of definability in terms of ordinals, i.e., definability by expressions containing names of ordinal numbers and logical constants, including quantification referring to sets. This concept should, I think, be investigated. It can be proved that it has the required closure property, by introducing the notion of truth for this whole transfinite language. I.e., by going over to the next language you will obtain no new definable sets (although you will obtain new definable properties of sets). The concept of constructible set I used in the consistency proof for the Continuum Hypothesis can be obtained in a very similar way, i.e. as a kind of definability in terms of ordinal numbers; but comparing constructibility with the concept of definability just outlined you will find that not all logical means of definition are admitted in the definition of constructible sets. Namely, quantification is admitted only with respect to constructible sets and not with respect to sets in general. This has the consequence that you can actually define sets and even sets of integers for which you cannot prove that they are constructible (although this can of course be consistently assumed) and, for this reason, I think constructibility cannot be considered as a satisfactory formulation of definability. But now coming back to the definition of definability, I suggested it might be objected that the introduction of all ordinals as primitive terms is too cheap a way out of the difficulty, and that the concept thus obtained completely fails to agree with the intuitive concept we set out to make precise, because there exist undenumerably many sets definable in this sense. There is certainly some justification in this objection. For it has some plausibility that all things conceivable by us are denumerable even if you disregard the question of expressibility in some language. But on the other hand there is much to be said in favor of the concept under consideration, namely, above all, it is clear that if the concept of mathematical definability is to be itself mathematically definable, it must necessarily be so that all ordinal numbers are definable, because otherwise you could define the first ordinal number not definable, and would thus obtain a contradiction. I think this does not mean that a concept of definability satisfying the postulate of denumerability is impossible but only that it would involve some extramathematical element concerning the psychology of the being which deals with mathematics. But irrespective of what the answer to this question may be, I would think that "definability in terms of ordinals" even if it is not an adequate formulation for "comprehensibility by our mind", is at least an adequate formulation in an absolute sense for a closely related property of sets, namely the property of "being formed according to a law" as opposed to "being formed by a random choice of the elements". For, in the ordinals, there is certainly no element of randomness and hence neither in sets defined in terms of them. This is particularly clear if you consider von Neumann's definition of ordinals because it is not based on any well ordering relations of sets which may very well involve some random element. Of course, you will have noticed that in both examples I gave, the concepts arrived at or envisaged were not absolute in the strictest sense, but only with respect to a certain system of things, namely the sets as described in axiomatic set theory, i.e. although there exist proofs and definitions not falling under these concepts, these definitions and proofs are to give nothing new within the domain of sets and of propositions expressible in terms of "set", "ε" and the logical constants. The question whether the two epistemological concepts considered or any other can be treated in a completely absolute way is of an entirely different nature. But, irrespective of whether this concept of definability corresponds to certain intuitive notions, I think it has some intrinsic mathematical interest and in particular there are questions arising in connection with it: (1) whether the sets definable in this sense satisfy the axioms of set theory. I think this question is to be answered in the affirmative and so will lead to another and probably simpler proof for the consistency of the axiom of choice. (2) It can be proved that the ordinals necessary to define all sets of integers which can be at all defined in this way will have an upper limit. I doubt that it will be possible to prove that this upper limit is ω_1 as in the case of the constructible sets.

AN UNSOLVABLE PROBLEM OF ELEMENTARY NUMBER THEORY

This paper is principally important for its explicit statement (since known as Church's thesis) that the functions which can be computed by a finite algorithm are precisely the recursive functions, and for the consequence that an explicit unsolvable problem can be given. Cf. also Church's abstract, Bulletin of the American Mathematical Society, vol. 41 (1935) p. 333.

Alonzo Church

AN UNSOLVABLE PROBLEM OF ELEMENTARY NUMBER THEORY.[1]

Reprinted from THE AMERICAN JOURNAL OF MATHEMATICS, vol. 58, pp. 345-363 (1936) with the kind permission of The Johns Hopkins Press.

1. Introduction. There is a class of problems of elementary number theory which can be stated in the form that it is required to find an effectively calculable function f of n positive integers, such that $f(x_1, x_2, \cdots, x_n) = 2$[2] is a necessary and sufficient condition for the truth of a certain proposition of elementary number theory involving $x_1, x_2, \cdots, x_n$ as free variables.

An example of such a problem is the problem to find a means of determining of any given positive integer n whether or not there exist positive integers x, y, z, such that $x^n + y^n = z^n$. For this may be interpreted, required to find an effectively calculable function f, such that $f(n)$ is equal to 2 if and only if there exist positive integers x, y, z, such that $x^n + y^n = z^n$. Clearly the condition that the function f be effectively calculable is an essential part of the problem, since without it the problem becomes trivial.

Another example of a problem of this class is, for instance, the problem of topology, to find a complete set of effectively calculable invariants of closed three-dimensional simplicial manifolds under homeomorphisms. This problem can be interpreted as a problem of elementary number theory in view of the fact that topological complexes are representable by matrices of incidence. In fact, as is well known, the property of a set of incidence matrices that it represent a closed three-dimensional manifold, and the property of two sets of incidence matrices that they represent homeomorphic complexes, can both be described in purely number-theoretic terms. If we enumerate, in a straightforward way, the sets of incidence matrices which represent closed three-dimensional manifolds, it will then be immediately provable that the problem under consideration (to find a complete set of effectively calculable invariants of closed three-dimensional manifolds) is equivalent to the problem, to find an effectively calculable function f of positive integers, such that $f(m, n)$ is equal to 2 if and only if the m-th set of incidence matrices and the n-th set of incidence matrices in the enumeration represent homeomorphic complexes.

Other examples will readily occur to the reader.

---

[1] Presented to the American Mathematical Society, April 19, 1935.

[2] The selection of the particular positive integer 2 instead of some other is, of course, accidental and non-essential.

The purpose of the present paper is to propose a definition of effective calculability [3] which is thought to correspond satisfactorily to the somewhat vague intuitive notion in terms of which problems of this class are often stated, and to show, by means of an example, that not every problem of this class is solvable.

2. Conversion and λ-definability. We select a particular list of symbols, consisting of the symbols { , }, (,), λ, [,], and an enumerably infinite set of symbols *a, b, c,* · · · to be called *variables.* And we define the word *formula* to mean any finite sequence of symbols out of this list. The terms *well-formed formula, free variable,* and *bound variable* are then defined by induction as follows. A variable $\boldsymbol{x}$ standing alone is a well-formed formula and the occurrence of $\boldsymbol{x}$ in it is an occurrence of $\boldsymbol{x}$ as a free variable in it; if the formulas $\boldsymbol{F}$ and $\boldsymbol{X}$ are well-formed, $\{\boldsymbol{F}\}(\boldsymbol{X})$ is well-formed, and an occurrence of $\boldsymbol{x}$ as a free (bound) variable in $\boldsymbol{F}$ or $\boldsymbol{X}$ is an occurrence of $\boldsymbol{x}$ as a free (bound) variable in $\{\boldsymbol{F}\}(\boldsymbol{X})$; if the formula $\boldsymbol{M}$ is well-formed and contains an occurrence of $\boldsymbol{x}$ as a free variable in $\boldsymbol{M}$, then $\lambda\boldsymbol{x}[\boldsymbol{M}]$ is well-formed, any occurrence of $\boldsymbol{x}$ in $\lambda\boldsymbol{x}[\boldsymbol{M}]$ is an occurrence of $\boldsymbol{x}$ as a bound variable in $\lambda\boldsymbol{x}[\boldsymbol{M}]$, and an occurrence of a variable $\boldsymbol{y}$, other than $\boldsymbol{x}$, as a free (bound) variable in $\boldsymbol{M}$ is an occurrence of $\boldsymbol{y}$ as a free (bound) variable in $\lambda\boldsymbol{x}[\boldsymbol{M}]$.

---

[3] As will appear, this definition of effective calculability can be stated in either of two equivalent forms, (1) that a function of positive integers shall be called effectively calculable if it is λ-definable in the sense of § 2 below, (2) that a function of positive integers shall be called effectively calculable if it is recursive in the sense of § 4 below. The notion of λ-definability is due jointly to the present author and S. C. Kleene, successive steps towards it having been taken by the present author in the *Annals of Mathematics,* vol. 34 (1933), p. 863, and by Kleene in the *American Journal of Mathematics,* vol. 57 (1935), p. 219. The notion of recursiveness in the sense of § 4 below is due jointly to Jacques Herbrand and Kurt Gödel, as is there explained. And the proof of equivalence of the two notions is due chiefly to Kleene, but also partly to the present author and to J. B. Rosser, as explained below. The proposal to identify these notions with the intuitive notion of effective calculability is first made in the present paper (but see the first footnote to § 7 below).

With the aid of the methods of Kleene (*American Journal of Mathematics,* 1935), the considerations of the present paper could, with comparatively slight modification, be carried through entirely in terms of λ-definability, without making use of the notion of recursiveness. On the other hand, since the results of the present paper were obtained, it has been shown by Kleene (see his forthcoming paper, "General recursive functions of natural numbers") that analogous results can be obtained entirely in terms of recursiveness, without making use of λ-definability. The fact, however, that two such widely different and (in the opinion of the author) equally natural definitions of effective calculability turn out to be equivalent adds to the strength of the reasons adduced below for believing that they constitute as general a characterization of this notion as is consistent with the usual intuitive understanding of it.

We shall use heavy type letters to stand for variable or undetermined formulas. And we adopt the convention that, unless otherwise stated, each heavy type letter shall represent a well-formed formula and each set of symbols standing apart which contains a heavy type letter shall represent a well-formed formula.

When writing particular well-formed formulas, we adopt the following abbreviations. A formula $\{\boldsymbol{F}\}(\boldsymbol{X})$ may be abbreviated as $\boldsymbol{F}(\boldsymbol{X})$ in any case where $\boldsymbol{F}$ is or is represented by a single symbol. A formula $\{\{\boldsymbol{F}\}(\boldsymbol{X})\}(\boldsymbol{Y})$ may be abbreviated as $\{\boldsymbol{F}\}(\boldsymbol{X}, \boldsymbol{Y})$, or, if $\boldsymbol{F}$ is or is represented by a single symbol, as $\boldsymbol{F}(\boldsymbol{X}, \boldsymbol{Y})$. And $\{\{\{\boldsymbol{F}\}(\boldsymbol{X})\}(\boldsymbol{Y})\}(\boldsymbol{Z})$ may be abbreviated as $\{\boldsymbol{F}\}(\boldsymbol{X}, \boldsymbol{Y}, \boldsymbol{Z})$, or as $\boldsymbol{F}(\boldsymbol{X}, \boldsymbol{Y}, \boldsymbol{Z})$, and so on. A formula $\lambda \boldsymbol{x}_1[\lambda \boldsymbol{x}_2[\cdots \lambda \boldsymbol{x}_n[\boldsymbol{M}]\cdots]]$ may be abbreviated as $\lambda \boldsymbol{x}_1 \boldsymbol{x}_2 \cdots \boldsymbol{x}_n \cdot \boldsymbol{M}$ or as $\lambda \boldsymbol{x}_1 \boldsymbol{x}_2 \cdots \boldsymbol{x}_n \boldsymbol{M}$.

We also allow ourselves at any time to introduce abbreviations of the form that a particular symbol α shall stand for a particular sequence of symbols $\boldsymbol{A}$, and indicate the introduction of such an abbreviation by the notation $\alpha \rightarrow \boldsymbol{A}$, to be read, "$\alpha$ stands for $\boldsymbol{A}$."

We introduce at once the following infinite list of abbreviations,

$$1 \rightarrow \lambda ab \cdot a(b),$$
$$2 \rightarrow \lambda ab \cdot a(a(b)),$$
$$3 \rightarrow \lambda ab \cdot a(a(a(b))),$$

and so on, each positive integer in Arabic notation standing for a formula of the form $\lambda ab \cdot a(a(\cdots a(b) \cdots))$.

The expression $\mathrm{S}_{\boldsymbol{N}}^{\boldsymbol{x}} \boldsymbol{M} \,|$ is used to stand for the result of substituting $\boldsymbol{N}$ for $\boldsymbol{x}$ throughout $\boldsymbol{M}$.

We consider the three following operations on well-formed formulas:

I. *To replace any part* $\lambda \boldsymbol{x}[\boldsymbol{M}]$ *of a formula by* $\lambda \boldsymbol{y}[\mathrm{S}_{\boldsymbol{y}}^{\boldsymbol{x}} \boldsymbol{M} \,|\,]$, *where* $\boldsymbol{y}$ *is a variable which does not occur in* $\boldsymbol{M}$.

II. *To replace any part* $\{\lambda \boldsymbol{x}[\boldsymbol{M}]\}(\boldsymbol{N})$ *of a formula by* $\mathrm{S}_{\boldsymbol{N}}^{\boldsymbol{x}} \boldsymbol{M} \,|$, *provided that the bound variables in* $\boldsymbol{M}$ *are distinct both from* $\boldsymbol{x}$ *and from the free variables in* $\boldsymbol{N}$.

III. *To replace any part* $\mathrm{S}_{\boldsymbol{N}}^{\boldsymbol{x}} \boldsymbol{M} \,|$ *(not immediately following* λ*) of a formula by* $\{\lambda \boldsymbol{x}[\boldsymbol{M}]\}(\boldsymbol{N})$, *provided that the bound variables in* $\boldsymbol{M}$ *are distinct both from* $\boldsymbol{x}$ *and from the free variables in* $\boldsymbol{N}$.

Any finite sequence of these operations is called a *conversion*, and if $\boldsymbol{B}$ is obtainable from $\boldsymbol{A}$ by a conversion we say that $\boldsymbol{A}$ is *convertible* into $\boldsymbol{B}$, or, "$\boldsymbol{A}$ conv $\boldsymbol{B}$." If $\boldsymbol{B}$ is identical with $\boldsymbol{A}$ or is obtainable from $\boldsymbol{A}$ by a single application of one of the operations I, II, III, we say that $\boldsymbol{A}$ is *immediately convertible* into $\boldsymbol{B}$.

A conversion which contains exactly one application of Operation II, and no application of Operation III, is called a *reduction*.

A formula is said to be *in normal form* if it is well-formed and contains no part of the form $\{\lambda \boldsymbol{x}[\boldsymbol{M}]\}(\boldsymbol{N})$. And $\boldsymbol{B}$ is said to be a *normal form of* $\boldsymbol{A}$ if $\boldsymbol{B}$ is in normal form and $\boldsymbol{A}$ conv $\boldsymbol{B}$.

The originally given order $a, b, c, \cdots$ of the variables is called their *natural order*. And a formula is said to be *in principal normal form* if it is in normal form, and no variable occurs in it both as a free variable and as a bound variable, and the variables which occur in it immediately following the symbol λ are, when taken in the order in which they occur in the formula, in natural order without repetitions, beginning with a and omitting only such variables as occur in the formula as free variables.[4] The formula $\boldsymbol{B}$ is said to be the *principal normal form of* $\boldsymbol{A}$ if $\boldsymbol{B}$ is in principal normal form and $\boldsymbol{A}$ conv $\boldsymbol{B}$.

Of the three following theorems, proof of the first is immediate, and the second and third have been proved by the present author and J. B. Rosser:[5]

THEOREM I. *If a formula is in normal form, no reduction of it is possible.*

THEOREM II. *If a formula has a normal form, this normal form is unique to within applications of Operation* I, *and any sequence of reductions of the formula must* (*if continued*) *terminate in the normal form.*

THEOREM III. *If a formula has a normal form, every well-formed part of it has a normal form.*

We shall call a function a *function of positive integers* if the range of each independent variable is the class of positive integers and the range of the dependent variable is contained in the class of positive integers. And when it is desired to indicate the number of independent variables we shall speak of a function of one positive integer, a function of two positive integers, and so on. Thus if F is a function of n positive integers, and $a_1, a_2, \cdots, a_n$ are positive integers, then $F(a_1, a_2, \cdots, a_n)$ must be a positive integer.

---

[4] For example, the formulas $\lambda ab \cdot b(a)$ and $\lambda a \cdot a(\lambda c \cdot b(c))$ are in principal normal form, and $\lambda ac \cdot c(a)$, and $\lambda bc \cdot c(b)$, and $\lambda a \cdot a(\lambda a \cdot b(a))$ are in normal form but not in principal normal form. Use of the principal normal form was suggested by S. C. Kleene as a means of avoiding the ambiguity of determination of the normal form of a formula, which is troublesome in certain connections.

Observe that the formulas $1, 2, 3, \cdots$ are all in principal normal form.

[5] Alonzo Church and J. B. Rosser, "Some properties of conversion," forthcoming (abstract in *Bulletin of the American Mathematical Society*, vol. 41, p. 332).

A function F of one positive integer is said to be λ-*definable* if it is possible to find a formula $\boldsymbol{F}$ such that, if $F(m) = r$ and $\boldsymbol{m}$ and $\boldsymbol{r}$ are the formulas for which the positive integers m and r (written in Arabic notation) stand according to our abbreviations introduced above, then $\{\boldsymbol{F}\}(\boldsymbol{m})$ conv $\boldsymbol{r}$.

Similarly, a function F of two positive integers is said to be λ-definable if it is possible to find a formula $\boldsymbol{F}$ such that, whenever $F(m, n) = r$, the formula $\{\boldsymbol{F}\}(\boldsymbol{m}, \boldsymbol{n})$ is convertible into $\boldsymbol{r}$ (m, n, r being positive integers and $\boldsymbol{m}$, $\boldsymbol{n}$, $\boldsymbol{r}$ the corresponding formulas). And so on for functions of three or more positive integers.[6]

It is clear that, in the case of any λ-definable function of positive integers, the process of reduction of formulas to normal form provides an algorithm for the effective calculation of particular values of the function.

3. The Gödel representation of a formula. Adapting to the formal notation just described a device which is due to Gödel,[7] we associate with every formula a positive integer to represent it, as follows. To each of the symbols {, (, [we let correspond the number 11, to each of the symbols },),] the number 13, to the symbol λ the number 1, and to the variables a, b, c, $\cdots$ the prime numbers 17, 19, 23, $\cdots$ respectively. And with a formula which is composed of the n symbols $\tau_1, \tau_2, \cdots, \tau_n$ in order we associate the number $2^{t_1}3^{t_2} \cdots p_n^{t_n}$, where t_i is the number corresponding to the symbol τ_i, and where p_n stands for the n-th prime number.

This number $2^{t_1}3^{t_2} \cdots p_n^{t_n}$ will be called the *Gödel representation* of the formula $\tau_1\tau_2 \cdots \tau_n$.

Two distinct formulas may sometimes have the same Gödel representation, because the numbers 11 and 13 each correspond to three different symbols, but it is readily proved that *no two distinct well-formed formulas can have the same Gödel representation.* It is clear, moreover, that there is an effective method by which, given any formula, its Gödel representation can be calculated; and likewise that there is an effective method by which, given any positive integer, it is possible to determine whether it is the Gödel representation of a well-formed formula and, if it is, to obtain that formula.

In this connection the Gödel representation plays a rôle similar to that

[6] Cf. S. C. Kleene, "A theory of positive integers in formal logic," *American Journal of Mathematics*, vol. 57 (1935), pp. 153-173 and 219-244, where the λ-definability of a number of familiar functions of positive integers, and of a number of important general classes of functions, is established. Kleene uses the term *definable,* or *formally definable,* in the sense in which we are here using λ-*definable*.

[7] Kurt Gödel, "Über formal unentscheidbare Sätze der Principia Mathematica und verwandter Systeme I," *Monatshefte für Mathematik und Physik*, vol. 38 (1931), pp. 173-198.[a]

of the matrix of incidence in combinatorial topology (cf. § 1 above). For there is, in the theory of well-formed formulas, an important class of problems, each of which is equivalent to a problem of elementary number theory obtainable by means of the Gödel representation.[8]

4. Recursive functions. We define a class of expressions, which we shall call *elementary expressions,* and which involve, besides parentheses and commas, the symbols 1, S, an infinite set of numerical variables $x, y, z, \cdots$, and, for each positive integer n, an infinite set $f_n, g_n, h_n, \cdots$ of functional variables with subscript n. This definition is by induction as follows. The symbol 1 or any numerical variable, standing alone, is an elementary expression. If A is an elementary expression, then $S(A)$ is an elementary expression. If $A_1, A_2, \cdots, A_n$ are elementary expressions and f_n is any functional variable with subscript n, then $f_n(A_1, A_2, \cdots, A_n)$ is an elementary expression.

The particular elementary expressions 1, $S(1)$, $S(S(1)), \cdots$ are called *numerals.* And the positive integers $1, 2, 3, \cdots$ are said to correspond to the numerals 1, $S(1)$, $S(S(1)), \cdots$.

An expression of the form $A = B$, where A and B are elementary expressions, is called an *elementary equation.*

The *derived equations* of a set E of elementary equations are defined by induction as follows. The equations of E themselves are derived equations. If $A = B$ is a derived equation containing a numerical variable x, then the result of substituting a particular numeral for all the occurrences of x in $A = B$ is a derived equation. If $A = B$ is a derived equation containing an elementary expression C (as part of either A or B), and if either $C = D$ or $D = C$ is a derived equation, then the result of substituting D for a particular occurrence of C in $A = B$ is a derived equation.

Suppose that no derived equation of a certain finite set E of elementary equations has the form $k = l$ where k and l are different numerals, that the functional variables which occur in E are $f_{n_1}{}^1, f_{n_2}{}^2, \cdots, f_{n_r}{}^r$ with subscripts $n_1, n_2, \cdots, n_r$ respectively, and that, for every value of i from 1 to r inclusive, and for every set of numerals $k_1{}^i, k_2{}^i, \cdots, k_{n_i}{}^i$, there exists a unique numeral k^i such that $f_{n_i}{}^i(k_1{}^i, k_2{}^i, \cdots, k_{n_i}{}^i) = k^i$ is a derived equation of E. And let $F^1, F^2, \cdots, F^r$ be the functions of positive integers defined by the con-

[8] This is merely a special case of the now familiar remark that, in view of the Gödel representation and the ideas associated with it, symbolic logic in general can be regarded, mathematically, as a branch of elementary number theory. This remark is essentially due to Hilbert (cf. for example, *Verhandlungen des dritten internationalen Mathematiker-Kongresses in Heidelberg,* 1904, p. 185; also Paul Bernays in *Die Naturwissenschaften,* vol. 10 (1922), pp. 97 and 98) but is most clearly formulated in terms of the Gödel representation.

dition that, in all cases, $F^i(m_1{}^i, m_2{}^i, \cdots, m_{n_i}{}^i)$ shall be equal to m^i, where $m_1{}^i, m_2{}^i, \cdots, m_{n_i}{}^i$, and m^i are the positive integers which correspond to the numerals $k_1{}^i, k_2{}^i, \cdots, k_{n_i}{}^i$, and k^i respectively. Then the set of equations E is said to *define*, or to be a set of *recursion equations* for, any one of the functions F^i, and the functional variable $f_{n_i}{}^i$ is said to *denote* the function F^i.

A function of positive integers for which a set of recursion equations can be given is said to be *recursive*.[9]

It is clear that for any recursive function of positive integers there exists an algorithm using which any required particular value of the function can be effectively calculated. For the derived equations of the set of recursion equations E are effectively enumerable, and the algorithm for the calculation of particular values of a function F^i, denoted by a functional variable $f_{n_i}{}^i$, consists in carrying out the enumeration of the derived equations of E until the required particular equation of the form $f_{n_i}{}^i(k_1{}^i, k_2{}^i, \cdots, k_{n_i}{}^i) = k^i$ is found.[10]

We call an infinite sequence of positive integers recursive if the function F such that $F(n)$ is the n-th term of the sequence is recursive.

We call a propositional function of positive integers recursive if the function whose value is 2 or 1, according to whether the propositional function is true or false, is recursive. By a recursive property of positive integers we shall mean a recursive propositional function of one positive integer, and by a recursive relation between positive integers we shall mean a recursive propositional function of two or more positive integers.

---

[9] This definition is closely related to, and was suggested by, a definition of recursive functions which was proposed by Kurt Gödel, in lectures at Princeton, N. J., 1934,[a] and credited by him in part to an unpublished suggestion of Jacques Herbrand. The principal features in which the present definition of recursiveness differs from Gödel's are due to S. C. Kleene.

In a forthcoming paper by Kleene to be entitled, "General recursive functions of natural numbers,"[a] (abstract in *Bulletin of the American Mathematical Society*, vol. 41), several definitions of recursiveness will be discussed and equivalences among them obtained. In particular, it follows readily from Kleene's results in that paper that every function recursive in the present sense is also recursive in the sense of Gödel (1934) and conversely.

[10] The reader may object that this algorithm cannot be held to provide an effective calculation of the required particular value of Fi unless the proof is constructive that the required equation $f_{n_i}{}^i(k_1{}^i, k_2{}^i, \cdots, k_{n_i}{}^i) = k^i$ will ultimately be found. But if so this merely means that he should take the existential quantifier which appears in our definition of a set of recursion equations in a constructive sense. What the criterion of constructiveness shall be is left to the reader.

The same remark applies in connection with the existence of an algorithm for calculating the values of a λ-definable function of positive integers.

A function F, for which the range of the dependent variable is contained in the class of positive integers and the range of the independent variable, or of each independent variable, is a subset (not necessarily the whole) of the class of positive integers, will be called *potentially recursive*, if it is possible to find a recursive function F' of positive integers (for which the range of the independent variable, or of each independent variable, is the whole of the class of positive integers), such that the value of F' agrees with the value of F in all cases where the latter is defined.

By an *operation on* well-formed formulas we shall mean a function for which the range of the dependent variable is contained in the class of well-formed formulas and the range of the independent variable, or of each independent variable, is the whole class of well-formed formulas. And we call such an operation recursive if the corresponding function obtained by replacing all formulas by their Gödel representations is potentially recursive.

Similarly any function for which the range of the dependent variable is contained either in the class of positive integers or in the class of well-formed formulas, and for which the range of each independent variable is identical either with the class of positive integers or with the class of well-formed formulas (allowing the case that some of the ranges are identical with one class and some with the other), will be said to be recursive if the corresponding function obtained by replacing all formulas by their Gödel representations is potentially recursive. We call an infinite sequence of well-formed formulas recursive if the corresponding infinite sequence of Gödel representations is recursive. And we call a property of, or relation between, well-formed formulas recursive if the corresponding property of, or relation between, their Gödel representations is potentially recursive. A set of well-formed formulas is said to be recursively enumerable if there exists a recursive infinite sequence which consists entirely of formulas of the set and contains every formula of the set at least once.[11]

In terms of the notion of recursiveness we may also define a *proposition of elementary number theory*, by induction as follows. If ϕ is a recursive propositional function of n positive integers (defined by giving a particular set of recursion equations for the corresponding function whose values are 2 and 1) and if $x_1, x_2, \cdots, x_n$ are variables which take on positive integers as values, then $\phi(x_1, x_2, \cdots, x_n)$ is a proposition of elementary number theory. If P is a proposition of elementary number theory involving x as a free

[11] It can be shown, in view of Theorem V below, that, if an infinite set of formulas is recursively enumerable in this sense, it is also recursively enumerable in the sense that there exists a recursive infinite sequence which consists entirely of formulas of the set and contains every formula of the set exactly once.

variable, then the result of substituting a particular positive integer for all occurrences of x as a free variable in P is a proposition of elementary number theory, and $(x)P$ and $(\exists x)P$ are propositions of elementary number theory, where (x) and $(\exists x)$ are respectively the universal and existential quantifiers of x over the class of positive integers.

It is then readily seen that the negation of a proposition of elementary number theory or the logical product or the logical sum of two propositions of elementary number theory is equivalent, in a simple way, to another proposition of elementary number theory.

5. Recursiveness of the Kleene $\mathfrak{p}$-function. We prove two theorems which establish the recursiveness of certain functions which are definable in words by means of the phrase, "The least positive integer such that," or, "The n-th positive integer such that."

THEOREM IV. *If F is a recursive function of two positive integers, and if for every positive integer x there exists a positive integer y such that $F(x, y) > 1$, then the function F^*, such that, for every positive integer x, $F^*(x)$ is equal to the least positive integer y for which $F(x, y) > 1$, is recursive.*

For a set of recursion equations for F^* consists of the recursion equations for F together with the equations,

$$
\begin{aligned}
&i_2(1, 2) = 2, && g_2(x, 1) = i_2(f_2(x, 1), 2),\\
&i_2(S(x), 2) = 1, && g_2(x, S(y)) = i_2(f_2(x, S(y)), g_2(x, y)),\\
&i_2(x, 1) = 3, && h_2(S(x), y) = x,\\
&i_2(x, S(S(y))) = 3, && h_2(g_2(x, y), x) = j_2(g_2(x, y), y),\\
&j_2(1, y) = y, && f_1(x) = h_2(1, x),\\
&j_2(S(x), y) = x,
\end{aligned}
$$

where the functional variables f_2 and f_1 denote the functions F and F^* respectively, and 2 and 3 are abbreviations for $S(1)$ and $S(S(1))$ respectively.[12]

THEOREM V. *If F is a recursive function of one positive integer, and if there exist an infinite number of positive integers x for which $F(x) > 1$, then the function F^0, such that, for every positive integer n, $F^0(n)$ is equal to the n-th positive integer x (in order of increasing magnitude) for which $F(x) > 1$, is recursive.*

---

[12] Since this result was obtained, it has been pointed out to the author by S. C. Kleene that it can be proved more simply by using the methods of the latter in *American Journal of Mathematics*, vol. 57 (1935), p. 231 *et seq.* His proof will be given in his forthcoming paper already referred to.

For a set of recursion equations for F^0 consists of the recursion equations for F together with the equations,

$$
\begin{aligned}
&g_2(1, y) = g_2(f_1(S(y)), S(y)),\\
&g_2(S(x), y) = y,\\
&g_1(1) = k,\\
&g_1(S(y)) = g_2(1, g_1(y)),
\end{aligned}
$$

where the functional variables g_1 and f_1 denote the functions F^0 and F respectively, and where k is the numeral to which corresponds the least positive integer x for which $F(x) > 1$.[13]

6. Recursiveness of certain functions of formulas. We list now a number of theorems which will be proved in detail in a forthcoming paper by S. C. Kleene [14] or follow immediately from considerations there given. We omit proofs here, except for brief indications in some instances.

Our statement of the theorems and our notation differ from Kleene's in that we employ the set of positive integers $(1, 2, 3, \cdots)$ in the rôle in which he employs the set of natural numbers $(0, 1, 2, \cdots)$. This difference is, of course, unessential. We have selected what is, from some points of view, the less natural alternative, in order to preserve the convenience and naturalness of the identification of the formula $\lambda ab \cdot a(b)$ with 1 rather than with 0.

THEOREM VI. *The property of a positive integer, that there exists a well-formed formula of which it is the Gödel representation is recursive.*

THEOREM VII. *The set of well-formed formulas is recursively enumerable.*

This follows from Theorems V and VI.

THEOREM VIII. *The function of two variables, whose value, when taken of the well-formed formulas* $\mathbf{F}$ *and* $\mathbf{X}$, *is the formula* $\{\mathbf{F}\}(\mathbf{X})$, *is recursive.*

THEOREM IX. *The function, whose value for each of the positive integers* $1, 2, 3, \cdots$ *is the corresponding formula* $\mathbf{1}, \mathbf{2}, \mathbf{3}, \cdots$, *is recursive.*

THEOREM X. *A function, whose value for each of the formulas* $\mathbf{1}, \mathbf{2}, \mathbf{3}, \cdots$ *is the corresponding positive integer, and whose value for other well-formed formulas is a fixed positive integer, is recursive. Likewise the function, whose value for each of the formulas* $\mathbf{1}, \mathbf{2}, \mathbf{3}, \cdots$ *is the corresponding positive integer*

[13] This proof is due to Kleene.

[14] S. C. Kleene, "λ-definability and recursiveness," forthcoming (abstract in *Bulletin of the American Mathematical Society*, vol. 41). In connection with many of the theorems listed, see also Kurt Gödel, *Monatshefte für Mathematik und Physik*, vol. 38 (1931), p. 181 *et seq.*, observing that every function which is recursive in the sense in which the word is there used by Gödel is also recursive in the present more general sense.

plus one, and whose value for other well-formed formulas is the positive integer 1, *is recursive.*

THEOREM XI. *The relation of immediate convertibility, between well-formed formulas, is recursive.*

THEOREM XII. *It is possible to associate simultaneously with every well-formed formula an enumeration of the formulas obtainable from it by conversion, in such a way that the function of two variables, whose value, when taken of a well-formed formula* ***A*** *and a positive integer n, is the n-th formula in the enumeration of the formulas obtainable from* ***A*** *by conversion, is recursive.*

THEOREM XIII. *The property of a well-formed formula, that it is in principal normal form, is recursive.*

THEOREM XIV. *The set of well-formed formulas which are in principal normal form is recursively enumerable.*

This follows from Theorems V, VII, XIII.

THEOREM XV. *The set of well-formed formulas which have a normal form is recursively enumerable.*[15]

For by Theorems XII and XIV this set can be arranged in an infinite square array which is recursively defined (i. e. defined by a recursive function of two variables). And the familiar process by which this square array is reduced to a single infinite sequence is recursive (i. e. can be expressed by means of recursive functions).

THEOREM XVI. *Every recursive function of positive integers is* λ*-definable.*[16]

THEOREM XVII. *Every* λ*-definable function of positive integers is recursive.*[17]

For functions of one positive integer this follows from Theorems IX, VIII, XII, XIII, IV, X. For functions of more than one positive integer

---

[15] This theorem was first proposed by the present author, with the outline of proof here indicated. Details of its proof are due to Kleene and will be given by him in his forthcoming paper, "λ-definability and recursiveness."

[16] This theorem can be proved as a straightforward application of the methods introduced by Kleene in the *American Journal of Mathematics* (*loc. cit.*). In the form here given it was first obtained by Kleene. The related result had previously been obtained by J. B. Rosser that, if we modify the definition of *well-formed* by omitting the requirement that $\boldsymbol{M}$ contain $\boldsymbol{x}$ as a free variable in order that $\lambda \boldsymbol{x}[\boldsymbol{M}]$ be well-formed, then every recursive function of positive integers is λ-definable in the resulting modified sense.

[17] This result was obtained independently by the present author and S. C. Kleene at about the same time.

it follows by the same method, using a generalization of Theorem IV to functions of more than two positive integers.

7. The notion of effective calculability. We now define the notion, already discussed, of an *effectively calculable* function of positive integers by identifying it with the notion of a recursive function of positive integers[18] (or of a λ-definable function of positive integers). This definition is thought to be justified by the considerations which follow, so far as positive justification can ever be obtained for the selection of a formal definition to correspond to an intuitive notion.

It has already been pointed out that, for every function of positive integers which is effectively calculable in the sense just defined, there exists an algorithm for the calculation of its values.

Conversely it is true, under the same definition of effective calculability, that every function, an algorithm for the calculation of the values of which exists, is effectively calculable. For example, in the case of a function F of one positive integer, an algorithm consists in a method by which, given any positive integer n, a sequence of expressions (in some notation) $E_{n1}, E_{n2}, \cdots, E_{nr_n}$ can be obtained; where E_{n1} is effectively calculable when n is given; where E_{ni} is effectively calculable when n and the expressions E_{nj}, $j < i$, are given; and where, when n and all the expressions E_{ni} up to and including E_{nr_n} are given, the fact that the algorithm has terminated becomes effectively known and the value of $F(n)$ is effectively calculable. Suppose that we set up a system of Gödel representations for the notation employed in the expressions E_{ni}, and that we then further adopt the method of Gödel of representing a finite sequence of expressions $E_{n1}, E_{n2}, \cdots, E_{ni}$ by the single positive integer $2^{e_{n1}}3^{e_{n2}} \cdots p_i^{e_{ni}}$ where $e_{n1}, e_{n2}, \cdots, e_{ni}$ are respectively the Gödel representations of $E_{n1}, E_{n2}, \cdots, E_{ni}$ (in particular representing a vacuous sequence of expressions by the positive integer 1). Then we may define a function G of two positive integers such that, if x represents the finite sequence $E_{n1}, E_{n2}, \cdots, E_{nk}$, then $G(n, x)$ is equal to the Gödel representation of E_{ni}, where $i = k + 1$, or is equal to 10 if $k = r_n$ (that is if the algorithm has terminated with E_{nk}), and in any other case $G(n, x)$ is equal to 1. And we may define a function H of two positive integers, such that the value of $H(n, x)$ is the same as that of $G(n, x)$, except in the case that $G(n, x) = 10$, in which case $H(n, x) = F(n)$. If the interpretation is allowed that the

[18] The question of the relationship betwen effective calculability and recursiveness (which it is here proposed to answer by identifying the two notions) was raised by Gödel in conversation with the author. The corresponding question of the relationship between effective calculability and λ-definability had previously been proposed by the author independently.

requirement of effective calculability which appears in our description of an algorithm means the effective calculability of the functions G and H,[19] and if we take the effective calculability of G and H to mean recursiveness (λ-definability), then the recursiveness (λ-definability) of F follows by a straightforward argument.

Suppose that we are dealing with some particular system of symbolic logic, which contains a symbol, $=$, for equality of positive integers, a symbol $\{\ \}(\ )$ for the application of a function of one positive integer to its argument, and expressions $1, 2, 3, \cdots$ to stand for the positive integers. The theorems of the system consist of a finite, or enumerably infinite, list of expressions, the *formal axioms*, together with all the expressions obtainable from them by a finite succession of applications of operations chosen out of a given finite, or enumerably infinite, list of operations, the *rules of procedure.* If the system is to serve at all the purposes for which a system of symbolic logic is usually intended, it is necessary that each rule of procedure be an effectively calculable operation, that the complete set of rules of procedure (if infinite) be effectively enumerable, that the complete set of formal axioms (if infinite) be effectively enumerable, and that the relation between a positive integer and the expression which stands for it be effectively determinable. Suppose that we interpret this to mean that, in terms of a system of Gödel representations for the expressions of the logic, each rule of procedure must be a recursive operation,[20] the complete set of rules of procedure must be recursively enumerable (in the sense that there exists a recursive function Φ such that $\Phi(n, x)$ is the representation of the result of applying the n-th rule of procedure to the ordered finite set of formulas represented by x), the complete set of formal axioms must be recursively enumerable, and the relation between a positive integer and the expression which stands for it must be recursive.[21] And let us call a function F of one positive integer[22] *calculable within* the logic if there exists an expression f in the logic such that $\{f\}(\mu) = \nu$ is a theorem when and only when $F(m) = n$ is true, μ and ν being the expressions which stand for the positive integers m and n. Then, since the

---

[19] If this interpretation or some similar one is not allowed, it is difficult to see how the notion of an algorithm can be given any exact meaning at all.

[20] As a matter of fact, in known systems of symbolic logic, e. g. in that of *Principia Mathematica*, the stronger statement holds, that the relation of *immediate consequence* (*unmittelbare Folge*) is recursive. Cf. Gödel, *loc. cit.*, p. 185. In any case where the relation of immediate consequence is recursive it is possible to find a set of rules of procedure, equivalent to the original ones, such that each rule is a (one-valued) recursive operation, and the complete set of rules is recursively enumerable.

[21] The author is here indebted to Gödel, who, in his 1934 lectures already referred to, proposed substantially these conditions, but in terms of the more restricted notion

complete set of theorems of the logic is recursively enumerable, it follows by Theorem IV above that every function of one positive integer which is calculable within the logic is also effectively calculable (in the sense of our definition).

Thus it is shown that no more general definition of effective calculability than that proposed above can be obtained by either of two methods which naturally suggest themselves (1) by defining a function to be effectively calculable if there exists an algorithm for the calculation of its values (2) by defining a function F (of one positive integer) to be effectively calculable if, for every positive integer m, there exists a positive integer n such that $F(m) = n$ is a provable theorem.

8. Invariants of conversion. The problem naturally suggests itself to find invariants of that transformation of formulas which we have called conversion. The only effectively calculable invariants at present known are the immediately obvious ones (e. g. the set of free variables contained in a formula). Others of importance very probably exist. But we shall prove (in Theorem XIX) that, under the definition of effective calculability proposed in § 7, *no complete set of effectively calculable invariants of conversion exists* (cf. § 1).

The results of Kleene (*American Journal of Mathematics*, 1935) make it clear that, if the problem of finding a complete set of effectively calculable invariants of conversion were solved, most of the familiar unsolved problems of elementary number theory would thereby also be solved. And from Theorem XVI above it follows further that to find a complete set of effectively calculable invariants of conversion would imply the solution of the Entscheidungsproblem for any system of symbolic logic whatever (subject to the very general restrictions of § 7). In the light of this it is hardly surprising that the problem to find such a set of invariants should be unsolvable.

It is to be remembered, however, that, if we consider only the statement of the problem (and ignore things which can be proved about it by more or less lengthy arguments), it appears to be a problem of the same class as the problems of number theory and topology to which it was compared in § 1, having no striking characteristic by which it can be distinguished from them. The temptation is strong to reason by analogy that other important problems of this class may also be unsolvable.

---

of recursiveness which he had employed in 1931, and using the condition that the relation of immediate consequence be recursive instead of the present conditions on the rules of procedure.

[22] We confine ourselves for convenience to the case of functions of one positive integer. The extension to functions of several positive integers is immediate.

LEMMA. *The problem, to find a recursive function of two formulas* $\boldsymbol{A}$ *and* $\boldsymbol{B}$ *whose value is* 2 *or* 1 *according as* $\boldsymbol{A}$ *conv* $\boldsymbol{B}$ *or not, is equivalent to the problem, to find a recursive function of one formula* $\boldsymbol{C}$ *whose value is* 2 *or* 1 *according as* $\boldsymbol{C}$ *has a normal form or not.*[23]

For, by Theorem X, the formula $\boldsymbol{a}$ (the formula $\boldsymbol{b}$), which stands for the positive integer which is the Gödel representation of the formula $\boldsymbol{A}$ (the formula $\boldsymbol{B}$), can be expressed as a recursive function of the formula $\boldsymbol{A}$ (the formula $\boldsymbol{B}$). Moreover, by Theorems VI and XII, there exists a recursive function F of two positive integers such that, if m is the Gödel representation of a well-formed formula $\boldsymbol{M}$, then $F(m, n)$ is the Gödel representation of the n-th formula in an enumeration of the formulas obtainable from $\boldsymbol{M}$ by conversion. And, by Theorem XVI, F is λ-definable, by a formula $\mathfrak{f}$. If we define,

$$Z_1 \rightarrow \mathfrak{Q}(\lambda x \cdot x(I), I),$$
$$Z_2 \rightarrow \mathfrak{Q}(\lambda xy \cdot S(x) - y, I),$$

where $\mathfrak{Q}$ is the formula defined by Kleene (*American Journal of Mathematics*, vol. 57 (1935), p. 226), then Z_1 and Z_2 λ-define the functions of one positive integer whose values, for a positive integer n, are the n-th terms respectively of the infinite sequences $1, 1, 2, 1, 2, 3, \cdots$ and $1, 2, 1, 3, 2, 1, \cdots$. By Theorem VIII the formula,

$$\{\lambda xy \cdot \mathfrak{p}(\lambda n \cdot \delta(\mathfrak{f}(x, Z_1(n)), \mathfrak{f}(y, Z_2(n))), 1)\}(\boldsymbol{a}, \boldsymbol{b}),$$

where $\mathfrak{p}$ and δ are defined as by Kleene (*loc. cit.*, p. 173 and p. 231), is a recursive function of $\boldsymbol{A}$ and $\boldsymbol{B}$, and this formula has a normal form if and only if $\boldsymbol{A}$ conv $\boldsymbol{B}$.

Again, by Theorem X, the formula $\boldsymbol{c}$, which stands for the positive integer which is the Gödel representation of the formula $\boldsymbol{C}$, can be expressed as a recursive function of the formula $\boldsymbol{C}$. By Theorems VI and XIII, there exists a recursive function G of one positive integer such that $G(m) = 2$ if m is the Gödel representation of a formula in principal normal form, and $G(m) = 1$ in any other case. And, by Theorem XVI, G is λ-definable, by a formula $\mathfrak{g}$. By Theorem VIII the formula,

$$\{\lambda x \cdot \mathfrak{p}(\lambda n \cdot \mathfrak{g}(\mathfrak{f}(x, n), 1, 1))\}(\boldsymbol{c})$$

---

[23] These two problems, in the forms, (1) to find an effective method of determining of any two formulas $\boldsymbol{A}$ and $\boldsymbol{B}$ whether $\boldsymbol{A}$ conv $\boldsymbol{B}$, (2) to find an effective method of determining of any formula $\boldsymbol{C}$ whether it has a normal form, were both proposed by Kleene to the author, in the course of a discussion of the properties of the $\mathfrak{p}$-function, about 1932. Some attempts towards solution of (1) by means of numerical invariants were actually made by Kleene at about that time.

where $\mathfrak{f}$ is the formula $\mathfrak{f}$ used in the preceding paragraph, is a recursive function of $\boldsymbol{C}$, and this formula is convertible into the formula 1 if and only if $\boldsymbol{C}$ has a normal form.

Thus we have proved that a formula $\boldsymbol{C}$ can be found as a recursive function of formulas $\boldsymbol{A}$ and $\boldsymbol{B}$, such that $\boldsymbol{C}$ has a normal form if and only if $\boldsymbol{A}$ conv $\boldsymbol{B}$; and that a formula $\boldsymbol{A}$ can be found as a recursive function of a formula $\boldsymbol{C}$, such that $\boldsymbol{A}$ conv 1 if and only if $\boldsymbol{C}$ has a normal form. From this the lemma follows.

THEOREM XVIII. *There is no recursive function of a formula* $\boldsymbol{C}$, *whose value is* 2 *or* 1 *according as* $\boldsymbol{C}$ *has a normal form or not.*

That is, the property of a well-formed formula, that it has a normal form, is not recursive.

For assume the contrary.

Then there exists a recursive function H of one positive integer such that $H(m) = 2$ if m is the Gödel representation of a formula which has a normal form, and $H(m) = 1$ in any other case. And, by Theorem XVI, H is λ-definable by a formula $\mathfrak{h}$.

By Theorem XV, there exists an enumeration of the well-formed formulas which have a normal form, and a recursive function A of one positive integer such that $A(n)$ is the Gödel representation of the n-th formula in this enumeration. And, by Theorem XVI, A is λ-definable, by a formula $\mathfrak{a}$.

By Theorems VI and VIII, there exists a recursive function B of two positive integers such that, if m and n are Gödel representations of well-formed formulas $\boldsymbol{M}$ and $\boldsymbol{N}$, then $B(m, n)$ is the Gödel representation of $\{\boldsymbol{M}\}(\boldsymbol{N})$. And, by Theorem XVI, B is λ-definable, by a formula $\mathfrak{b}$.

By Theorems VI and X, there exists a recursive function C of one positive integer such that, if m is the Gödel representation of one of the formulas $1, 2, 3, \cdots$, then $C(m)$ is the corresponding positive integer plus one, and in any other case $C(m) = 1$. And, by Theorem XVI, C is λ-definable, by a formula $\mathfrak{c}$.

By Theorem IX there exists a recursive function Z^{-1} of one positive integer, whose value for each of the positive integers $1, 2, 3, \cdots$ is the Gödel representation of the corresponding formula $1, 2, 3, \cdots$. And, by Theorem XVI, Z^{-1} is λ-definable, by a formula $\mathfrak{z}$.

Let $\mathfrak{f}$ and $\mathfrak{g}$ be the formulas $\mathfrak{f}$ and $\mathfrak{g}$ used in the proof of the Lemma. By Kleene 15III Cor. (*loc. cit.*, p. 220), a formula $\mathfrak{d}$ can be found such that,

$$\mathfrak{d}(1) \text{ conv } \lambda x \cdot x(1)$$
$$\mathfrak{d}(2) \text{ conv } \lambda u \cdot \mathfrak{c}(\mathfrak{f}(u, \mathfrak{p}(\lambda m \cdot \mathfrak{g}(\mathfrak{f}(u, m)), 1))).$$

We define,

$$\mathfrak{e} \to \lambda n \cdot \mathfrak{d}(\mathfrak{h}(\mathfrak{b}(\mathfrak{a}(n), \mathfrak{z}(n))), \mathfrak{b}(\mathfrak{a}(n), \mathfrak{z}(n))).$$

Then if $\boldsymbol{n}$ is one of the formulas $1, 2, 3, \cdots$, $\mathfrak{e}(\boldsymbol{n})$ is convertible into one of the formulas $1, 2, 3, \cdots$ in accordance with the following rules: (1) if $\mathfrak{b}(\mathfrak{a}(\boldsymbol{n}), \mathfrak{z}(\boldsymbol{n}))$ conv a formula which stands for the Gödel representation of a formula which has no normal form, $\mathfrak{e}(\boldsymbol{n})$ conv 1, (2) if $\mathfrak{b}(\mathfrak{a}(\boldsymbol{n}), \mathfrak{z}(\boldsymbol{n}))$ conv a formula which stands for the Gödel representation of a formula which has a principal normal form which is not one of the formulas $1, 2, 3, \cdots$, $\mathfrak{e}(\boldsymbol{n})$ conv 1, (3) if $\mathfrak{b}(\mathfrak{a}(\boldsymbol{n}), \mathfrak{z}(\boldsymbol{n}))$ conv a formula which stands for the Gödel representation of a formula which has a principal normal form which is one of the formulas $1, 2, 3, \cdots$, $\mathfrak{e}(\boldsymbol{n})$ conv the next following formula in the list $1, 2, 3, \cdots$.

By Theorem III, since $\mathfrak{e}(1)$ has a normal form, the formula $\mathfrak{e}$ has a normal form. Let $\mathfrak{E}$ be the formula which stands for the Gödel representation of $\mathfrak{e}$. Then, if $\boldsymbol{n}$ is any one of the formulas $1, 2, 3, \cdots$, $\mathfrak{E}$ is not convertible into the formula $\mathfrak{a}(\boldsymbol{n})$, because $\mathfrak{b}(\mathfrak{E}, \mathfrak{z}(\boldsymbol{n}))$ is, by the definition of $\mathfrak{b}$, convertible into the formula which stands for the Gödel representation of $\mathfrak{e}(\boldsymbol{n})$, while $\mathfrak{b}(\mathfrak{a}(\boldsymbol{n}), \mathfrak{z}(\boldsymbol{n}))$ is, by the preceding paragraph, convertible into the formula stands for the Gödel representation of a formula definitely not convertible into $\mathfrak{e}(\boldsymbol{n})$ (Theorem II). But, by our definition of $\mathfrak{a}$, it must be true of one of the formulas $\boldsymbol{n}$ in the list $1, 2, 3, \cdots$ that $\mathfrak{a}(\boldsymbol{n})$ conv $\mathfrak{E}$.

Thus, since our assumption to the contrary has led to a contradiction, the theorem must be true.

In order to present the essential ideas without any attempt at exact statement, the preceding proof may be outlined as follows. We are to deduce a contradiction from the assumption that it is effectively determinable of every well-formed formula whether or not it has a normal form. If this assumption holds, it is effectively determinable of every well-formed formula whether or not it is convertible into one of the formulas $1, 2, 3, \cdots$; for, given a well formed formula $\boldsymbol{R}$, we can first determine whether or not it has a normal form, and if it has we can obtain the principal normal form by enumerating the formulas into which $\boldsymbol{R}$ is convertible (Theorem XII) and picking out the first formula in principal normal form which occurs in the enumeration, and we can then determine whether the principal normal form is one of the formulas $1, 2, 3, \cdots$. Let $\boldsymbol{A}_1, \boldsymbol{A}_2, \boldsymbol{A}_3, \cdots$ be an effective enumeration of the well-formed formulas which have a normal form (Theorem XV). Let E be a function of one positive integer, defined by the rule that, where $\boldsymbol{m}$ and $\boldsymbol{n}$ are the formulas which stand for the positive integers m and n respectively, $E(n) = 1$ if $\{\boldsymbol{A}_n\}(\boldsymbol{n})$ is not convertible into one of the formulas $1, 2, 3, \cdots$, and $E(n) = m + 1$ if $\{\boldsymbol{A}_n\}(\boldsymbol{n})$ conv $\boldsymbol{m}$ and $\boldsymbol{m}$ is one of the formulas $1, 2, 3, \cdots$. The function E is effectively calculable and is therefore λ-definable, by a formula $\mathfrak{e}$. The formula $\mathfrak{e}$ has a normal form, since $\mathfrak{e}(1)$ has a normal form. But $\mathfrak{e}$ is not any one of the formulas $\boldsymbol{A}_1, \boldsymbol{A}_2, \boldsymbol{A}_3, \cdots$, because, for every n, $\mathfrak{e}(\boldsymbol{n})$ is a formula which is not convertible into $\{\boldsymbol{A}_n\}(\boldsymbol{n})$. And this contradicts the property of the enumeration $\boldsymbol{A}_1, \boldsymbol{A}_2, \boldsymbol{A}_3, \cdots$ that it contains all well-formed formulas which have a normal form.

Corollary 1. *The set of well-formed formulas which have no normal form is not recursively enumerable.*[24]

For, to outline the argument, the set of well-formed formulas which have a normal form is recursively enumerable, by Theorem XV. If the set of those which do not have a normal form were aslo recursively enumerable, it would be possible to tell effectively of any well-formed formula whether it had a normal form, by the process of searching through the two enumerations until it was found in one or the other. This, however, is contrary to Theorem XVIII.

This corollary gives us an example of an effectively enumerable set (the set of well-formed formulas) which is divided into two non-overlapping subsets of which one is effectively enumerable and the other not. Indeed, in view of the difficulty of attaching any reasonable meaning to the assertion that a set is enumerable but not effectively enumerable, it may even be permissible to go a step further and say that here is an example of an enumerable set which is divided into two non-overlapping subsets of which one is enumerable and the other non-enumerable.[25]

Corollary 2. *Let a function F of one positive integer be defined by the rule that $F(n)$ shall equal 2 or 1 according as n is or is not the Gödel representation of a formula which has a normal form. Then F (if its definition be admitted as valid at all) is an example of a non-recursive function of positive integers.*[26]

This follows at once from Theorem XVIII.

---

[24] This corollary was proposed by J. B. Rosser.

The outline of proof here given for it is open to the objection, recently called to the author's attention by Paul Bernays, that it ostensibly requires a non-constructive use of the principle of excluded middle. This objection is met by a revision of the proof, the revised proof to consist in taking any recursive enumeration of formulas which have no normal form and showing that this enumeration is not a complete enumeration of such formulas, by constructing a formula $\mathfrak{e}(\boldsymbol{n})$ such that (1) the supposition that $\mathfrak{e}(\boldsymbol{n})$ occurs in the enumeration leads to contradiction (2) the supposition that $\mathfrak{e}(\boldsymbol{n})$ has a normal form leads to contradiction.

[25] Cf. the remarks of the author in *The American Mathematical Monthly*, vol. 41 (1934), pp. 356-361.

[26] Other examples of non-recursive functions have since been obtained by S. C. Kleene in a different connection. See his forthcoming paper, "General recursive functions of natural numbers."

Consider the infinite sequence of positive integers, $F(1), F(2), F(3), \cdots$ It is impossible to specify effectively a method by which, given any n, the n-th term of this sequence could be calculated. But it is also impossible ever to select a particular term of this sequence and prove about that term that its value cannot be calculated (because of the obvious theorem that if this sequence has terms whose values cannot be calculated then the value of each of those terms 1). Therefore it is natural to raise the question whether, in spite of the fact that there is no systematic method of effectively calculating the terms of this sequence, it might not be true of each term individually that there existed a method of calculating its value. To this question perhaps the best answer is that the question itself has no meaning, on the ground that the universal quantifier which it contains is intended to express a mere infinite succession of accidents rather than anything systematic.

There is in consequence some room for doubt whether the assertion that the function F exists can be given a reasonable meaning.

THEOREM XIX. *There is no recursive function of two formulas* ***A*** *and* ***B****, whose value is* 2 *or* 1 *according as* ***A*** *conv* ***B*** *or not.*

This follows at once from Theorem XVIII and the Lemma preceding it.

As a corollary of Theorem XIX, it follows that the Entscheidungsproblem is unsolvable in the case of any system of symbolic logic which is ω-consistent (ω-widerspruchsfrei) in the sense of Gödel (*loc. cit.*, p. 187)[i] and is strong enough to allow certain comparatively simple methods of definition and proof. For in any such system the proposition will be expressible about two positive integers a and b that they are Gödel representations of formulas ***A*** and ***B*** such that ***A*** is immediately convertible into ***B***. Hence, utilizing the fact that a conversion is a finite sequence of immediate conversions, the proposition $\Psi(a, b)$ will be expressible that a and b are Gödel representations of formulas ***A*** and ***B*** such that ***A*** conv ***B***. Moreover if ***A*** conv ***B***, and a and b are the Gödel representations of ***A*** and ***B*** respectively, the proposition $\Psi(a, b)$ will be provable in the system, by a proof which amounts to exhibiting, in terms of Gödel representations, a particular finite sequence of immediate conversions, leading from ***A*** to ***B***; and if ***A*** is not convertible into ***B***, the ω-consistency of the system means that $\Psi(a, b)$ will not be provable. If the Entscheidungsproblem for the system were solved, there would be a means of determining effectively of every proposition $\Psi(a, b)$ whether it was provable, and hence a means of determining effectively of every pair of formulas ***A*** and ***B*** whether ***A*** conv ***B***, contrary to Theorem XIX.

In particular, if the system of *Principia Mathematica* be ω-consistent, its Entscheidungsproblem is unsolvable.

[i] 23, 24

Hilbert characterized the problem of determining whether or not a given formula of his "engere Funktionenkalkül" (otherwise known as: first order functional calculus, predicate calculus, or quantification theory) is valid as the fundamental problem of mathematical logic. This was because it seemed clear to Hilbert that with the solution of this problem, the Entscheidungsproblem, it should be possible at least in principle to settle all mathematical questions in a purely mechanical manner. Hence, given unsolvable problems at all, if Hilbert was correct, then the Entscheidungsproblem itself should be unsolvable.

Church's paper sketches a proof that this is the case. The paper as originally published contained an easily repaired technical error; a correction was published as an additional note in the same volume of the Journal of Symbolic Logic as the original. At the author's suggestion, we have incorporated the content of the "correction" into the original paper. The few connecting phrases it was found necessary to add to the author's own words are included in square brackets.

What the author actually shows is that there is no algorithm which can test a formula to determine whether or not it can be derived from the rules laid down by Hilbert and Ackermann. (cf. the author's footnote 2). The inference that no algorithm is possible for determining validity then depends (as Church points out) on the Gödel completeness theorem according to which validity is in fact equivalent to derivability from these rules. Because the proof (actually the very statement!) of the Gödel completeness theorem is non-constructive, the unsolvability of the Entscheidungsproblem is likewise obtained only non-constructively. Church's own feelings (at least at the time of his writing) on this matter were so strong that he concluded:

"The unsolvability of this second form of the Entscheidungsproblem of the engere Funktionenkalkül cannot, therefore, be regarded as established beyond question." Most mathematicians would probably have greater confidence in non-constructive mathematics than that expressed here.

Turing's "On computable numbers, with an application to the Entscheidungsproblem" paper (this anthology, page 145) also outlines an independent proof of the unsolvability of the Entscheidungsproblem.

Still another proof is an immediate consequence of Theorem X of Gödel's "On formally undecidable propositions of Principia Mathematica and related systems I." (this anthology, page 32) using the result that the problem is unsolvable to determine whether or not a given primitive recursive function (given by a sequence of applications of composition and primitive recursion to initial functions) is identically 0. This last unsolvability result is immediate from Kleene's "General recursive functions of natural numbers" (this anthology, page 251). It may be noted also that this argument gives a direct proof of the unsolvability of what Church called the "second form of the Entscheidungsproblem".

A NOTE ON THE ENTSCHEIDUNGSPROBLEM*,**

Alonzo Church

In a recent paper[1] the author has proposed a definition of the commonly used term "effectively calculable" and has shown on the basis of this definition that the general case of the Entscheidungsproblem is unsolvable in any system of symbolic logic which is adequate to a certain portion of arithmetic and is ω-consistent. The purpose of the present note is to outline an extension of this result to the engere Funktionenkalkül of Hilbert and Ackermann.[2]

In the author's cited paper it is pointed out that there can be associated recursively with every well-formed formula[3] a recursive enumeration of the formulas into which it is convertible.[3] This means the existence of a recursively defined function a of two positive integers such that, if y is the Gödel representation of a well-formed formula Y then $a(x,y)$ is the Gödel representation of the xth formula in the enumeration of the formulas into which Y is convertible.

Consider the system L of symbolic logic which arises from the engere Funktionenkalkül by adding to it: as additional undefined symbols, a symbol 1 for the number 1 (regarded as an individual), a symbol = for the propositional function = (equality of individuals), a symbol s for the arithmetic function $x+1$,

*Reprinted from the Journal of Symbolic Logic vol. 1 no. 1 (1936) and vol. 1 no. 3 (1936), by kind permission of the author, the Journal of Symbolic Logic and the Association for Symbolic Logic Inc.

**Received April 15, 1936. Correction received August 13, 1936.

1. "An unsolvable problem of elementary number theory." American Journal of Mathematics. vol. 58 (1936).[a]

2. Grundzüge der Theoretischen Logik. Berlin, 1928.

3. Definitions of the terms "well-formed formula" and "convertible" are given in the cited paper.

a symbol a for the arithmetic function a described in the preceding paragraph, and symbols $b_1, b_2, \ldots, b_k$ for the auxiliary arithmetic functions which are employed in the recursive definition of a; and as additional axioms, [the expressions obtained from] the recursion equations for the functions $a, b_1, b_2, \ldots, b_k$ (expressed with free individual variables, the class of individuals being taken as identical with the class of positive integers), by quantifying all the individual variables by means of universal quantifiers initially placed, [and the] axioms of equality:

$$(x)[x = x]$$

$$(x)(y)(z)[x=y \rightarrow [x=z \rightarrow y=z]],$$

$$(x)(y)[x=y \rightarrow s(x)=s(y)],$$

$$(x)(y)(z)[x=y \rightarrow a(x,z)=a(y,z)],$$

$$(x)(y)(z)[x=y \rightarrow a(z,x)=a(z,y)],$$

and similar axioms for each of the functions $b_1, b_2, \ldots, b_k$.[4]

4. The proof in the April, 1936 version of this paper contained an error in that the axioms for equality of L were stated with free (individual and functional) variables.

When a formula containing free variables is asserted, these free variables may be thought of as having been bound by suppressed universal quantifiers. And on combining several such formulas (or negating such a formula) it may be necessary to restore the suppressed quantifiers in order to avoid confusions of scope. Thus, if U contains free variables, the proposition meant when U→R is asserted is not an implication between the proposition meant when U is asserted and that meant when R is asserted (this observation, and the consequent technique of restoring suppressed quantifiers in such cases, are, of course, a familiar matter to users of the functional calculus). It is in this, or, more strictly, in formal matters which parallel it, that the error lies in the present instance.

The author is indebted to Paul Bernays for pointing out this error and suggesting the method of correcting it, as also for calling attention to the desirability of distinguishing in this connection (as is done below) between proofs which are constructive (finite) and those which are not.

In the presence of the axioms and rules of the engere Funktionenkalkül, the above axioms suffice for the derivation of any expression which may be obtained from $x=y \rightarrow [F(x) \rightarrow F(y)$ by replacing F by a propositional function, which is expressible in the notation of L, and which does not involve free propositional function variables. Cf. Hilbert and Bernays, Grundlagen der Mathematik, Berlin, 1934, vol. 1 pp. 373-375.

The consistency of the system L follows by the methods of existing proofs.[5] The ω-consistency of L is a matter of more difficulty, but for our present purpose the following weaker property of L is sufficient:

> If P contains no quantifiers and (Ex)P is provable in L then some one of $P_1, P_2, P_3, \ldots$ is provable in L (where $P_1, P_2, P_3, \ldots$ are respectively the results of substituting for x the symbols for 1, 2, 3, ... throughout P).

This property has been proved by Paul Bernays[6] for any one of a class of systems of which L is one. Hence, by the argument of the author's cited paper, follows:

> The general case of the Entscheidungsproblem[7] of the system L is unsolvable.

5. Cf. Wilhelm Ackermann, "Begrundung des 'tertium non datur' mittels der Hilbertschen Theorie der Widerspruchsfreiheit", Mathematische Annalen. vol. 93 (1924-5) pp. 1-136; J. v. Neumann, "Zur Hilbertschen Beweistheorie", Mathematische Zeitschrift, vol. 26 (1927) pp. 1-46; Jacques Herbrand, "Sur la non-contradiction de l'arithmétique". Journal für die reine und angewandte Mathematik, vol. 166 (1931-2) pp. 1-8.

6. In lectures at Princeton, N. J., 1936. The methods employed are those of existing consistency proofs. Bernays's proof is adequate to establish the property constructively in this positive form; see the mimeographed notes of his lectures at The Institute for Advanced Study, p. 122.

[In the original (April, 1936) version of this paper] the property of L proved by Bernays [was stated] as follows:

> If P contains no quantifiers and (Ex)P is provable in L then not all of $P_1, P_2, P_3, \ldots$ are provable in L.

[The present restatement is given] in order to avoid all question over the inference from "not all" to "some not". Similarly the condition of ω-consistency, on the last page of "An unsolvable problem of elementary number theory", (American Journal of Mathematics, vol. 58 (1936) pp. 345-363)[a] should be replaced by the condition: Where E is the existential quantifier over the class of positive integers, if (Ex)P is provable then of some one of the formulas $P_1, P_2, P_3, \ldots$ it is true that the negation is not provable. This condition we may call "strong ω-consistency".

7. The Entscheidungsproblem of a system of symbolic logic is here understood

Now by a device which is well known, it is possible to replace the system L by an equivalent system L′ which contains no symbols for arithmetic functions. This is done by replacing $s, a, b_1, b_2, \ldots, b_k$ by the symbols $S, A, B_1, B_2, \ldots, B_k$ for the propositional functions $x = s(y)$, $x = a(y, z)$, etc., and making corresponding alterations in the formal axioms of L.

In particular, [L′ should include, as additional axioms,]

$$(y)(Ex)S(x, y),$$
$$(x)(y)(z)[S(x, z) \& S(y, z) \rightarrow x = y],$$

and similar axioms for each of the propositional functions A, $B_1, B_2, \ldots, B_k$; [so the axioms of equality of L′ may be taken to be]:

$$(x)[x = x],$$
$$(x)(y)(z)[x = y \rightarrow [x = z \rightarrow y = z]],$$
$$(x)(y)(z)[x = y \rightarrow [S(x, z) \rightarrow S(y, z)]],$$
$$(x)(y)(z)[x = y \rightarrow [S(z, x) \rightarrow S(z, y)]],$$
$$(x)(y)(z)(t)[x = y \rightarrow [A(x, z, t) \rightarrow A(y, z, t)]],$$
$$(x)(y)(z)(t)[x = y \rightarrow [A(z, x, t) \rightarrow A(z, y, t)]],$$
$$(x)(y)(z)(t)[x = y \rightarrow [A(z, t, x) \rightarrow A(z, t, y)]],$$

and similar axioms for each of the propositional functions B_1, $B_2, \ldots, B_k$.

[in the sense of what is called the deducibility problem below, that is, as] the problem to find an effective method by which, given any expression Q in the notation of the system, it can be determined whether or not Q is provable in the system. Hilbert and Ackermann (loc. cit.) understand the Entscheidungsproblem of the engere Funktionenkalkül in a slightly different sense. But the two senses are equivalent in view of the proof by Kurt Gödel of the completeness of the engere Funktionenkalkül (Monatshefte für Mathematik und Physik, vol. 37 (1930), pp. 349-360).

The system L′ differs from the engere Funktionenkalkül by the additional undefined terms $1, =, S, A, B_1, B_2, \ldots, B_k$ and a number of formal expressions [containing no free variables] introduced as additional axioms. Let T be the logical product of these additional axioms, let z be an individual variable which does not occur in any of the formal axioms of L′, and let $G_1, G_2, \ldots, G_{k+3}$ be propositional function variables which do not occur in any of the formal axioms of L′. Let U be the result of substituting throughout T the symbols $z, G_1, G_2, \ldots, G_{k+3}$ for the symbols $1, =, S, A, B_1, B_2, \ldots, B_k$ respectively.

Let Q be a formal expression in the notation of L′. We may suppose without loss of generality that Q contains none of the variables $z, G_1, G_2, \ldots, G_{k+3}$. Let R be the result of substituting throughout Q the symbols $z, G_1, G_2, \ldots, G_{k+3}$ for the symbols $1, =, S, A, B_1, B_2, \ldots, B_k$ respectively. Then Q is provable in L′ if and only if $U \rightarrow R$ is provable in the engere Funktionenkalkül.

Thus a solution of the general case of the Entscheidungsproblem of the engere Funktionenkalkül would lead to a solution of the general case of the Entscheidungsproblem of L′ and hence of L. Therefore:

The general case of the Entscheidungsproblem of the engere Funktionenkalkül is unsolvable.[8]

[This proof] is thought to be correct by accepted mathematical standards.

8. From this follows further the unsolvability of the particular case of the Entscheidungsproblem of the engere Funktionenkalkül which concerns the provability of expressions of the form $(Ex_1)(Ex_2)(Ex_3)(y_1)(y_2)\ldots(y_n)P$, where P contains no quantifiers and no individual variables except $x_1, x_2, x_3, y_1, y_2, \ldots, y_n$. Cf. Kurt Gödel, "Zum Entscheidungsproblem des logischen Funktionenkalküls". Monatshefte für Mathematik und Physik, vol. 40 (1933), pp. 433-443.

It is desirable, however, to distinguish between constructive and non-constructive proofs, and then the argument given [is seen to provide] a constructive proof of the unsolvability of what we may call the deducibility problem of the engere Funktionenkalkül, that is the problem to find an effective procedure which is capable of determining, about any given expression in the notation of the engere Funktionenkalkül, whether it is deducible in that system. The inference, however, to the unsolvability of the other form of the Entscheidungsproblem, which concerns a procedure for determining universal validity, depends on the non-constructively proved theorem of Gödel that every universally valid expression is deducible in the engere Funktionenkalkül, as well as on the assumption of the converse of this, that every deducible expression is universally valid. The unsolvability of this second form of the Entscheidungsproblem of the engere Funktionenkalkül cannot, therefore, be regarded as established beyond question.

For the system L′, however, the unsolvability of both forms of the Entscheidungsproblem follows constructively.

ON COMPUTABLE NUMBERS, WITH AN APPLICATION TO THE ENTSCHEIDUNGSPROBLEM

This is a brilliant paper, but the reader should be warned that many of the technical details are incorrect as given. A careful critique is given in a special appendix to a paper of Emil Post's, this anthology, pp. 299-303. In any case, it may well be found most instructive to read this paper for its general sweep, ignoring the petty technical details. (In an up-to-date treatment these would be handled quite differently anyhow.)

ON COMPUTABLE NUMBERS, WITH AN APPLICATION TO THE ENTSCHEIDUNGSPROBLEM

By A. M. TURING.

Reprinted with the kind permission of the London Mathematical Society from the Proceedings of the London Mathematical Society, ser. 2, vol. 42 (1936-7), pp. 230-265; corrections, Ibid, vol 43 (1937) pp. 544-546.

The "computable" numbers may be described briefly as the real numbers whose expressions as a decimal are calculable by finite means. Although the subject of this paper is ostensibly the computable *numbers*, it is almost equally easy to define and investigate computable functions of an integral variable or a real or computable variable, computable predicates, and so forth. The fundamental problems involved are, however, the same in each case, and I have chosen the computable numbers for explicit treatment as involving the least cumbrous technique. I hope shortly to give an account of the relations of the computable numbers, functions, and so forth to one another. This will include a development of the theory of functions of a real variable expressed in terms of computable numbers. According to my definition, a number is computable if its decimal can be written down by a machine.

In §§ 9, 10 I give some arguments with the intention of showing that the computable numbers include all numbers which could naturally be regarded as computable. In particular, I show that certain large classes of numbers are computable. They include, for instance, the real parts of all algebraic numbers, the real parts of the zeros of the Bessel functions, the numbers π, e, etc. The computable numbers do not, however, include all definable numbers, and an example is given of a definable number which is not computable.

Although the class of computable numbers is so great, and in many ways similar to the class of real numbers, it is nevertheless enumerable. In § 8 I examine certain arguments which would seem to prove the contrary. By the correct application of one of these arguments, conclusions are reached which are superficially similar to those of Gödel†. These results

† Gödel, "Über formal unentscheidbare Sätze der Principia Mathematica und verwant der Systeme, I", *Monatshefte Math. Phys.*, 38 (1931), 173–198.[a]

have valuable applications. In particular, it is shown (§ 11) that the Hilbertian Entscheidungsproblem can have no solution.

In a recent paper Alonzo Church† has introduced an idea of "effective calculability", which is equivalent to my "computability", but is very differently defined. Church also reaches similar conclusions about the Entscheidungsproblem‡. The proof of equivalence between "computability" and "effective calculability" is outlined in an appendix to the present paper.

1. *Computing machines.*

We have said that the computable numbers are those whose decimals are calculable by finite means. This requires rather more explicit definition. No real attempt will be made to justify the definitions given until we reach § 9. For the present I shall only say that the justification lies in the fact that the human memory is necessarily limited.

We may compare a man in the process of computing a real number to a machine which is only capable of a finite number of conditions $q_1, q_2, \ldots, q_R$ which will be called "m-configurations". The machine is supplied with a "tape" (the analogue of paper) running through it, and divided into sections (called "squares") each capable of bearing a "symbol". At any moment there is just one square, say the r-th, bearing the symbol $\mathfrak{S}(r)$ which is "in the machine". We may call this square the "scanned square". The symbol on the scanned square may be called the "scanned symbol". The "scanned symbol" is the only one of which the machine is, so to speak, "directly aware". However, by altering its m-configuration the machine can effectively remember some of the symbols which it has "seen" (scanned) previously. The possible behaviour of the machine at any moment is determined by the m-configuration q_n and the scanned symbol $\mathfrak{S}(r)$. This pair q_n, $\mathfrak{S}(r)$ will be called the "configuration": thus the configuration determines the possible behaviour of the machine. In some of the configurations in which the scanned square is blank (*i.e.* bears no symbol) the machine writes down a new symbol on the scanned square: in other configurations it erases the scanned symbol. The machine may also change the square which is being scanned, but only by shifting it one place to right or left. In addition to any of these operations the m-configuration may be changed. Some of the symbols written down

---

† Alonzo Church, "An unsolvable problem of elementary number theory", *American J. of Math.*, 58 (1936), 345–363.[a]

‡ Alonzo Church, "A note on the Entscheidungsproblem", *J. of Symbolic Logic*, 1 (1936), 40–41.[a]

will form the sequence of figures which is the decimal of the real number which is being computed. The others are just rough notes to "assist the memory". It will only be these rough notes which will be liable to erasure.

It is my contention that these operations include all those which are used in the computation of a number. The defence of this contention will be easier when the theory of the machines is familiar to the reader. In the next section I therefore proceed with the development of the theory and assume that it is understood what is meant by "machine", "tape", "scanned", etc.

2. *Definitions.*

Automatic machines.

If at each stage the motion of a machine (in the sense of § 1) is *completely* determined by the configuration, we shall call the machine an "automatic machine" (or *a*-machine).

For some purposes we might use machines (choice machines or *c*-machines) whose motion is only partially determined by the configuration (hence the use of the word "possible" in § 1). When such a machine reaches one of these ambiguous configurations, it cannot go on until some arbitrary choice has been made by an external operator. This would be the case if we were using machines to deal with axiomatic systems. In this paper I deal only with automatic machines, and will therefore often omit the prefix *a*-.

Computing machines.

If an *a*-machine prints two kinds of symbols, of which the first kind (called figures) consists entirely of 0 and 1 (the others being called symbols of the second kind), then the machine will be called a computing machine. If the machine is supplied with a blank tape and set in motion, starting from the correct initial *m*-configuration, the subsequence of the symbols printed by it which are of the first kind will be called the *sequence computed by the machine*. The real number whose expression as a binary decimal is obtained by prefacing this sequence by a decimal point is called the *number computed by the machine*.

At any stage of the motion of the machine, the number of the scanned square, the complete sequence of all symbols on the tape, and the *m*-configuration will be said to describe the *complete configuration* at that stage. The changes of the machine and tape between successive complete configurations will be called the *moves* of the machine.

Circular and circle-free machines.

If a computing machine never writes down more than a finite number of symbols of the first kind, it will be called *circular*. Otherwise it is said to be *circle-free*.

A machine will be circular if it reaches a configuration from which there is no possible move, or if it goes on moving, and possibly printing symbols of the second kind, but cannot print any more symbols of the first kind. The significance of the term "circular" will be explained in § 8.

Computable sequences and numbers.

A sequence is said to be computable if it can be computed by a circle-free machine. A number is computable if it differs by an integer from the number computed by a circle-free machine.

We shall avoid confusion by speaking more often of computable sequences than of computable numbers.

3. *Examples of computing machines.*

I. A machine can be constructed to compute the sequence 010101.... The machine is to have the four m-configurations "𝔟", "𝔠", "𝔣", "𝔢" and is capable of printing "0" and "1". The behaviour of the machine is described in the following table in which "R" means "the machine moves so that it scans the square immediately on the right of the one it was scanning previously". Similarly for "L". "E" means "the scanned symbol is erased" and "P" stands for "prints". This table (and all succeeding tables of the same kind) is to be understood to mean that for a configuration described in the first two columns the operations in the third column are carried out successively, and the machine then goes over into the m-configuration described in the last column. When the second column is left blank, it is understood that the behaviour of the third and fourth columns applies for any symbol and for no symbol. The machine starts in the m-configuration 𝔟 with a blank tape.

Configuration		*Behaviour*	
m-config.	*symbol*	*operations*	*final m-config.*
𝔟	None	$P0, R$	𝔠
𝔠	None	R	𝔢
𝔢	None	$P1, R$	𝔣
𝔣	None	R	𝔟

If (contrary to the description in § 1) we allow the letters L, R to appear more than once in the operations column we can simplify the table considerably.

m-config.	*symbol*	*operations*	*final m-config.*
$\mathfrak{b}$	None	$P0$	$\mathfrak{b}$
	0	$R, R, P1$	$\mathfrak{b}$
	1	$R, R, P0$	$\mathfrak{b}$

II. As a slightly more difficult example we can construct a machine to compute the sequence 001011011101111011111.... The machine is to be capable of five m-configurations, viz. "$\mathfrak{o}$", "$\mathfrak{q}$", "$\mathfrak{p}$", "$\mathfrak{f}$", "$\mathfrak{b}$" and of printing "ə", "x", "0", "1". The first three symbols on the tape will be "əə0"; the other figures follow on alternate squares. On the intermediate squares we never print anything but "x". These letters serve to "keep the place" for us and are erased when we have finished with them. We also arrange that in the sequence of figures on alternate squares there shall be no blanks.

Configuration		*Behaviour*	
m-config.	*symbol*	*operations*	*final m-config.*
$\mathfrak{b}$		Pə, R, Pə, R, $P0$, R, R, $P0$, L, L	$\mathfrak{o}$
$\mathfrak{o}$	1	R, Px, L, L, L	$\mathfrak{o}$
	0		$\mathfrak{q}$
$\mathfrak{q}$	Any (0 or 1)	R, R	$\mathfrak{q}$
	None	$P1, L$	$\mathfrak{p}$
$\mathfrak{p}$	x	E, R	$\mathfrak{q}$
	ə	R	$\mathfrak{f}$
	None	L, L	$\mathfrak{p}$
$\mathfrak{f}$	Any	R, R	$\mathfrak{f}$
	None	$P0, L, L$	$\mathfrak{o}$

To illustrate the working of this machine a table is given below of the first few complete configurations. These complete configurations are described by writing down the sequence of symbols which are on the tape, with the m-configuration written below the scanned symbol. The successive complete configurations are separated by colons.

: ə ə 0   0 : ə ə 0   0 : ə ə 0   0 : ə ə 0   0   : ə ə 0   0   1 :

$\mathfrak{b}$ $\mathfrak{o}$ $\mathfrak{q}$ $\mathfrak{q}$ $\mathfrak{q}$ $\mathfrak{p}$

ə ə 0   0   1 : ə ə 0   0   1 : ə ə 0   0   1 : ə ə 0   0   1 :

$\mathfrak{p}$ $\mathfrak{p}$ $\mathfrak{f}$ $\mathfrak{f}$

ə ə 0   0   1 : ə ə 0   0   1   : ə ə 0   0   1   0 :

$\mathfrak{f}$ $\mathfrak{f}$ $\mathfrak{o}$

ə ə 0   0   1 x 0 : ....

$\mathfrak{o}$

This table could also be written in the form

$\mathfrak{b}$: ə ə $\mathfrak{o}$ 0   0 : ə ə $\mathfrak{q}$ 0   0 : ...,   (C)

in which a space has been made on the left of the scanned symbol and the m-configuration written in this space. This form is less easy to follow, but we shall make use of it later for theoretical purposes.

The convention of writing the figures only on alternate squares is very useful: I shall always make use of it. I shall call the one sequence of alternate squares F-squares and the other sequence E-squares. The symbols on E-squares will be liable to erasure. The symbols on F-squares form a continuous sequence. There are no blanks until the end is reached. There is no need to have more than one E-square between each pair of F-squares: an apparent need of more E-squares can be satisfied by having a sufficiently rich variety of symbols capable of being printed on E-squares. If a symbol β is on an F-square S and a symbol α is on the E-square next on the right of S, then S and β will be said to be *marked* with α. The process of printing this α will be called marking β (or S) with α.

4. *Abbreviated tables.*

There are certain types of process used by nearly all machines, and these, in some machines, are used in many connections. These processes include copying down sequences of symbols, comparing sequences, erasing all symbols of a given form, etc. Where such processes are concerned we can abbreviate the tables for the m-configurations considerably by the use of "skeleton tables". In skeleton tables there appear capital German letters and small Greek letters. These are of the nature of "variables". By replacing each capital German letter throughout by an m-configuration and each small Greek letter by a symbol, we obtain the table for an m-configuration.

The skeleton tables are to be regarded as nothing but abbreviations: they are not essential. So long as the reader understands how to obtain the complete tables from the skeleton tables, there is no need to give any exact definitions in this connection.

Let us consider an example:

m-config.	*Symbol*	*Behaviour*	*Final m-config.*
$\mathfrak{f}(\mathfrak{C}, \mathfrak{B}, \alpha)$	ə	L	$\mathfrak{f}_1(\mathfrak{C}, \mathfrak{B}, \alpha)$
	not ə	L	$\mathfrak{f}(\mathfrak{C}, \mathfrak{B}, \alpha)$
$\mathfrak{f}_1(\mathfrak{C}, \mathfrak{B}, \alpha)$	α		$\mathfrak{C}$
	not α	R	$\mathfrak{f}_1(\mathfrak{C}, \mathfrak{B}, \alpha)$
	None	R	$\mathfrak{f}_2(\mathfrak{C}, \mathfrak{B}, \alpha)$
$\mathfrak{f}_2(\mathfrak{C}, \mathfrak{B}, \alpha)$	α		$\mathfrak{C}$
	not α	R	$\mathfrak{f}_1(\mathfrak{C}, \mathfrak{B}, \alpha)$
	None	R	$\mathfrak{B}$

From the m-configuration $\mathfrak{f}(\mathfrak{C}, \mathfrak{B}, \alpha)$ the machine finds the symbol of form α which is farthest to the left (the "first α") and the m-configuration then becomes $\mathfrak{C}$. If there is no α then the m-configuration becomes $\mathfrak{B}$.

If we were to replace $\mathfrak{C}$ throughout by $\mathfrak{q}$ (say), $\mathfrak{B}$ by $\mathfrak{r}$, and α by x, we should have a complete table for the m-configuration $\mathfrak{f}(\mathfrak{q}, \mathfrak{r}, x)$. $\mathfrak{f}$ is called an "m-configuration function" or "m-function".

The only expressions which are admissible for substitution in an m-function are the m-configurations and symbols of the machine. These have to be enumerated more or less explicitly: they may include expressions such as $\mathfrak{p}(\mathfrak{e}, x)$; indeed they must if there are any m-functions used at all. If we did not insist on this explicit enumeration, but simply stated that the machine had certain m-configurations (enumerated) and all m-configurations obtainable by substitution of m-configurations in certain m-functions, we should usually get an infinity of m-configurations; *e.g.*, we might say that the machine was to have the m-configuration $\mathfrak{q}$ and all m-configurations obtainable by substituting an m-configuration for $\mathfrak{C}$ in $\mathfrak{p}(\mathfrak{C})$. Then it would have $\mathfrak{q}$, $\mathfrak{p}(\mathfrak{q})$, $\mathfrak{p}\big(\mathfrak{p}(\mathfrak{q})\big)$, $\mathfrak{p}\Big(\mathfrak{p}\big(\mathfrak{p}(\mathfrak{q})\big)\Big)$, ... as m-configurations.

Our interpretation rule then is this. We are given the names of the m-configurations of the machine, mostly expressed in terms of m-functions. We are also given skeleton tables. All we want is the complete table for the m-configurations of the machine. This is obtained by repeated substitution in the skeleton tables.

Further examples.

(In the explanations the symbol "$\rightarrow$" is used to signify "the machine goes into the *m*-configuration. . . .")

$\mathfrak{e}(\mathfrak{C}, \mathfrak{B}, \alpha)$			$\mathfrak{f}\left(\mathfrak{e}_1(\mathfrak{C}, \mathfrak{B}, \alpha), \mathfrak{B}, \alpha\right)$	From $\mathfrak{e}(\mathfrak{C}, \mathfrak{B}, \alpha)$ the first α is erased and $\rightarrow \mathfrak{C}$. If there is no $\alpha \rightarrow \mathfrak{B}$.
$\mathfrak{e}_1(\mathfrak{C}, \mathfrak{B}, \alpha)$		E	$\mathfrak{C}$	
$\mathfrak{e}(\mathfrak{B}, \alpha)$			$\mathfrak{e}\left(\mathfrak{e}(\mathfrak{B}, \alpha), \mathfrak{B}, \alpha\right)$	From $\mathfrak{e}(\mathfrak{B}, \alpha)$ all letters α are erased and $\rightarrow \mathfrak{B}$.

The last example seems somewhat more difficult to interpret than most. Let us suppose that in the list of *m*-configurations of some machine there appears $\mathfrak{e}(\mathfrak{b}, x)$ ($= \mathfrak{q}$, say). The table is

	$\mathfrak{e}(\mathfrak{b}, x)$	$\mathfrak{e}\left(\mathfrak{e}(\mathfrak{b}, x), \mathfrak{b}, x\right)$
or	$\mathfrak{q}$	$\mathfrak{e}(\mathfrak{q}, \mathfrak{b}, x)$.

Or, in greater detail:

$\mathfrak{q}$		$\mathfrak{e}(\mathfrak{q}, \mathfrak{b}, x)$
$\mathfrak{e}(\mathfrak{q}, \mathfrak{b}, x)$		$\mathfrak{f}\left(\mathfrak{e}_1(\mathfrak{q}, \mathfrak{b}, x), \mathfrak{b}, x\right)$
$\mathfrak{e}_1(\mathfrak{q}, \mathfrak{b}, x)$	E	$\mathfrak{q}$.

In this we could replace $\mathfrak{e}_1(\mathfrak{q}, \mathfrak{b}, x)$ by $\mathfrak{q}'$ and then give the table for $\mathfrak{f}$ (with the right substitutions) and eventually reach a table in which no *m*-functions appeared.

$\mathfrak{pe}(\mathfrak{C}, \beta)$			$\mathfrak{f}\left(\mathfrak{pe}_1(\mathfrak{C}, \beta), \mathfrak{C}, \text{ə}\right)$	From $\mathfrak{pe}(\mathfrak{C}, \beta)$ the machine prints β at the end of the sequence of symbols and $\rightarrow \mathfrak{C}$.
$\mathfrak{pe}_1(\mathfrak{C}, \beta)$	Any	R, R	$\mathfrak{pe}_1(\mathfrak{C}, \beta)$	
	None	$P\beta$	$\mathfrak{C}$	
$\mathfrak{l}(\mathfrak{C})$		L	$\mathfrak{C}$	From $\mathfrak{f}'(\mathfrak{C}, \mathfrak{B}, \alpha)$ it does the same as for $\mathfrak{f}(\mathfrak{C}, \mathfrak{B}, \alpha)$ but moves to the left before $\rightarrow \mathfrak{C}$.
$\mathfrak{r}(\mathfrak{C})$		R	$\mathfrak{C}$	
$\mathfrak{f}'(\mathfrak{C}, \mathfrak{B}, \alpha)$			$\mathfrak{f}\left(\mathfrak{l}(\mathfrak{C}), \mathfrak{B}, \alpha\right)$	
$\mathfrak{f}''(\mathfrak{C}, \mathfrak{B}, \alpha)$			$\mathfrak{f}\left(\mathfrak{r}(\mathfrak{C}), \mathfrak{B}, \alpha\right)$	
$\mathfrak{c}(\mathfrak{C}, \mathfrak{B}, \alpha)$			$\mathfrak{f}'\left(\mathfrak{c}_1(\mathfrak{C}), \mathfrak{B}, \alpha\right)$	$\mathfrak{c}(\mathfrak{C}, \mathfrak{B}, \alpha)$. The machine writes at the end the first symbol marked α and $\rightarrow \mathfrak{C}$.
$\mathfrak{c}_1(\mathfrak{C})$	β		$\mathfrak{pe}(\mathfrak{C}, \beta)$	

The last line stands for the totality of lines obtainable from it by replacing β by any symbol which may occur on the tape of the machine concerned.

$\mathfrak{ce}(\mathfrak{C}, \mathfrak{B}, \alpha)$		$\mathfrak{c}\big(\mathfrak{e}(\mathfrak{C}, \mathfrak{B}, \alpha), \mathfrak{B}, \alpha\big)$	
$\mathfrak{ce}(\mathfrak{B}, \alpha)$		$\mathfrak{ce}\big(\mathfrak{ce}(\mathfrak{B}, \alpha), \mathfrak{B}, \alpha\big)$	

$\mathfrak{ce}(\mathfrak{B}, \alpha)$. The machine copies down in order at the end all symbols marked α and erases the letters α; $\rightarrow \mathfrak{B}$.

$\mathfrak{re}(\mathfrak{C}, \mathfrak{B}, \alpha, \beta)$			$\mathfrak{f}\big(\mathfrak{re}_1(\mathfrak{C}, \mathfrak{B}, \alpha, \beta), \mathfrak{B}, \alpha\big)$
$\mathfrak{re}_1(\mathfrak{C}, \mathfrak{B}, \alpha, \beta)$		$E, P\beta$	$\mathfrak{C}$
$\mathfrak{re}(\mathfrak{B}, \alpha, \beta)$			$\mathfrak{re}\big(\mathfrak{re}(\mathfrak{B}, \alpha, \beta), \mathfrak{B}, \alpha, \beta\big)$

$\mathfrak{re}(\mathfrak{C}, \mathfrak{B}, \alpha, \beta)$. The machine replaces the first α by β and $\rightarrow \mathfrak{C} \rightarrow \mathfrak{B}$ if there is no α.

$\mathfrak{re}(\mathfrak{B}, \alpha, \beta)$. The machine replaces all letters α by β; $\rightarrow \mathfrak{B}$.

$\mathfrak{cr}(\mathfrak{C}, \mathfrak{B}, \alpha)$			$\mathfrak{c}\big(\mathfrak{re}(\mathfrak{C}, \mathfrak{B}, \alpha, a), \mathfrak{B}, \alpha\big)$
$\mathfrak{cr}(\mathfrak{B}, \alpha)$			$\mathfrak{cr}\big(\mathfrak{cr}(\mathfrak{B}, \alpha), \mathfrak{re}(\mathfrak{B}, a, \alpha), \alpha\big)$

$\mathfrak{cr}(\mathfrak{B}, \alpha)$ differs from $\mathfrak{ce}(\mathfrak{B}, \alpha)$ only in that the letters α are not erased. The m-configuration $\mathfrak{cr}(\mathfrak{B}, \alpha)$ is taken up when no letters "a" are on the tape.

$\mathfrak{cp}(\mathfrak{C}, \mathfrak{A}, \mathfrak{E}, \alpha, \beta)$			$\mathfrak{f}'\big(\mathfrak{cp}_1(\mathfrak{C}, \mathfrak{A}, \beta), \mathfrak{f}(\mathfrak{A}, \mathfrak{E}, \beta), \alpha\big)$
$\mathfrak{cp}_1(\mathfrak{C}, \mathfrak{A}, \beta)$	γ		$\mathfrak{f}'\big(\mathfrak{cp}_2(\mathfrak{C}, \mathfrak{A}, \gamma), \mathfrak{A}, \beta\big)$
$\mathfrak{cp}_2(\mathfrak{C}, \mathfrak{A}, \gamma)$	γ		$\mathfrak{C}$
	not γ		$\mathfrak{A}$.

The first symbol marked α and the first marked β are compared. If there is neither α nor β, $\rightarrow \mathfrak{E}$. If there are both and the symbols are alike, $\rightarrow \mathfrak{C}$. Otherwise $\rightarrow \mathfrak{A}$.

$$\mathfrak{cpe}(\mathfrak{C}, \mathfrak{A}, \mathfrak{E}, \alpha, \beta) \qquad \mathfrak{cp}\Big(\mathfrak{e}\big(\mathfrak{e}(\mathfrak{C}, \mathfrak{C}, \beta), \mathfrak{C}, \alpha\big), \mathfrak{A}, \mathfrak{E}, \alpha, \beta\Big)$$

$\mathfrak{cpe}(\mathfrak{C}, \mathfrak{A}, \mathfrak{E}, \alpha, \beta)$ differs from $\mathfrak{cp}(\mathfrak{C}, \mathfrak{A}, \mathfrak{E}, \alpha, \beta)$ in that in the case when there is similarity the first α and β are erased.

$$\mathfrak{cpe}(\mathfrak{A}, \mathfrak{E}, \alpha, \beta) \qquad \mathfrak{cpe}\big(\mathfrak{cpe}(\mathfrak{A}, \mathfrak{E}, \alpha, \beta), \mathfrak{A}, \mathfrak{E}, \alpha, \beta\big).$$

$\mathfrak{cpe}(\mathfrak{A}, \mathfrak{E}, \alpha, \beta)$. The sequence of symbols marked α is compared with the sequence marked β. $\rightarrow \mathfrak{E}$ if they are similar. Otherwise $\rightarrow \mathfrak{A}$. Some of the symbols α and β are erased.

$\mathfrak{q}(\mathfrak{C})$	Any	R	$\mathfrak{q}(\mathfrak{C})$	$\mathfrak{q}(\mathfrak{C}, \alpha)$. The machine finds the last symbol of form α. $\rightarrow \mathfrak{C}$.
	None	R	$\mathfrak{q}_1(\mathfrak{C})$	
$\mathfrak{q}_1(\mathfrak{C})$	Any	R	$\mathfrak{q}(\mathfrak{C})$	
	None		$\mathfrak{C}$	
$\mathfrak{q}(\mathfrak{C}, \alpha)$			$\mathfrak{q}(\mathfrak{q}_1(\mathfrak{C}, \alpha))$	
$\mathfrak{q}_1(\mathfrak{C}, \alpha)$	α		$\mathfrak{C}$	
	not α	L	$\mathfrak{q}_1(\mathfrak{C}, \alpha)$	
$\mathfrak{pe}_2(\mathfrak{C}, \alpha, \beta)$			$\mathfrak{pe}(\mathfrak{pe}(\mathfrak{C}, \beta), \alpha)$	$\mathfrak{pe}_2(\mathfrak{C}, \alpha, \beta)$. The machine prints $\alpha\ \beta$ at the end.
$\mathfrak{ce}_2(\mathfrak{B}, \alpha, \beta)$			$\mathfrak{ce}(\mathfrak{ce}(\mathfrak{B}, \beta), \alpha)$	$\mathfrak{ce}_3(\mathfrak{B}, \alpha, \beta, \gamma)$. The machine copies down at the end first the symbols marked α, then those marked β, and finally those marked γ; it erases the symbols α, β, γ.
$\mathfrak{ce}_3(\mathfrak{B}, \alpha, \beta, \gamma)$			$\mathfrak{ce}(\mathfrak{ce}_2(\mathfrak{B}, \beta, \gamma), \alpha)$	
$\mathfrak{e}(\mathfrak{C})$	ə	R	$\mathfrak{e}_1(\mathfrak{C})$	From $\mathfrak{e}(\mathfrak{C})$ the marks are erased from all marked symbols. $\rightarrow \mathfrak{C}$.
	Not ə	L	$\mathfrak{e}(\mathfrak{C})$	
$\mathfrak{e}_1(\mathfrak{C})$	Any	R, E, R	$\mathfrak{e}_1(\mathfrak{C})$	
	None		$\mathfrak{C}$	

5. *Enumeration of computable sequences.*

A computable sequence γ is determined by a description of a machine which computes γ. Thus the sequence 001011011101111... is determined by the table on p.120, and, in fact, any computable sequence is capable of being described in terms of such a table.

It will be useful to put these tables into a kind of standard form. In the first place let us suppose that the table is given in the same form as the first table, for example, I on p.119. That is to say, that the entry in the operations column is always of one of the forms $E : E$, $R : E$, $L : Pa : Pa$, $R : Pa$, $L : R : L :$ or no entry at all. The table can always be put into this form by introducing more m-configurations. Now let us give numbers to the m-configurations, calling them $q_1, \ldots, q_R$, as in § 1. The initial m-configuration is always to be called q_1. We also give numbers to the symbols $S_1, \ldots, S_m$

and, in particular, blank $= S_0$, $0 = S_1$, $1 = S_2$. The lines of the table are now of form

m-config.	*Symbol*	*Operations*	*Final m-config.*	
q_i	S_j	PS_k, L	q_m	(N_1)
q_i	S_j	PS_k, R	q_m	(N_2)
q_i	S_j	PS_k	q_m	(N_3)

Lines such as

q_i	S_j	E, R	q_m

are to be written as

q_i	S_j	PS_0, R	q_m

and lines such as

q_i	S_j	R	q_m

to be written as

q_i	S_j	PS_j, R	q_m

In this way we reduce each line of the table to a line of one of the forms (N_1), (N_2), (N_3).

From each line of form (N_1) let us form an expression $q_i S_j S_k L q_m$; from each line of form (N_2) we form an expression $q_i S_j S_k R q_m$; and from each line of form (N_3) we form an expression $q_i S_j S_k N q_m$.

Let us write down all expressions so formed from the table for the machine and separate them by semi-colons. In this way we obtain a complete description of the machine. In this description we shall replace q_i by the letter "*D*" followed by the letter "*A*" repeated *i* times, and S_j by "*D*" followed by "*C*" repeated *j* times. This new description of the machine may be called the *standard description* (S.D). It is made up entirely from the letters "*A*", "*C*", "*D*", "*L*", "*R*", "*N*", and from ";".

If finally we replace "*A*" by "1", "*C*" by "2", "*D*" by "3", "*L*" by "4", "*R*" by "5", "*N*" by "6", and ";" by "7" we shall have a description of the machine in the form of an arabic numeral. The integer represented by this numeral may be called a *description number* (D.N) of the machine. The D.N determine the S.D and the structure of the machine uniquely. The machine whose D.N is n may be described as $\mathcal{M}(n)$.

To each computable sequence there corresponds at least one description number, while to no description number does there correspond more than one computable sequence. The computable sequences and numbers are therefore enumerable.

Let us find a description number for the machine I of § 3. When we rename the m-configurations its table becomes:

q_1	S_0	PS_1, R	q_2
q_2	S_0	PS_0, R	q_3
q_3	S_0	PS_2, R	q_4
q_4	S_0	PS_0, R	q_1

Other tables could be obtained by adding irrelevant lines such as

q_1	S_1	PS_1, R	q_2

Our first standard form would be

$$q_1 S_0 S_1 R q_2;\ q_2 S_0 S_0 R q_3;\ q_3 S_0 S_2 R q_4;\ q_4 S_0 S_0 R q_1;.$$

The standard description is

DADDCRDAA ;DAADDRDAAA ;

DAAADDCCRDAAAA ;DAAAADDRDA ;

A description number is

31332531173113353111731113322531111731111335317

and so is

3133253117311335311173111332253111173111133531731323253117

A number which is a description number of a circle-free machine will be called a *satisfactory* number. In § 8 it is shown that there can be no general process for determining whether a given number is satisfactory or not.

6. *The universal computing machine.*

It is possible to invent a single machine which can be used to compute any computable sequence. If this machine $\mathcal{U}$ is supplied with a tape on the beginning of which is written the S.D of some computing machine $\mathcal{M}$, then $\mathcal{U}$ will compute the same sequence as $\mathcal{M}$. In this section I explain in outline the behaviour of the machine. The next section is devoted to giving the complete table for $\mathcal{U}$.

Let us first suppose that we have a machine $\mathcal{M}'$ which will write down on the F-squares the successive complete configurations of $\mathcal{M}$. These might be expressed in the same form as on p. 121, using the second description, (C), with all symbols on one line. Or, better, we could transform this description (as in § 5) by replacing each m-configuration by "D" followed by "A" repeated the appropriate number of times, and by replacing each symbol by "D" followed by "C" repeated the appropriate number of times. The numbers of letters "A" and "C" are to agree with the numbers chosen in § 5, so that, in particular, "0" is replaced by "DC", "1" by "DCC", and the blanks by "D". These substitutions are to be made after the complete configurations have been put together, as in (C). Difficulties arise if we do the substitution first. In each complete configuration the blanks would all have to be replaced by "D", so that the complete configuration would not be expressed as a finite sequence of symbols.

If in the description of the machine II of § 3 we replace "$\mathfrak{o}$" by "DAA", "ə" by "$DCCC$", "$\mathfrak{q}$" by "$DAAA$", then the sequence (C) becomes:

$$DA:DCCCDCCCDAADCDDC:DCCCDCCCDAAADCDDC:\ldots \quad (C_1)$$

(This is the sequence of symbols on F-squares.)

It is not difficult to see that if $\mathcal{M}$ can be constructed, then so can $\mathcal{M}'$. The manner of operation of $\mathcal{M}'$ could be made to depend on having the rules of operation (*i.e.*, the S.D) of $\mathcal{M}$ written somewhere within itself (*i.e.* within $\mathcal{M}'$); each step could be carried out by referring to these rules. We have only to regard the rules as being capable of being taken out and exchanged for others and we have something very akin to the universal machine.

One thing is lacking: at present the machine $\mathcal{M}'$ prints no figures. We may correct this by printing between each successive pair of complete configurations the figures which appear in the new configuration but not in the old. Then (C_1) becomes

$$DDA:0:0:DCCCDCCCDAADCDDC:DCCC.\ldots \quad (C_2)$$

It is not altogether obvious that the E-squares leave enough room for the necessary "rough work", but this is, in fact, the case.

The sequences of letters between the colons in expressions such as (C_1) may be used as standard descriptions of the complete configurations. When the letters are replaced by figures, as in § 5, we shall have a numerical description of the complete configuration, which may be called its description number.

7. *Detailed description of the universal machine.*

A table is given below of the behaviour of this universal machine. The m-configurations of which the machine is capable are all those occurring in the first and last columns of the table, together with all those which occur when we write out the unabbreviated tables of those which appear in the table in the form of m-functions. *E.g.*, $\mathfrak{e}(\mathfrak{anf})$ appears in the table and is an m-function. Its unabbreviated table is (see p. 125)

$\mathfrak{e}(\mathfrak{anf})$	ə	R	$\mathfrak{e}_1(\mathfrak{anf})$
	not ə	L	$\mathfrak{e}(\mathfrak{anf})$
$\mathfrak{e}_1(\mathfrak{anf})$	Any	R, E, R	$\mathfrak{e}_1(\mathfrak{anf})$
	None		$\mathfrak{anf}$

Consequently $\mathfrak{e}_1(\mathfrak{anf})$ is an m-configuration of $\mathfrak{U}$.

When $\mathfrak{U}$ is ready to start work the tape running through it bears on it the symbol ə on an F-square and again ə on the next E-square; after this, on F-squares only, comes the S.D of the machine followed by a double colon "::" (a single symbol, on an F-square). The S.D consists of a number of instructions, separated by semi-colons.

Each instruction consists of five consecutive parts

(i) "D" followed by a sequence of letters "A". This describes the relevant m-configuration.

(ii) "D" followed by a sequence of letters "C". This describes the scanned symbol.

(iii) "D" followed by another sequence of letters "C". This describes the symbol into which the scanned symbol is to be changed.

(iv) "L", "R", or "N", describing whether the machine is to move to left, right, or not at all.

(v) "D" followed by a sequence of letters "A". This describes the final m-configuration.

The machine $\mathfrak{U}$ is to be capable of printing "A", "C", "D", "0", "1", "u", "v", "w", "x", "y", "z". The S.D is formed from ";", "A", "C", "D", "L", "R", "N".

Subsidiary skeleton table.

$\mathfrak{con}(\mathfrak{C}, \alpha)$	Not A	R, R	$\mathfrak{con}(\mathfrak{C}, \alpha)$
	A	$L, P\alpha, R$	$\mathfrak{con}_1(\mathfrak{C}, \alpha)$
$\mathfrak{con}_1(\mathfrak{C}, \alpha)$	A	$R, P\alpha, R$	$\mathfrak{con}_1(\mathfrak{C}, \alpha)$
	D	$R, P\alpha, R$	$\mathfrak{con}_2(\mathfrak{C}, \alpha)$
$\mathfrak{con}_2(\mathfrak{C}, \alpha)$	C	$R, P\alpha, R$	$\mathfrak{con}_2(\mathfrak{C}, \alpha)$
	Not C	R, R	$\mathfrak{C}$

$\mathfrak{con}(\mathfrak{C}, \alpha)$. Starting from an F-square, S say, the sequence C of symbols describing a configuration closest on the right of S is marked out with letters α. $\rightarrow \mathfrak{C}$.

$\mathfrak{con}(\mathfrak{C}, \ )$. In the final configuration the machine is scanning the square which is four squares to the right of the last square of C. C is left unmarked.

The table for $\mathfrak{U}$.

$\mathfrak{b}$		$\mathfrak{f}(\mathfrak{b}_1, \mathfrak{b}_1, ::)$
$\mathfrak{b}_1$	$R, R, P:, R, R, PD, R, R, PA$	$\mathfrak{anf}$

$\mathfrak{b}$. The machine prints $:DA$ on the F-squares after $::$ $\rightarrow \mathfrak{anf}$.

$\mathfrak{anf}$		$\mathfrak{g}(\mathfrak{anf}_1, :)$
$\mathfrak{anf}_1$		$\mathfrak{con}(\mathfrak{fom}, y)$

$\mathfrak{anf}$. The machine marks the configuration in the last complete configuration with y. $\rightarrow \mathfrak{fom}$.

$\mathfrak{fom}$	;	R, Pz, L	$\mathfrak{con}(\mathfrak{fmp}, x)$
	z	L, L	$\mathfrak{fom}$
	not z nor ;	L	$\mathfrak{fom}$

$\mathfrak{fom}$. The machine finds the last semi-colon not marked with z. It marks this semi-colon with z and the configuration following it with x.

$\mathfrak{fmp}$	$\mathfrak{cpe}\big(\mathfrak{e}(\mathfrak{fom}, x, y), \mathfrak{sim}, x, y\big)$

$\mathfrak{fmp}$. The machine compares the sequences marked x and y. It erases all letters x and y. $\rightarrow \mathfrak{sim}$ if they are alike. Otherwise $\rightarrow \mathfrak{fom}$.

$\mathfrak{anf}$. Taking the long view, the last instruction relevant to the last configuration is found. It can be recognised afterwards as the instruction following the last semi-colon marked z. $\rightarrow \mathfrak{sim}$.

$\mathfrak{sim}$			$\mathfrak{f}'(\mathfrak{sim}_1, \mathfrak{sim}_1, z)$
$\mathfrak{sim}_1$			$\mathfrak{con}(\mathfrak{sim}_2, \ )$
$\mathfrak{sim}_2$	A		$\mathfrak{sim}_3$
	not A	R, Pu, R, R, R	$\mathfrak{sim}_2$
$\mathfrak{sim}_3$	not A	L, Py	$\mathfrak{e}(\mathfrak{mk}, z)$
	A	L, Py, R, R, R	$\mathfrak{sim}_3$

$\mathfrak{sim}$. The machine marks out the instructions. That part of the instructions which refers to operations to be carried out is marked with u, and the final m-configuration with y. The letters z are erased.

$\mathfrak{mk}$			$\mathfrak{g}(\mathfrak{mk}, :)$
$\mathfrak{mk}_1$	not A	R, R	$\mathfrak{mk}_1$
	A	L, L, L, L	$\mathfrak{mk}_2$
$\mathfrak{mk}_2$	C	R, Px, L, L, L	$\mathfrak{mk}_2$
	:		$\mathfrak{mk}_4$
	D	R, Px, L, L, L	$\mathfrak{mk}_3$
$\mathfrak{mk}_3$	not :	R, Pv, L, L, L	$\mathfrak{mk}_3$
	:		$\mathfrak{mk}_4$
$\mathfrak{mk}_4$			$\mathfrak{con}\Big(\mathfrak{l}\big(\mathfrak{l}(\mathfrak{mk}_5)\big), \ \Big)$
$\mathfrak{mk}_5$	Any	R, Pw, R	$\mathfrak{mk}_5$
	None	$P:$	$\mathfrak{sh}$

$\mathfrak{mk}$. The last complete configuration is marked out into four sections. The configuration is left unmarked. The symbol directly preceding it is marked with x. The remainder of the complete configuration is divided into two parts, of which the first is marked with v and the last with w. A colon is printed after the whole. $\to \mathfrak{sh}$.

$\mathfrak{sh}$			$\mathfrak{f}(\mathfrak{sh}_1, \mathfrak{inst}, u)$
$\mathfrak{sh}_1$		L, L, L	$\mathfrak{sh}_2$
$\mathfrak{sh}_2$	D	R, R, R, R	$\mathfrak{sh}_2$
	not D		$\mathfrak{inst}$
$\mathfrak{sh}_3$	C	R, R	$\mathfrak{sh}_4$
	not C		$\mathfrak{inst}$
$\mathfrak{sh}_4$	C	R, R	$\mathfrak{sh}_5$
	not C		$\mathfrak{pe}_2(\mathfrak{inst}, 0, :)$
$\mathfrak{sh}_5$	C		$\mathfrak{inst}$
	not C		$\mathfrak{pe}_2(\mathfrak{inst}, 1, :)$

$\mathfrak{sh}$. The instructions (marked u) are examined. If it is found that they involve "Print 0" or "Print 1", then 0: or 1: is printed at the end.

$\mathfrak{inst}$			$\mathfrak{g}\big(\mathfrak{l}(\mathfrak{inst}_1), u\big)$	$\mathfrak{inst}$. The next complete configuration is written down, carrying out the marked instructions. The letters u, v, w, x, y are erased. $\rightarrow \mathfrak{anf}$.
$\mathfrak{inst}_1$	α	R, E	$\mathfrak{inst}_1(\alpha)$	
$\mathfrak{inst}_1(L)$			$\mathfrak{ce}_5(\mathfrak{ov}, v, y, x, u, w)$	
$\mathfrak{inst}_1(R)$			$\mathfrak{ce}_5(\mathfrak{ov}, v, x, u, y, w)$	
$\mathfrak{inst}_1(N)$			$\mathfrak{ec}_5(\mathfrak{ov}, v, x, y, u, w)$	
$\mathfrak{ov}$			$\mathfrak{e}(\mathfrak{anf})$	

8. *Application of the diagonal process.*

It may be thought that arguments which prove that the real numbers are not enumerable would also prove that the computable numbers and sequences cannot be enumerable*. It might, for instance, be thought that the limit of a sequence of computable numbers must be computable. This is clearly only true if the sequence of computable numbers is defined by some rule.

Or we might apply the diagonal process. "If the computable sequences are enumerable, let α_n be the n-th computable sequence, and let $\phi_n(m)$ be the m-th figure in α_n. Let β be the sequence with $1-\phi_n(n)$ as its n-th figure. Since β is computable, there exists a number K such that $1-\phi_n(n)=\phi_K(n)$ all n. Putting $n=K$, we have $1=2\phi_K(K)$, *i.e.* 1 is even. This is impossible. The computable sequences are therefore not enumerable".

The fallacy in this argument lies in the assumption that β is computable. It would be true if we could enumerate the computable sequences by finite means, but the problem of enumerating computable sequences is equivalent to the problem of finding out whether a given number is the D.N of a circle-free machine, and we have no general process for doing this in a finite number of steps. In fact, by applying the diagonal process argument correctly, we can show that there cannot be any such general process.

The simplest and most direct proof of this is by showing that, if this general process exists, then there is a machine which computes β. This proof, although perfectly sound, has the disadvantage that it may leave the reader with a feeling that "there must be something wrong". The proof which I shall give has not this disadvantage, and gives a certain insight into the significance of the idea "circle-free". It depends not on constructing β, but on constructing β', whose n-th figure is $\phi_n(n)$.

---

* *Cf.* Hobson, *Theory of functions of a real variable* (2nd ed., 1921), 87, 88.

Let us suppose that there is such a process; that is to say, that we can invent a machine $\mathfrak{D}$ which, when supplied with the S.D of any computing machine $\mathcal{M}$ will test this S.D and if $\mathcal{M}$ is circular will mark the S.D with the symbol "u" and if it is circle-free will mark it with "s". By combining the machines $\mathfrak{D}$ and $\mathfrak{U}$ we could construct a machine $\mathfrak{H}$ to compute the sequence β'. The machine $\mathfrak{D}$ may require a tape. We may suppose that it uses the E-squares beyond all symbols on F-squares, and that when it has reached its verdict all the rough work done by $\mathfrak{D}$ is erased.

The machine $\mathfrak{H}$ has its motion divided into sections. In the first $N-1$ sections, among other things, the integers 1, 2, ..., $N-1$ have been written down and tested by the machine $\mathfrak{D}$. A certain number, say $R(N-1)$, of them have been found to be the D.N's of circle-free machines. In the N-th section the machine $\mathfrak{D}$ tests the number N. If N is satisfactory, *i.e.*, if it is the D.N of a circle-free machine, then $R(N) = 1+R(N-1)$ and the first $R(N)$ figures of the sequence of which a D.N is N are calculated. The $R(N)$-th figure of this sequence is written down as one of the figures of the sequence β' computed by $\mathfrak{H}$. If N is not satisfactory, then $R(N) = R(N-1)$ and the machine goes on to the $(N+1)$-th section of its motion.

From the construction of $\mathfrak{H}$ we can see that $\mathfrak{H}$ is circle-free. Each section of the motion of $\mathfrak{H}$ comes to an end after a finite number of steps. For, by our assumption about $\mathfrak{D}$, the decision as to whether N is satisfactory is reached in a finite number of steps. If N is not satisfactory, then the N-th section is finished. If N is satisfactory, this means that the machine $\mathfrak{M}(N)$ whose D.N is N is circle-free, and therefore its $R(N)$-th figure can be calculated in a finite number of steps. When this figure has been calculated and written down as the $R(N)$-th figure of β', the N-th section is finished. Hence $\mathfrak{H}$ is circle-free.

Now let K be the D.N of $\mathfrak{H}$. What does $\mathfrak{H}$ do in the K-th section of its motion? It must test whether K is satisfactory, giving a verdict "s" or "u". Since K is the D.N of $\mathfrak{H}$ and since $\mathfrak{H}$ is circle-free, the verdict cannot be "u". On the other hand the verdict cannot be "s". For if it were, then in the K-th section of its motion $\mathfrak{H}$ would be bound to compute the first $R(K-1)+1 = R(K)$ figures of the sequence computed by the machine with K as its D.N and to write down the $R(K)$-th as a figure of the sequence computed by $\mathfrak{H}$. The computation of the first $R(K)-1$ figures would be carried out all right, but the instructions for calculating the $R(K)$-th would amount to "calculate the first $R(K)$ figures computed by H and write down the $R(K)$-th". This $R(K)$-th figure would never be found. *I.e.*, $\mathfrak{H}$ is circular, contrary both to what we have found in the last paragraph and to the verdict "s". Thus both verdicts are impossible and we conclude that there can be no machine $\mathfrak{D}$.

We can show further that *there can be no machine* $\mathcal{E}$ *which, when supplied with the S.D of an arbitrary machine* $\mathcal{M}$, *will determine whether* $\mathcal{M}$ *ever prints a given symbol* (0 *say*).

We will first show that, if there is a machine $\mathcal{E}$, then there is a general process for determining whether a given machine $\mathcal{M}$ prints 0 infinitely often. Let $\mathcal{M}_1$ be a machine which prints the same sequence as $\mathcal{M}$, except that in the position where the first 0 printed by $\mathcal{M}$ stands, $\mathcal{M}_1$ prints $\bar{0}$. $\mathcal{M}_2$ is to have the first two symbols 0 replaced by $\bar{0}$, and so on. Thus, if $\mathcal{M}$ were to print

$$A\,B\,A\,0\,1\,A\,A\,B\,0\,0\,1\,0\,A\,B\ldots,$$

then $\mathcal{M}_1$ would print

$$A\,B\,A\,\bar{0}\,1\,A\,A\,B\,0\,0\,1\,0\,A\,B\ldots$$

and $\mathcal{M}_2$ would print

$$A\,B\,A\,\bar{0}\,1\,A\,A\,B\,\bar{0}\,0\,1\,0\,A\,B\ldots.$$

Now let $\mathcal{F}$ be a machine which, when supplied with the S.D of $\mathcal{M}$, will write down successively the S.D of $\mathcal{M}$, of $\mathcal{M}_1$, of $\mathcal{M}_2$, ... (there is such a machine). We combine $\mathcal{F}$ with $\mathcal{E}$ and obtain a new machine, $\mathcal{G}$. In the motion of $\mathcal{G}$ first $\mathcal{F}$ is used to write down the S.D of $\mathcal{M}$, and then $\mathcal{E}$ tests it, : 0 : is written if it is found that $\mathcal{M}$ never prints 0; then $\mathcal{F}$ writes the S.D of $\mathcal{M}_1$, and this is tested, : 0 : being printed if and only if $\mathcal{M}_1$ never prints 0, and so on. Now let us test $\mathcal{G}$ with $\mathcal{E}$. If it is found that $\mathcal{G}$ never prints 0, then $\mathcal{M}$ prints 0 infinitely often; if $\mathcal{G}$ prints 0 sometimes, then $\mathcal{M}$ does not print 0 infinitely often.

Similarly there is a general process for determining whether $\mathcal{M}$ prints 1 infinitely often. By a combination of these processes we have a process for determining whether $\mathcal{M}$ prints an infinity of figures, *i.e.* we have a process for determining whether $\mathcal{M}$ is circle-free. There can therefore be no machine $\mathcal{E}$.

The expression "there is a general process for determining ..." has been used throughout this section as equivalent to "there is a machine which will determine ...". This usage can be justified if and only if we can justify our definition of "computable". For each of these "general process" problems can be expressed as a problem concerning a general process for determining whether a given integer n has a property $G(n)$ [*e.g.* $G(n)$ might mean "n is satisfactory" or "n is the Gödel representation of a provable formula"], and this is equivalent to computing a number whose n-th figure is 1 if $G(n)$ is true and 0 if it is false.

9. *The extent of the computable numbers.*

No attempt has yet been made to show that the "computable" numbers include all numbers which would naturally be regarded as computable. All arguments which can be given are bound to be, fundamentally, appeals to intuition, and for this reason rather unsatisfactory mathematically. The real question at issue is "What are the possible processes which can be carried out in computing a number?"

The arguments which I shall use are of three kinds.

(*a*) A direct appeal to intuition.

(*b*) A proof of the equivalence of two definitions (in case the new definition has a greater intuitive appeal).

(*c*) Giving examples of large classes of numbers which are computable.

Once it is granted that computable numbers are all "computable", several other propositions of the same character follow. In particular, it follows that, if there is a general process for determining whether a formula of the Hilbert function calculus is provable, then the determination can be carried out by a machine.

I. [Type (*a*)]. This argument is only an elaboration of the ideas of § 1.

Computing is normally done by writing certain symbols on paper. We may suppose this paper is divided into squares like a child's arithmetic book. In elementary arithmetic the two-dimensional character of the paper is sometimes used. But such a use is always avoidable, and I think that it will be agreed that the two-dimensional character of paper is no essential of computation. I assume then that the computation is carried out on one-dimensional paper, *i.e.* on a tape divided into squares. I shall also suppose that the number of symbols which may be printed is finite. If we were to allow an infinity of symbols, then there would be symbols differing to an arbitrarily small extent†. The effect of this restriction of the number of symbols is not very serious. It is always possible to use sequences of symbols in the place of single symbols. Thus an Arabic numeral such as

---

† If we regard a symbol as literally printed on a square we may suppose that the square is $0 \leqslant x \leqslant 1$, $0 \leqslant y \leqslant 1$. The symbol is defined as a set of points in this square, viz. the set occupied by printer's ink. If these sets are restricted to be measurable, we can define the "distance" between two symbols as the cost of transforming one symbol into the other if the cost of moving unit area of printer's ink unit distance is unity, and there is an infinite supply of ink at $x = 2$, $y = 0$. With this topology the symbols form a conditionally compact space.

17 or 999999999999999 is normally treated as a single symbol. Similarly in any European language words are treated as single symbols (Chinese, however, attempts to have an enumerable infinity of symbols). The differences from our point of view between the single and compound symbols is that the compound symbols, if they are too lengthy, cannot be observed at one glance. This is in accordance with experience. We cannot tell at a glance whether 9999999999999999 and 999999999999999 are the same.

The behaviour of the computer at any moment is determined by the symbols which he is observing, and his "state of mind" at that moment. We may suppose that there is a bound B to the number of symbols or squares which the computer can observe at one moment. If he wishes to observe more, he must use successive observations. We will also suppose that the number of states of mind which need be taken into account is finite. The reasons for this are of the same character as those which restrict the number of symbols. If we admitted an infinity of states of mind, some of them will be "arbitrarily close" and will be confused. Again, the restriction is not one which seriously affects computation, since the use of more complicated states of mind can be avoided by writing more symbols on the tape.

Let us imagine the operations performed by the computer to be split up into "simple operations" which are so elementary that it is not easy to imagine them further divided. Every such operation consists of some change of the physical system consisting of the computer and his tape. We know the state of the system if we know the sequence of symbols on the tape, which of these are observed by the computer (possibly with a special order), and the state of mind of the computer. We may suppose that in a simple operation not more than one symbol is altered. Any other changes can be split up into simple changes of this kind. The situation in regard to the squares whose symbols may be altered in this way is the same as in regard to the observed squares. We may, therefore, without loss of generality, assume that the squares whose symbols are changed are always "observed" squares.

Besides these changes of symbols, the simple operations must include changes of distribution of observed squares. The new observed squares must be immediately recognisable by the computer. I think it is reasonable to suppose that they can only be squares whose distance from the closest of the immediately previously observed squares does not exceed a certain fixed amount. Let us say that each of the new observed squares is within L squares of an immediately previously observed square.

In connection with "immediate recognisability", it may be thought that there are other kinds of square which are immediately recognisable. In particular, squares marked by special symbols might be taken as immediately recognisable. Now if these squares are marked only by single symbols there can be only a finite number of them, and we should not upset our theory by adjoining these marked squares to the observed squares. If, on the other hand, they are marked by a sequence of symbols, we cannot regard the process of recognition as a simple process. This is a fundamental point and should be illustrated. In most mathematical papers the equations and theorems are numbered. Normally the numbers do not go beyond (say) 1000. It is, therefore, possible to recognise a theorem at a glance by its number. But if the paper was very long, we might reach Theorem 157767733443477; then, further on in the paper, we might find "... hence (applying Theorem 157767733443477) we have ...". In order to make sure which was the relevant theorem we should have to compare the two numbers figure by figure, possibly ticking the figures off in pencil to make sure of their not being counted twice. If in spite of this it is still thought that there are other "immediately recognisable" squares, it does not upset my contention so long as these squares can be found by some process of which my type of machine is capable. This idea is developed in III below.

The simple operations must therefore include:

(*a*) Changes of the symbol on one of the observed squares.

(*b*) Changes of one of the squares observed to another square within L squares of one of the previously observed squares.

It may be that some of these changes necessarily involve a change of state of mind. The most general single operation must therefore be taken to be one of the following:

(*A*) A possible change (*a*) of symbol together with a possible change of state of mind.

(*B*) A possible change (*b*) of observed squares, together with a possible change of state of mind.

The operation actually performed is determined, as has been suggested on p. 136, by the state of mind of the computer and the observed symbols. In particular, they determine the state of mind of the computer after the operation is carried out.

We may now construct a machine to do the work of this computer. To each state of mind of the computer corresponds an "*m*-configuration" of the machine. The machine scans B squares corresponding to the B squares observed by the computer. In any move the machine can change a symbol on a scanned square or can change any one of the scanned squares to another square distant not more than L squares from one of the other scanned squares. The move which is done, and the succeeding configuration, are determined by the scanned symbol and the m-configuration. The machines just described do not differ very essentially from computing machines as defined in § 2, and corresponding to any machine of this type a computing machine can be constructed to compute the same sequence, that is to say the sequence computed by the computer.

II. [Type (b)].

If the notation of the Hilbert functional calculus† is modified so as to be systematic, and so as to involve only a finite number of symbols, it becomes possible to construct an automatic‡ machine $\mathfrak{K}$, which will find all the provable formulae of the calculus§.

Now let α be a sequence, and let us denote by $G_\alpha(x)$ the proposition "The x-th figure of α is 1", so that|| $-G_\alpha(x)$ means "The x-th figure of α is 0". Suppose further that we can find a set of properties which define the sequence α and which can be expressed in terms of $G_\alpha(x)$ and of the propositional functions $N(x)$ meaning "x is a non-negative integer" and $F(x, y)$ meaning "$y = x+1$". When we join all these formulae together conjunctively, we shall have a formula, $\mathfrak{A}$ say, which defines α. The terms of $\mathfrak{A}$ must include the necessary parts of the Peano axioms, viz.,

$$(\exists u) N(u) \,\&\, (x)\Big(N(x) \to (\exists y) F(x, y)\Big) \,\&\, \Big(F(x, y) \to N(y)\Big),$$

which we will abbreviate to P.

When we say "$\mathfrak{A}$ defines α", we mean that $-\mathfrak{A}$ is not a provable formula, and also that, for each n, one of the following formulae (A_n) or (B_n) is provable.

$$\mathfrak{A} \,\&\, F^{(n)} \to G_\alpha(u^{(n)}), \qquad (A_n)¶$$

$$\mathfrak{A} \,\&\, F^{(n)} \to \Big(-G_\alpha(u^{(n)})\Big), \qquad (B_n),$$

where $F^{(n)}$ stands for $F(u, u') \,\&\, F(u', u'') \,\&\, \ldots F(u^{(n-1)}, u^{(n)})$.

---

† The expression "the functional calculus" is used throughout to mean the *restricted* Hilbert functional calculus.

‡ It is most natural to construct first a choice machine (§ 2) to do this. But it is then easy to construct the required automatic machine. We can suppose that the choices are always choices between two possibilities 0 and 1. Each proof will then be determined by a sequence of choices $i_1, i_2, \ldots, i_n$ ($i_1 = 0$ or 1, $i_2 = 0$ or 1, ..., $i_n = 0$ or 1), and hence the number $2^n + i_1 2^{n-1} + i_2 2^{n-2} + \ldots + i_n$ completely determines the proof. The automatic machine carries out successively proof 1, proof 2, proof 3, ...

§ The author has found a description of such a machine.

|| The negation sign is written before an expression and not over it.

¶ A sequence of r primes is denoted by $^{(r)}$.

I say that α is then a computable sequence: a machine $\mathcal{K}_\alpha$ to compute α can be obtained by a fairly simple modification of $\mathcal{K}$.

We divide the motion of $\mathcal{K}_\alpha$ into sections. The n-th section is devoted to finding the n-th figure of α. After the $(n-1)$-th section is finished a double colon :: is printed after all the symbols, and the succeeding work is done wholly on the squares to the right of this double colon. The first step is to write the letter "A" followed by the formula (A_n) and then "B" followed by (B_n). The machine $\mathcal{K}_\alpha$ then starts to do the work of $\mathcal{K}$, but whenever a provable formula is found, this formula is compared with (A_n) and with (B_n). If it is the same formula as (A_n), then the figure "1" is printed, and the n-th section is finished. If it is (B_n), then "0" is printed and the section is finished. If it is different from both, then the work of $\mathcal{K}$ is continued from the point at which it had been abandoned. Sooner or later one of the formulae (A_n) or (B_n) is reached; this follows from our hypotheses about α and $\mathfrak{A}$, and the known nature of $\mathcal{K}$. Hence the n-th section will eventually be finished. $\mathcal{K}_\alpha$ is circle-free; α is computable.

It can also be shown that the numbers α definable in this way by the use of axioms include all the computable numbers. This is done by describing computing machines in terms of the function calculus.

It must be remembered that we have attached rather a special meaning to the phrase "$\mathfrak{A}$ defines α". The computable numbers do not include all (in the ordinary sense) definable numbers. Let δ be a sequence whose n-th figure is 1 or 0 according as n is or is not satisfactory. It is an immediate consequence of the theorem of § 8 that δ is not computable. It is (so far as we know at present) possible that any assigned number of figures of δ can be calculated, but not by a uniform process. When sufficiently many figures of δ have been calculated, an essentially new method is necessary in order to obtain more figures.

III. This may be regarded as a modification of I or as a corollary of II.

We suppose, as in I, that the computation is carried out on a tape; but we avoid introducing the "state of mind" by considering a more physical and definite counterpart of it. It is always possible for the computer to break off from his work, to go away and forget all about it, and later to come back and go on with it. If he does this he must leave a note of instructions (written in some standard form) explaining how the work is to be continued. This note is the counterpart of the "state of mind". We will suppose that the computer works in such a desultory manner that he never does more than one step at a sitting. The note of instructions must enable him to carry out one step and write the next note. Thus the state of progress of the computation at any stage is completely determined by the note of instructions and the symbols on the tape. That is, the state of the system may be described by a single expression (sequence of symbols), consisting of the symbols on the tape followed by Δ (which we suppose not to appear elsewhere) and then by the note of instructions. This expression may be called the "state formula". We know that the state formula at any given stage is determined by the state formula before the last step was made, and we assume that the relation of these two formulae is expressible in the functional calculus. In other words, we assume that there is an axiom $\mathfrak{A}$ which expresses the rules governing the behaviour of the computer, in terms of the relation of the state formula at any stage to the state formula at the preceding stage. If this is so, we can construct a machine to write down the successive state formulae, and hence to compute the required number.

10. *Examples of large classes of numbers which are computable.*

It will be useful to begin with definitions of a computable function of an integral variable and of a computable variable, etc. There are many equivalent ways of defining a computable function of an integral variable. The simplest is, possibly, as follows. If γ is a computable sequence in which 0 appears infinitely† often, and n is an integer, then let us define $\xi(\gamma, n)$ to be the number of figures 1 between the n-th and the $(n+1)$-th figure 0 in γ. Then $\phi(n)$ is computable if, for all n and some γ, $\phi(n) = \xi(\gamma, n)$. An equivalent definition is this. Let $H(x, y)$ mean $\phi(x) = y$. Then, if we can find a contradiction-free axiom $\mathfrak{A}_\phi$, such that $\mathfrak{A}_\phi \to P$, and if for each integer n there exists an integer N, such that

$$\mathfrak{A}_\phi \ \& \ F^{(N)} \to H(u^{(n)}, u^{(\phi(n))}),$$

and such that, if $m \neq \phi(n)$, then, for some N',

$$\mathfrak{A}_\phi \ \& \ F^{(N')} \to \left(-H(u^{(n)}, u^{(m)} \right),$$

then ϕ may be said to be a computable function.

We cannot define general computable functions of a real variable, since there is no general method of describing a real number, but we can define a computable function of a computable variable. If n is satisfactory, let γ_n be the number computed by $\mathcal{M}(n)$, and let

$$a_n = \tan\left(\pi(\gamma_n - \tfrac{1}{2})\right),$$

† If $\mathcal{M}$ computes γ, then the problem whether $\mathcal{M}$ prints 0 infinitely often is of the same character as the problem whether $\mathcal{M}$ is circle-free.

unless $\gamma_n = 0$ or $\gamma_n = 1$, in either of which cases $\alpha_n = 0$. Then, as n runs through the satisfactory numbers, α_n runs through the computable numbers†. Now let $\phi(n)$ be a computable function which can be shown to be such that for any satisfactory argument its value is satisfactory‡. Then the function f, defined by $f(\alpha_n) = \alpha_{\phi(n)}$, is a computable function and all computable functions of a computable variable are expressible in this form.

Similar definitions may be given of computable functions of several variables, computable-valued functions of an integral variable, etc.

I shall enunciate a number of theorems about computability, but I shall prove only (ii) and a theorem similar to (iii).

(i) A computable function of a computable function of an integral or computable variable is computable.

(ii) Any function of an integral variable defined recursively in terms of computable functions is computable. *I.e.* if $\phi(m, n)$ is computable, and r is some integer, then $\eta(n)$ is computable, where

$$\eta(0) = r,$$

$$\eta(n) = \phi\big(n, \eta(n-1)\big).$$

(iii) If $\phi\ (m, n)$ is a computable function of two integral variables, then $\phi(n, n)$ is a computable function of n.

(iv) If $\phi(n)$ is a computable function whose value is always 0 or 1, then the sequence whose n-th figure is $\phi(n)$ is computable.

Dedekind's theorem does not hold in the ordinary form if we replace "real" throughout by "computable". But it holds in the following form:

(v) If $G(\alpha)$ is a propositional function of the computable numbers and

$$(a) \quad (\exists\alpha)(\exists\beta)\big\{G(\alpha)\ \&\ \big(-G(\beta)\big)\big\},$$

$$(b) \quad G(\alpha)\ \&\ \big(-G(\beta)\big) \rightarrow (\alpha < \beta),$$

and there is a general process for determining the truth value of $G(\alpha)$, then

† A function α_n may be defined in many other ways so as to run through the computable numbers.

‡ Although it is not possible to find a general process for determining whether a given number is satisfactory, it is often possible to show that certain classes of numbers are satisfactory.

there is a computable number ξ such that

$$G(a) \rightarrow a \leqslant \xi,$$
$$-G(a) \rightarrow a \geqslant \xi.$$

In other words, the theorem holds for any section of the computables such that there is a general process for determining to which class a given number belongs.

Owing to this restriction of Dedekind's theorem, we cannot say that a computable bounded increasing sequence of computable numbers has a computable limit. This may possibly be understood by considering a sequence such as

$$-1, -\tfrac{1}{2}, -\tfrac{1}{4}, -\tfrac{1}{8}, -\tfrac{1}{16}, \tfrac{1}{2}, \ldots.$$

On the other hand, (v) enables us to prove

(vi) If a and β are computable and $a < \beta$ and $\phi(a) < 0 < \phi(\beta)$, where $\phi(a)$ is a computable increasing continuous function, then there is a unique computable number γ, satisfying $a < \gamma < \beta$ and $\phi(\gamma) = 0$.

Computable convergence.

We shall say that a sequence β_n of computable numbers *converges computably* if there is a computable integral valued function $N(\epsilon)$ of the computable variable ϵ, such that we can show that, if $\epsilon > 0$ and $n > N(\epsilon)$ and $m > N(\epsilon)$, then $|\beta_n - \beta_m| < \epsilon$.

We can then show that

(vii) A power series whose coefficients form a computable sequence of computable numbers is computably convergent at all computable points in the interior of its interval of convergence.

(viii) The limit of a computably convergent sequence is computable.

And with the obvious definition of "uniformly computably convergent":

(ix) The limit of a uniformly computably convergent computable sequence of computable functions is a computable function. Hence

(x) The sum of a power series whose coefficients form a computable sequence is a computable function in the interior of its interval of convergence.

From (viii) and $\pi = 4(1 - \tfrac{1}{3} + \tfrac{1}{5} - \ldots)$ we deduce that π is computable.

From $e = 1 + 1 + \dfrac{1}{2!} + \dfrac{1}{3!} + \ldots$ we deduce that e is computable.

From (vi) we deduce that all real algebraic numbers are computable.

From (vi) and (x) we deduce that the real zeros of the Bessel functions are computable.

Proof of (ii).

Let $H(x, y)$ mean "$\eta(x) = y$", and let $K(x, y, z)$ mean "$\phi(x, y) = z$". $\mathfrak{A}_\phi$ is the axiom for $\phi(x, y)$. We take $\mathfrak{A}_\eta$ to be

$$\mathfrak{A}_\phi \,\&\, P \,\&\, \big(F(x, y) \to G(x, y)\big) \,\&\, \big(G(x, y) \,\&\, G(y, z) \to G(x, z)\big)$$

$$\&\, \big(F^{(r)} \to H(u, u^{(r)})\big) \,\&\, \big(F(v, w) \,\&\, H(v, x) \,\&\, K(w, x, z) \to H(w, z)\big)$$

$$\&\, \big[H(w, z) \,\&\, G(z, t) \vee G(t, z) \to \big(-H(w, t)\big)\big].$$

I shall not give the proof of consistency of $\mathfrak{A}_\eta$. Such a proof may be constructed by the methods used in Hilbert and Bernays, *Grundlagen der Mathematik* (Berlin, 1934), p. 209 *et seq.* The consistency is also clear from the meaning.

Suppose that, for some n, N, we have shown

$$\mathfrak{A}_\eta \,\&\, F^{(N)} \to H(u^{(n-1)}, u^{(\eta(n-1))}),$$

then, for some M,

$$\mathfrak{A}_\phi \,\&\, F^{(M)} \to K(u^{(n)}, u^{(\eta(n-1))}, u^{(\eta(n))}),$$

$$\mathfrak{A}_\eta \,\&\, F^{(M)} \to F(u^{(n-1)}, u^{(n)}) \,\&\, H(u^{(n-1)}, u^{(\eta(n-1))})$$

$$\&\, K(u^{(n)}, u^{(\eta(n-1))}, u^{(\eta(n))}),$$

and

$$\mathfrak{A}_\eta \,\&\, F^{(M)} \to [F(u^{(n-1)}, u^{(n)}) \,\&\, H(u^{(n-1)}, u^{(\eta(n-1))})$$

$$\&\, K(u^{(n)}, u^{(\eta(n-1))}, u^{(\eta(n))}) \to H(u^{(n)}, u^{(\eta(n))})].$$

Hence $$\mathfrak{A}_\eta \,\&\, F^{(M)} \to H(u^{(n)}, u^{(\eta(n))}).$$

Also $$\mathfrak{A}_\eta \,\&\, F^{(r)} \to H(u, u^{(\eta(0))}).$$

Hence for each n some formula of the form

$$\mathfrak{A}_\eta \,\&\, F^{(M)} \to H(u^{(n)}, u^{(\eta(n))})$$

is provable. Also, if $M' \geqslant M$ and $M' \geqslant m$ and $m \neq \eta(u)$, then

$$\mathfrak{A}_\eta \,\&\, F^{(M')} \to G(u^{\eta((n))}, u^{(m)}) \vee G(u^{(m)}, u^{(\eta(n))})$$

and

$$\mathfrak{A}_\eta \,\&\, F^{(M')} \to \Big[\{G(u^{(\eta(n))}, u^{(m)}) \vee G(u^{(m)}, u^{(\eta(n))}) \\ \& \, H(u^{(n)}, u^{(\eta(n))})\} \to \big(-H(u^{(n)}, u^{(m)})\big)\Big].$$

Hence $$\mathfrak{A}_\eta \,\&\, F^{(M')} \to \big(-H(u^{(n)}, u^{(m)})\big).$$

The conditions of our second definition of a computable function are therefore satisfied. Consequently η is a computable function.

Proof of a modified form of (iii).

Suppose that we are given a machine $\mathfrak{N}$, which, starting with a tape bearing on it əə followed by a sequence of any number of letters "F" on F-squares and in the m-configuration $\mathfrak{b}$, will compute a sequence γ_n depending on the number n of letters "F". If $\phi_n(m)$ is the m-th figure of γ_n, then the sequence β whose n-th figure is $\phi_n(n)$ is computable.

We suppose that the table for $\mathfrak{N}$ has been written out in such a way that in each line only one operation appears in the operations column. We also suppose that Ξ, Θ, $\bar{0}$, and $\bar{1}$ do not occur in the table, and we replace ə throughout by Θ, 0 by $\bar{0}$, and 1 by $\bar{1}$. Further substitutions are then made. Any line of form

$\mathfrak{A}$	α	$P\bar{0}$	$\mathfrak{B}$

we replace by

$\mathfrak{A}$	α	$P\bar{0}$	$\mathfrak{re}(\mathfrak{B}, \mathfrak{u}, h, k)$

and any line of the form

$\mathfrak{A}$	α	$P\bar{1}$	$\mathfrak{B}$

by

$\mathfrak{A}$	α	$P\bar{1}$	$\mathfrak{re}(\mathfrak{B}, \mathfrak{v}, h, k)$

and we add to the table the following lines:

$\mathfrak{u}$		$\mathfrak{pe}(\mathfrak{u}_1, 0)$
$\mathfrak{u}_1$	$R, Pk, R, P\Theta, R, P\Theta$	$\mathfrak{u}_2$
$\mathfrak{u}_2$		$\mathfrak{re}(\mathfrak{u}_3, \mathfrak{u}_3, k, h)$
$\mathfrak{u}_3$		$\mathfrak{pe}(\mathfrak{u}_2, F)$

and similar lines with $\mathfrak{v}$ for $\mathfrak{u}$ and 1 for 0 together with the following line

$\mathfrak{c}$	$R, P\Xi, R, Ph$	$\mathfrak{b}$.

We then have the table for the machine $\mathfrak{N}'$ which computes β. The initial m-configuration is $\mathfrak{c}$, and the initial scanned symbol is the second ə.

11. *Application to the Entscheidungsproblem.*

The results of § 8 have some important applications. In particular, they can be used to show that the Hilbert Entscheidungsproblem can have no solution. For the present I shall confine myself to proving this particular theorem. For the formulation of this problem I must refer the reader to Hilbert and Ackermann's *Grundzüge der Theoretischen Logik* (Berlin, 1931), chapter 3.

I propose, therefore, to show that there can be no general process for determining whether a given formula $\mathfrak{A}$ of the functional calculus **K** is provable, *i.e.* that there can be no machine which, supplied with any one $\mathfrak{A}$ of these formulae, will eventually say whether $\mathfrak{A}$ is provable.

It should perhaps be remarked that what I shall prove is quite different from the well-known results of Gödel†. Gödel has shown that (in the formalism of Principia Mathematica) there are propositions $\mathfrak{A}$ such that neither $\mathfrak{A}$ nor $-\mathfrak{A}$ is provable. As a consequence of this, it is shown that no proof of consistency of Principia Mathematica (or of **K**) can be given within that formalism. On the other hand, I shall show that there is no general method which tells whether a given formula $\mathfrak{A}$ is provable in **K**, or, what comes to the same, whether the system consisting of **K** with $-\mathfrak{A}$ adjoined as an extra axiom is consistent.

If the negation of what Gödel has shown had been proved, *i.e.* if, for each $\mathfrak{A}$, either $\mathfrak{A}$ or $-\mathfrak{A}$ is provable, then we should have an immediate solution of the Entscheidungsproblem. For we can invent a machine $\mathcal{K}$ which will prove consecutively all provable formulae. Sooner or later $\mathcal{K}$ will reach either $\mathfrak{A}$ or $-\mathfrak{A}$. If it reaches $\mathfrak{A}$, then we know that $\mathfrak{A}$ is provable. If it reaches $-\mathfrak{A}$, then, since **K** is consistent (Hilbert and Ackermann, p. 65), we know that $\mathfrak{A}$ is not provable.

Owing to the absence of integers in **K** the proofs appear somewhat lengthy. The underlying ideas are quite straightforward.

Corresponding to each computing machine $\mathcal{M}$ we construct a formula Un ($\mathcal{M}$) and we show that, if there is a general method for determining whether Un ($\mathcal{M}$) is provable, then there is a general method for determining whether $\mathcal{M}$ ever prints 0.

The interpretations of the propositional functions involved are as follows :

$R_{S_l}(x, y)$ is to be interpreted as "in the complete configuration x (of $\mathcal{M}$) the symbol on the square y is S".

† *Loc. cit.*[a]

$I(x, y)$ is to be interpreted as "in the complete configuration x the square y is scanned".

$K_{q_m}(x)$ is to be interpreted as "in the complete configuration x the m-configuration is q_m.

$F(x, y)$ is to be interpreted as "y is the immediate successor of x".

Inst $\{q_i S_j S_k L q_l\}$ is to be an abbreviation for

$$(x, y, x', y')\Big\{\Big(R_{S_j}(x, y) \,\&\, I(x, y) \,\&\, K_{q_i}(x) \,\&\, F(x, x') \,\&\, F(y', y)\Big)$$
$$\rightarrow \Big(I(x', y') \,\&\, R_{S_k}(x', y) \,\&\, K_{q_l}(x')$$
$$\&\, (z)\Big[F(y', z) \vee \Big(R_{S_j}(x, z) \rightarrow R_{S_k}(x', z)\Big)\Big]\Big)\Big\}.$$

Inst $\{q_i S_j S_k R q_l\}$ and Inst $\{q_i S_j S_k N q_l\}$

are to be abbreviations for other similarly constructed expressions.

Let us put the description of $\mathcal{M}$ into the first standard form of § 6. This description consists of a number of expressions such as "$q_i \; S_j \; S_k \; L q_l$" (or with R or N substituted for L). Let us form all the corresponding expressions such as Inst $\{q_i S_j S_k L q_l\}$ and take their logical sum. This we call Des $(\mathcal{M})$.

The formula Un $(\mathcal{M})$ is to be

$$(\exists u)\Big[N(u) \,\&\, (x)\Big(N(x) \rightarrow (\exists x') F(x, x')\Big)$$
$$\&\, (y, z)\Big(F(y, z) \rightarrow N(y) \,\&\, N(z)\Big) \,\&\, (y)\, R_{S_0}(u, y)$$
$$\&\, I(u, u) \,\&\, K_{q_1}(u) \,\&\, \mathrm{Des}(\mathcal{M})\Big]$$
$$\rightarrow (\exists s)(\exists t)\,[N(s) \,\&\, N(t) \,\&\, R_{S_1}(s, t)].$$

$[N(u) \,\&\, \ldots \,\&\, \mathrm{Des}\,(\mathcal{M})]$ may be abbreviated to $A(\mathcal{M})$.

When we substitute the meanings suggested on p.146–46 we find that Un $(\mathcal{M})$ has the interpretation "in some complete configuration of $\mathcal{M}$, S_1 (*i.e.* 0) appears on the tape". Corresponding to this I prove that

(a) If S_1 appears on the tape in some complete configuration of $\mathcal{M}$, then Un $(\mathcal{M})$ is provable.

(b) If Un $(\mathcal{M})$ is provable, then S_1 appears on the tape in some complete configuration of $\mathcal{M}$.

When this has been done, the remainder of the theorem is trivial.

LEMMA 1. *If S_1 appears on the tape in some complete configuration of $\mathcal{M}$, then* Un ($\mathcal{M}$) *is provable.*

We have to show how to prove Un ($\mathcal{M}$). Let us suppose that in the n-th complete configuration the sequence of symbols on the tape is $S_{r(n,0)}$, $S_{r(n,1)}$, ..., $S_{r(n,n)}$, followed by nothing but blanks, and that the scanned symbol is the $i(n)$-th, and that the m-configuration is $q_{k(n)}$. Then we may form the proposition

$$R_{S_{r(n,0)}}(u^{(n)}, u) \,\&\, R_{S_{r(n,1)}}(u^{(n)}, u') \,\&\, \ldots \,\&\, R_{S_{r(n,n)}}(u^{(n)}, u^{(n)})$$
$$\&\, I(u^{(n)}, u^{(i(n))}) \,\&\, K_{q_{k(n)}}(u^{(n)})$$
$$\&\, (y) F\Big((y, u') \vee F(u, y) \vee F(u', y) \vee \ldots \vee F(u^{(n-1)}, y) \vee R_{S_0}(u^{(n)}, y)\Big),$$

which we may abbreviate to CC_n.

As before, $F(u, u') \,\&\, F(u', u'') \,\&\, \ldots \,\&\, F(u^{(r-1)}, u^{(r)})$ is abbreviated to $F^{(r)}$.

I shall show that all formulae of the form $A(\mathcal{M}) \,\&\, F^{(n)} \to CC_n$ (abbreviated to CF_n) are provable. The meaning of CF_n is "The n-th complete configuration of $\mathcal{M}$ is so and so", where "so and so" stands for the actual n-th complete configuration of $\mathcal{M}$. That CF_n should be provable is therefore to be expected.

CF_0 is certainly provable, for in the complete configuration the symbols are all blanks, the m-configuration is q_1, and the scanned square is u, *i.e.* CC_0 is

$$(y)\, R_{S_0}(u, y) \,\&\, I(u, u) \,\&\, K_{q_1}(u).$$

$A(\mathcal{M}) \to CC_0$ is then trivial.

We next show that $CF_n \to CF_{n+1}$ is provable for each n. There are three cases to consider, according as in the move from the n-th to the $(n+1)$-th configuration the machine moves to left or to right or remains stationary. We suppose that the first case applies, *i.e.* the machine moves to the left. A similar argument applies in the other cases. If $r\big(n, i(n)\big) = a$, $r\big(n+1, i(n+1)\big) = c$, $k\big(i(n)\big) = b$, and $k\big(i(n+1)\big) = d$, then Des ($\mathcal{M}$) must include Inst $\{q_a S_b S_d L q_c\}$ as one of its terms, *i.e.*

$$\text{Des}(\mathcal{M}) \to \text{Inst}\{q_a S_b S_d L q_c\}.$$

Hence $$A(\mathcal{M}) \,\&\, F^{(n+1)} \to \text{Inst}\{q_a S_b S_d L q_c\} \,\&\, F^{(n+1)}.$$

But $$\text{Inst}\{q_a S_b S_d L q_c\} \,\&\, F^{(n+1)} \to (CC_n \to CC_{n+1})$$

is provable, and so therefore is

$$A(\mathcal{M}) \,\&\, F^{(n+1)} \to (CC_n \to CC_{n+1})$$

and $\left(A(\mathcal{M}) \;\&\; F^{(n)} \rightarrow CC_n\right) \rightarrow \left(A(\mathcal{M}) \;\&\; F^{(n+1)} \rightarrow CC_{n+1}\right)$,

i.e. $$CF_n \rightarrow CF_{n+1}.$$

CF_n is provable for each n. Now it is the assumption of this lemma that S_1 appears somewhere, in some complete configuration, in the sequence of symbols printed by $\mathcal{M}$; that is, for some integers N, K, CC_N has $R_{S_1}(u^{(N)}, u^{(K)})$ as one of its terms, and therefore $CC_N \rightarrow R_{S_1}(u^{(N)}, u^{(K)})$ is provable. We have then

$$CC_N \rightarrow R_{S_1}(u^{(N)}, u^{(K)})$$

and $$A(\mathcal{M}) \;\&\; F^{(N)} \rightarrow CC^N.$$

We also have

$$(\exists u) A(\mathcal{M}) \rightarrow (\exists u)(\exists u') \ldots (\exists u^{(N')}) \left(A(\mathcal{M}) \;\&\; F^{(N)}\right),$$

where $N' = \max(N, K)$. And so

$$(\exists u) A(\mathcal{M}) \rightarrow (\exists u)(\exists u') \ldots (\exists u^{(N')}) R_{S_1}(u^{(N)}, u^{(K)}),$$

$$(\exists u) A(\mathcal{M}) \rightarrow (\exists u^{(N)})(\exists u^{(K)}) R_{S_1}(u^{(N)}, u^{(K)}),$$

$$(\exists u) A(\mathcal{M}) \rightarrow (\exists s)(\exists t) R_{S_1}(s, t),$$

i.e. Un($\mathcal{M}$) is provable.

This completes the proof of Lemma 1.

LEMMA 2. *If* Un($\mathcal{M}$) *is provable, then* S_1 *appears on the tape in some complete configuration of* $\mathcal{M}$.

If we substitute any propositional functions for function variables in a provable formula, we obtain a true proposition. In particular, if we substitute the meanings tabulated on pp. 145–46 in Un($\mathcal{M}$), we obtain a true proposition with the meaning "S_1 appears somewhere on the tape in some complete configuration of $\mathcal{M}$".

We are now in a position to show that the Entscheidungsproblem cannot be solved. Let us suppose the contrary. Then there is a general (mechanical) process for determining whether Un($\mathcal{M}$) is provable. By Lemmas 1 and 2, this implies that there is a process for determining whether $\mathcal{M}$ ever prints 0, and this is impossible, by § 8. Hence the Entscheidungsproblem cannot be solved.

In view of the large number of particular cases of solutions of the Entscheidungsproblem for formulae with restricted systems of quantors, it is interesting to express $\text{Un}(\mathcal{M})$ in a form in which all quantors are at the beginning. $\text{Un}(\mathcal{M})$ is, in fact, expressible in the form

$$(u)(\exists x)(w)(\exists u_1)\dots(\exists u_n)\mathfrak{B}, \tag{I}$$

where $\mathfrak{B}$ contains no quantors, and $n = 6$. By unimportant modifications we can obtain a formula, with all essential properties of $\text{Un}(\mathcal{M})$, which is of form (I) with $n = 5$.

Added 28 *August*, 1936.

Appendix.

Computability and effective calculability

The theorem that all effectively calculable (λ-definable) sequences are computable and its converse are proved below in outline. It is assumed that the terms "well-formed formula" (W.F.F.) and "conversion" as used by Church and Kleene are understood. In the second of these proofs the existence of several formulae is assumed without proof; these formulae may be constructed straightforwardly with the help of, *e.g.*, the results of Kleene in "A theory of positive integers in formal logic", *American Journal of Math.*, 57 (1935), 153-173, 219-244.

The W.F.F. representing an integer n will be denoted by N_n. We shall say that a sequence γ whose n-th figure is $\phi_\gamma(n)$ is λ-definable or effectively calculable if $1+\phi_\gamma(u)$ is a λ-definable function of n, *i.e.* if there is a W.F.F. M_γ such that, for all integers n,

$$\{M_\gamma\}(N_n) \text{ conv } N_{\phi_\gamma(n)+1},$$

i.e. $\{M_\gamma\}(N_n)$ is convertible into $\lambda xy.x\big(x(y)\big)$ or into $\lambda xy.x(y)$ according as the n-th figure of λ is 1 or 0.

To show that every λ-definable sequence γ is computable, we have to show how to construct a machine to compute γ. For use with machines it is convenient to make a trivial modification in the calculus of conversion. This alteration consists in using $x, x', x'', \dots$ as variables instead of $a, b, c, \dots$. We now construct a machine $\mathcal{L}$ which, when supplied with the formula M_γ, writes down the sequence γ. The construction of $\mathcal{L}$ is somewhat similar to that of the machine $\mathcal{K}$ which proves all provable formulae of the functional calculus. We first construct a choice machine $\mathcal{L}_1$, which, if supplied with a W.F.F., M say, and suitably manipulated, obtains any formula into which M is convertible. $\mathcal{L}_1$ can then be modified so as to yield an automatic machine $\mathcal{L}_2$ which obtains successively all the formulae into which M is convertible (cf. foot-note p. 138). The machine $\mathcal{L}$ includes $\mathcal{L}_2$ as a part. The motion of the machine $\mathcal{L}$ when supplied with the formula M_γ is divided into sections of which the n-th is devoted to finding the n-th figure of γ. The first stage in this n-th section is the formation of $\{M_\gamma\}(N_n)$. This formula is then supplied to the machine $\mathcal{L}_2$, which converts it successively into various other formulae. Each formula into which it is convertible eventually appears, and each, as it is found, is compared with

$$\lambda x\left[\lambda x'\left[\{x\}\left(\{x\}(x')\right)\right]\right], \quad i.e.\ N_2,$$

and with
$$\lambda x\left[\lambda x'[\{x\}(x')]\right], \quad i.e.\ N_1.$$

If it is identical with the first of these, then the machine prints the figure 1 and the n-th section is finished. If it is identical with the second, then 0 is printed and the section is finished. If it is different from both, then the work of $\mathcal{L}_2$ is resumed. By hypothesis, $\{M_\gamma\}(N_n)$ is convertible into one of the formulae N_2 or N_1; consequently the n-th section will eventually be finished, *i.e.* the n-th figure of γ will eventually be written down.

To prove that every computable sequence γ is λ-definable, we must show how to find a formula M_γ such that, for all integers n,

$$\{M_\gamma\}(N_n) \text{ conv } N_{1+\phi_\gamma(n)}.$$

Let $\mathcal{M}$ be a machine which computes γ and let us take some description of the complete configurations of $\mathcal{M}$ by means of numbers, *e.g.* we may take the D.N of the complete configuration as described in § 6. Let $\xi(n)$ be the D.N of the n-th complete configuration of $\mathcal{M}$. The table for the machine $\mathcal{M}$ gives us a relation between $\xi(n+1)$ and $\xi(n)$ of the form

$$\xi(n+1) = \rho_\gamma\left(\xi(n)\right),$$

where ρ_γ is a function of very restricted, although not usually very simple, form: it is determined by the table for $\mathcal{M}$. ρ_γ is λ-definable (I omit the proof of this), *i.e.* there is a W.F.F. A_γ such that, for all integers n,

$$\{A_\gamma\}(N_{\xi(n)}) \text{ conv } N_{\xi(n+1)}.$$

Let U_γ stand for

$$\lambda u\left[\left\{\{u\}(A_\gamma)\right\}(N_r)\right],$$

where $r = \xi(0)$; then, for all integers n,

$$\{U_\gamma\}(N_n) \text{ conv } N_{\xi(n)}.$$

It may be proved that there is a formula V such that

$$\Big\{\{V\}(N_{\xi(n+1)})\Big\}(N_{\xi(n)}) \begin{cases} \text{conv } N_1 & \text{if, in going from the } n\text{-th to the } (n+1)\text{-th complete configuration, the figure 0 is printed.} \\ \text{conv } N_2 & \text{if the figure 1 is printed.} \\ \text{conv } N_3 & \text{otherwise.} \end{cases}$$

Let W_γ stand for

$$\lambda u\Big[\Big\{\{V\}\Big(\{A_\gamma\}\big(\{U_\gamma\}(u)\big)\Big)\Big\}\big(\{U_\gamma\}(u)\big)\Big],$$

so that, for each integer n,

$$\Big\{\{V\}(N_{\xi(n+1)})\Big\}(N_{\xi(n)}) \text{ conv } \{W_\gamma\}(N_n),$$

and let Q be a formula such that

$$\Big\{\{Q\}(W_\gamma)\Big\}(N_s) \text{ conv } N_{r(s)},$$

where $r(s)$ is the s-th integer q for which $\{W_\gamma\}(N_q)$ is convertible into either N_1 or N_2. Then, if M_γ stands for

$$\lambda w\Big[\{W_\gamma\}\Big(\Big\{\{Q\}(W_\gamma)\Big\}(w)\Big)\Big],$$

it will have the required property†.

The Graduate College,
Princeton University,
New Jersey, U.S.A.

---

† In a complete proof of the λ-definability of computable sequences it would be best to modify this method by replacing the numerical description of the complete configurations by a description which can be handled more easily with our apparatus. Let us choose certain integers to represent the symbols and the m-configurations of the machine. Suppose that in a certain complete configuration the numbers representing the successive symbols on the tape are $s_1 s_2 \dots s_n$, that the m-th symbol is scanned, and that the m-configuration has the number t; then we may represent this complete configuration by the formula

$$\big[[N_{s_1}, N_{s_2}, \dots, N_{s_{m-1}}],\ [N_t, N_{s_m}],\ [N_{s_{m+1}}, \dots, N_{s_n}]\big],$$

where $\qquad [a, b]$ stands for $\lambda u\Big[\Big\{\{u\}(a)\Big\}(b)\Big]$,

$$[a, b, c] \text{ stands for } \lambda u\Big[\Big\{\Big\{\{u\}(a)\Big\}(b)\Big\}(c)\Big],$$

etc.

ON COMPUTABLE NUMBERS, WITH AN APPLICATION TO THE ENTSCHEIDUNGSPROBLEM. A CORRECTION

By A. M. Turing.

In a paper entitled "On computable numbers, with an application to the Entscheidungsproblem"* the author gave a proof of the insolubility of the Entscheidungsproblem of the "engere Funktionenkalkül". This proof contained some formal errors† which will be corrected here: there are also some other statements in the same paper which should be modified, although they are not actually false as they stand.

The expression for Inst $\{q_i S_j S_k L q_l\}$ on p. 146 of the paper quoted should read

$$(x, y, x', y')\Big\{\Big(R_{S_j}(x, y) \,\&\, I(x, y) \,\&\, K_{q_i}(x) \,\&\, F(x, x') \,\&\, F(y', y)\Big)$$

$$\to \Big(I(x', y') \,\&\, R_{S_k}(x', y) \,\&\, K_{q_l}(x') \,\&\, F(y', z) \vee \Big[\Big(R_{S_0}(x, z) \to R_{S_0}(x', z)\Big)$$

$$\&\, \Big(R_{S_1}(x, z) \to R_{S_1}(x', z)\Big) \,\&\, \ldots \,\&\, \Big(R_{S_M}(x, z) \to R_{S_M}(x', z)\Big)\Big]\Big)\Big\},$$

$S_0, S_1, \ldots, S_M$ being the symbols which $\mathcal{M}$ can print. The statement on p. 147, line 33, viz.

$$\text{“Inst}\,\{q_a S_b S_d L q_c\} \,\&\, F^{(n+1)} \to (CC_n \to CC_{n+1})$$

is provable" is false (even with the new expression for Inst $\{q_a S_b S_d L q_c\}$): we are unable for example to deduce $F^{(n+1)} \to \big(-F(u, u'')\big)$ and therefore can never use the term

$$F(y', z) \vee \Big[\Big(R_{S_0}(x, z) \to R_{S_0}(x', z)\Big) \,\&\, \ldots \,\&\, \Big(R_{S_M}(x, z) \to R_{S_M}(x', z)\Big)\Big]$$

* *Proc. London Math. Soc.* (2), 42 (1936–7), 230–265.

† The author is indebted to P. Bernays for pointing out these errors.

in Inst $\{q_a S_b S_d L q_c\}$. To correct this we introduce a new functional variable G [$G(x, y)$ to have the interpretation "x precedes y"]. Then, if Q is an abbreviation for

$$(x)(\exists w)(y, z)\Big\{F(x, w) \,\&\, \big(F(x, y) \to G(x, y)\big) \,\&\, \big(F(x, z) \,\&\, G(z, y) \to G(x, y)\big)$$
$$\&\Big[G(z, x) \vee \big(G(x, y) \,\&\, F(y, z)\big) \vee \big(F(x, y) \,\&\, F(z, y)\big) \to \big(-F(x, z)\big)\Big]\Big\}$$

the corrected formula $\mathrm{Un}(\mathfrak{M})$ is to be

$$(\exists u) A(\mathfrak{M}) \to (\exists s)(\exists t) R_{S_1}(s, t),$$

where $A(\mathfrak{M})$ is an abbreviation for

$$Q \,\&\, (y) R_{S_0}(u, y) \,\&\, I(u, u) \,\&\, K_{q_1}(u) \,\&\, \mathrm{Des}(\mathfrak{M}).$$

The statement on page 147 (line 33) must then read

$$\mathrm{Inst}\{q_a S_b S_d L q_c\} \,\&\, Q \,\&\, F^{(n+1)} \to (CC_n \to CC_{n+1}),$$

and line 29 should read

$$r\big(n, i(n)\big) = b, \quad r\big(n+1, i(n)\big) = d, \quad k(n) = a, \quad k(n+1) = c.$$

For the words "logical sum" on p. 146, line 15, read "conjunction". With these modifications the proof is correct. $\mathrm{Un}(\mathfrak{M})$ may be put in the form (I) (p. 149) with $n = 4$.

Some difficulty arises from the particular manner in which "computable number" was defined (p. 119). If the computable numbers are to satisfy intuitive requirements we should have:

If we can give a rule which associates with each positive integer n two rationals a_n, b_n satisfying $a_n \leqslant a_{n+1} < b_{n+1} \leqslant b_n$, $b_n - a_n < 2^{-n}$, then there is a computable number α for which $a_n \leqslant \alpha \leqslant b_n$ each n. (A)

A proof of this may be given, valid by ordinary mathematical standards, but involving an application of the principle of excluded middle. On the other hand the following is false:

There is a rule whereby, given the rule of formation of the sequences a_n, b_n in (A) *we can obtain a* D.N. *for a machine to compute α.* (B)

That (B) is false, at least if we adopt the convention that the decimals of numbers of the form $m/2^n$ shall always terminate with zeros, can be seen in this way. Let $\mathfrak{N}$ be some machine, and define c_n as follows: $c_n = \frac{1}{2}$ if $\mathfrak{N}$ has not printed a figure 0 by the time the n-th complete configuration is reached $c_n = \frac{1}{2} - 2^{-m-3}$ if 0 had first been printed at the m-th complete configuration ($m \leqslant n$). Put $a_n = c_n - 2^{-n-2}$, $b_n = c_n + 2^{-n-2}$. Then the inequalities of (A) are satisfied, and the first figure of α is 0 if $\mathfrak{N}$ ever prints 0 and is 1 otherwise. If (B) were true we should have a means of finding the first figure of α given the D.N. of $\mathfrak{N}$: *i.e.* we should be able to determine whether $\mathfrak{N}$ ever prints 0, contrary to the results of § 8 of the paper quoted. Thus although (A) shows that there must be machines which compute the Euler constant (for example) we cannot at present describe any such machine, for we do not yet know whether the Euler constant is of the form $m/2^n$.

This disagreeable situation can be avoided by modifying the manner in which computable numbers are associated with computable sequences, the totality of computable numbers being left unaltered. It may be done in many ways* of which this is an example. Suppose that the first figure of a computable sequence γ is i and that this is followed by 1 repeated n times, then by 0 and finally by the sequence whose r-th figure is c_r; then the sequence γ is to correspond to the real number

$$(2i-1)n + \sum_{r=1}^{\infty} (2c_r - 1)\left(\tfrac{2}{3}\right)^r.$$

If the machine which computes γ is regarded as computing also this real number then (B) holds. The uniqueness of representation of real numbers by sequences of figures is now lost, but this is of little theoretical importance, since the D.N.'s are not unique in any case.

SYSTEMS OF LOGIC BASED ON ORDINALS

This is a profound and difficult paper in which Turing explores the possibility of avoiding the Gödel incompleteness theorem by replacing a single given logic by a system of logics obtained from one another by transfinite iterations. What is iterated is the addition of principles which are intuitively as evident as the given system, but which yield stronger logics; e.g. one iterates the adjunction of a sentence which asserts that the given logic is consistent.

This paper also introduces the idea of an "oracle" for a possibly non-recursive set being made available to a Turing machine.

* This use of overlapping intervals for the definition of real numbers is due originally to Brouwer.

SYSTEMS OF LOGIC BASED ON ORDINALS†

By A. M. TURING.

Reprinted with the kind permission of the London Mathematical Society from the Proceedings of the London Mathematical Society, ser. 2, vol. 45 (1939) pp. 161-228.

The well-known theorem of Gödel (Gödel [1], [2])[a] shows that every system of logic is in a certain sense incomplete, but at the same time it indicates means whereby from a system L of logic a more complete system L' may be obtained. By repeating the process we get a sequence L, $L_1 = L'$, $L_2 = L_1'$, ... each more complete than the preceding. A logic L_ω may then be constructed in which the provable theorems are the totality of theorems provable with the help of the logics $L, L_1, L_2, \ldots$. We may then form $L_{2\omega}$ related to L_ω in the same way as L_ω was related to L. Proceeding in this way we can associate a system of logic with any constructive ordinal‡. It may be asked whether a sequence of logics of this kind is complete in the sense that to any problem A there corresponds

† This paper represents work done while a Jane Eliza Procter Visiting Fellow at Princeton University, where the author received most valuable advice and assistance from Prof. Alonzo Church.

‡ The situation is not quite so simple as is suggested by this crude argument. See pages 183–187, 196, 197.

an ordinal α such that A is solvable by means of the logic L_α. I propose to investigate this question in a rather more general case, and to give some other examples of ways in which systems of logic may be associated with constructive ordinals.

1. *The calculus of conversion. Gödel representations.*

It will be convenient to be able to use the "conversion calculus" of Church for the description of functions and for some other purposes. This will make greater clarity and simplicity of expression possible. I give a short account of this calculus. For detailed descriptions see Church **[3]**,[a] **[2]**, Kleene **[1]**, Church and Rosser **[1]**.

The formulae of the calculus are formed from the symbols {, }, (,), [,], λ, δ, and an infinite list of others called variables; we shall take for our infinite list $a, b, \ldots, z, x', x'', \ldots$. Certain finite sequences of such symbols are called *well-formed formulae* (abbreviated to W.F.F.); we define this class inductively, and define simultaneously the free and the bound variables of a W.F.F. Any variable is a W.F.F.; it is its only free variable, and it has no bound variables.* δ is a W.F.F. and has no free or bound variables. If **M** and **N** are W.F.F. then {**M**}(**N**) is a W.F.F., whose free variables are the free variables of **M** together with the free variables of **N**, and whose bound variables are the bound variables of **M** together with those of **N**. If **M** is a W.F.F. and **V** is one of its free variables, then λ**V**[**M**] is a W.F.F. whose free variables are those of **M** with the exception of **V**, and whose bound variables are those of **M** together with **V**. No sequence of symbols is a W.F.F. except in consequence of these three statements.

In metamathematical statements we use heavy type letters to stand for variable or undetermined formulae, as was done in the last paragraph, and in future such letters will stand for well-formed formulae unless otherwise stated. Small letters in heavy type will stand for formulae representing undetermined positive integers (see below).

A W.F.F. is said to be in normal form if it has no parts of the form {λ**V**[**M**]}(**N**) and none of the form {{δ}(**M**)}(**N**), where **M** and **N** have no free variables.

We say that one W.F.F. is *immediately convertible* into another if it is obtained from it either by:

(i) Replacing one occurrence of a well-formed part λ**V**[**M**] by λ**U**[**N**], where the variable **U** does not occur in **M**, and **N** is obtained from **M** by replacing the variable **V** by **U** throughout.

(ii) Replacing a well-formed part $\{\lambda \mathbf{V}[\mathbf{M}]\}(\mathbf{N})$ by the formula which is obtained from **M** by replacing **V** by **N** throughout, provided that the bound variables of **M** are distinct both from **V** and from the free variables of **N**.

(iii) The process inverse to (ii).

(iv) Replacing a well-formed part $\{\{\delta\}(\mathbf{M})\}(\mathbf{M})$ by

$$\lambda f\Big[\lambda x\big[\{f\}\big(\{f\}(x)\big)\big]\Big]$$

if **M** is in normal form and has no free variables.

(v) Replacing a well-formed part $\{\{\delta\}(\mathbf{M})\}(\mathbf{N})$ by

$$\lambda f\big[\lambda x[\{f\}(x)]\big]$$

if **M** and **N** are in normal form, are not transformable into one another by repeated application of (i), and have no free variables.

(vi) The process inverse to (iv).

(vii) The process inverse to (v).

These rules could have been expressed in such a way that in no case could there be any doubt about the admissibility or the result of the transformation [in particular this can be done in the case of process (v)].

A formula **A** is said to be *convertible* into another **B** (abbreviated to "**A** conv **B**") if there is a finite chain of immediate conversions leading from one formula to the other. It is easily seen that the relation of convertibility is an equivalence relation, *i.e.* it is symmetric, transitive, and reflexive.

Since the formulae are liable to be very lengthy, we need means for abbreviating them. If we wish to introduce a particular letter as an abbreviation for a particular lengthy formula we write the letter followed by "$\rightarrow$" and then by the formula, thus

$$I \rightarrow \lambda x[x]$$

indicates that I is an abbreviation for $\lambda x[x]$. We also use the arrow in less sharply defined senses, but never so as to cause any real confusion. In these cases the meaning of the arrow may be rendered by the words "stands for".

If a formula $\mathbf{F}$ is, or is represented by, a single symbol we abbreviate $\{\mathbf{F}\}(\mathbf{X})$ to $\mathbf{F}(\mathbf{X})$. A formula $\big\{\{\mathbf{F}\}(\mathbf{X})\big\}(\mathbf{Y})$ may be abbreviated to

$$\{\mathbf{F}\}(\mathbf{X}, \mathbf{Y}),$$

or to $\mathbf{F}(\mathbf{X}, \mathbf{Y})$ if $\mathbf{F}$ is, or is represented by, a single symbol. Similarly for $\Big\{\big\{\{\mathbf{F}\}(\mathbf{X})\big\}(\mathbf{Y})\Big\}(\mathbf{Z})$, etc. A formula $\lambda\mathbf{V}_1\Big[\lambda\mathbf{V}_2\ldots\big[\lambda\mathbf{V}_r[\mathbf{M}]\big]\ldots\Big]$ may be abbreviated to $\lambda\mathbf{V}_1\mathbf{V}_2\ldots\mathbf{V}_r.\mathbf{M}$.

We have not as yet assigned any meanings to our formulae, and we do not intend to do so in general. An exception may be made for the case of the positive integers, which are very conveniently represented by the formulae $\lambda fx.f(x)$, $\lambda fx.f\big(f(x)\big)$, ... In fact we introduce the abbreviations

$$1 \to \lambda fx.f(x)$$

$$2 \to \lambda fx.f\big(f(x)\big)$$

$$3 \to \lambda fx.f\Big(f\big(f(x)\big)\Big), \quad \text{etc.,}$$

and we also say, for example, that $\lambda fx.f\big(f(x)\big)$, or in full

$$\lambda f\Big[\lambda x\Big[\{f\}\big(\{f\}(x)\big)\Big]\Big],$$

represents the positive integer 2. Later we shall allow certain formulae to represent ordinals, but otherwise we leave them without explicit meaning; an implicit meaning may be suggested by the abbreviations used. In any case where any meaning is assigned to formulae it is desirable that the meaning should be invariant under conversion. Our definitions of the positive integers do not violate this requirement, since it may be proved that no two formulae representing different positive integers are convertible the one into the other.

In connection with the positive integers we introduce the abbreviation

$$S \to \lambda ufx.f\big(u(f, x)\big).$$

This formula has the property that, if $\mathbf{n}$ represents a positive integer, $S(\mathbf{n})$ is convertible to a formula representing its successor†.

Formulae representing undetermined positive integers will be represented by small letters in heavy type, and we adopt once for all the

† This follows from (A) below.

convention that, if a small letter, n say, stands for a positive integer, then the same letter in heavy type, $\mathbf{n}$, stands for the formula representing the positive integer. When no confusion arises from so doing, we shall not trouble to distinguish between an integer and the formula which represents it.

Suppose that $f(n)$ is a function of positive integers taking positive integers as values, and that there is a W.F.F. $\mathbf{F}$ not containing δ such that, for each positive integer n, $\mathbf{F}(\mathbf{n})$ is convertible to the formula representing $f(n)$. We shall then say that $f(n)$ is *λ-definable* or *formally definable*, and that $\mathbf{F}$ *formally defines* $f(n)$. Similar conventions are used for functions of more than one variable. The sum function is, for instance, formally defined by $\lambda abfx \,.\, a\big(f, b(f, x)\big)$; in fact, for any positive integers m, n, p for which $m+n=p$, we have

$$\big\{\lambda abfx \,.\, a\big(f, b(f, x)\big)\big\}(\mathbf{m}, \mathbf{n}) \text{ conv } \mathbf{p}.$$

In order to emphasize this relation we introduce the abbreviation

$$\mathbf{X}+\mathbf{Y} \rightarrow \big\{\lambda abfx \,.\, a\big(f, b(f, x)\big)\big\}(\mathbf{X}, \mathbf{Y})$$

and we shall use similar notations for sums of three or more terms, products, etc.

For any W.F.F. $\mathbf{G}$ we shall say that $\mathbf{G}$ *enumerates* the sequence $\mathbf{G}(1)$, $\mathbf{G}(2)$, ... and any other sequence whose terms are convertible to those of this sequence.

When a formula is convertible to another which is in normal form, the second is described as a *normal form* of the first, which is then said to *have a normal form.* I quote here some of the more important theorems concerning normal forms.

(A) *If a formula has two normal forms they are convertible into one another by the use of* (i) *alone.* (Church and Rosser [**1**], 479, 481.)

(B) *If a formula has a normal form then every well-formed part of it has a normal form.* (Church and Rosser [**1**], 480–481.)

(C) *There is (demonstrably) no process whereby it can be said of a formula whether it has a normal form.* (Church [**3**], 360;[i] Theorem XVIII.)

We often need to be able to describe formulae by means of positive integers. The method used here is due to Gödel (Gödel [**1**]).[a] To each single symbol s of the calculus we assign an integer $r[s]$ as in the table below.

s	{, (, or [	},), or]	λ	δ	a	...	z	x'	x''	x'''	...
$r[s]$	1	2	3	4	5	...	30	31	32	33	...

[i]104

If $s_1, s_2, \ldots, s_k$ is a sequence of symbols, then $2^{r[s_1]} 3^{r[s_2]} \ldots p_k^{r[s_k]}$ (where p_k is the k-th prime number) is called the *Gödel representation* (G.R.) of that sequence of symbols. No two W.F.F. have the same G.R.

Two theorems on G.R. of W.F.F. are quoted here.

(D) *There is a* W.F.F. "form" *such that if* a *is the* G.R. *of a* W.F.F. **A** *without free variables, then* form (a) conv **A**. (This follows from a similar theorem to be found in Church [**2**], 53 66. Metads are used there in place of G.R.)

(E) *There is a* W.F.F. Gr *such that, if* **A** *is a* W.F.F. *with a normal form without free variables, then* Gr(**A**) conv **a**, *where* a *is the* G.R. *of a normal form of* **A**. [Church [**2**], 53, 66, as (D).]

2. *Effective calculability. Abbreviation of treatment.*

A function is said to be "effectively calculable" if its values can be found by some purely mechanical process. Although it is fairly easy to get an intuitive grasp of this idea, it is nevertheless desirable to have some more definite, mathematically expressible definition. Such a definition was first given by Gödel at Princeton in 1934 (Gödel [**2**], 26),[i] following in part an unpublished suggestion of Herbrand, and has since been developed by Kleene [**2**]).[a] These functions were described as "general recursive" by Gödel. We shall not be much concerned here with this particular definition. Another definition of effective calculability has been given by Church (Church [**3**], 356–358),[ii] who identifies it with λ-definability. The author has recently suggested a definition corresponding more closely to the intuitive idea (Turing [**1**],[a] see also Post [**1**][a]). It was stated above that "a function is effectively calculable if its values can be found by some purely mechanical process". We may take this statement literally, understanding by a purely mechanical process one which could be carried out by a machine. It is possible to give a mathematical description, in a certain normal form, of the structures of these machines. The development of these ideas leads to the author's definition of a computable function, and to an identification of computability† with effective calculability. It is not difficult, though somewhat laborious, to prove that these three definitions are equivalent (Kleene [**3**], Turing [**2**]).

† We shall use the expression "computable function" to mean a function calculable by a machine, and we let "effectively calculable" refer to the intuitive idea without particular identification with any one of these definitions. We do not restrict the values taken by a computable function to be natural numbers; we may for instance have computable propositional functions.

[i] 70 [ii] 100–102

In the present paper we shall make considerable use of Church's identification of effective calculability with λ-definability, or, what comes to the same thing, of the identification with computability and one of the equivalence theorems. In most cases where we have to deal with an effectively calculable function, we shall introduce the corresponding W.F.F. with some such phrase as "the function f is effectively calculable, let F be a formula λ defining it", or "let F be a formula such that $F(\mathbf{n})$ is convertible to . . . whenever $\mathbf{n}$ represents a positive integer". In such cases there is no difficulty in seeing how a machine could in principle be designed to calculate the values of the function concerned; and, assuming this done, the equivalence theorem can be applied. A statement of what the formula F actually is may be omitted. We may immediately introduce on this basis a W.F.F. ϖ with the property that

$$\varpi(\mathbf{m}, \mathbf{n}) \text{ conv } \mathbf{r},$$

if r is the greatest positive integer, if any, for which m^r divides n and r is 1 if there is none. We also introduce Dt with the properties

$$\mathrm{Dt}(\mathbf{n}, \mathbf{n}) \text{ conv } 3,$$

$$\mathrm{Dt}(\mathbf{n}+\mathbf{m}, \mathbf{n}) \text{ conv } 2,$$

$$\mathrm{Dt}(\mathbf{n}, \mathbf{n}+\mathbf{m}) \text{ conv } 1.$$

There is another point to be made clear in connection with the point of view that we are adopting. It is intended that all proofs that are given should be regarded no more critically than proofs in classical analysis. The subject matter, roughly speaking, is constructive systems of logic, but since the purpose is directed towards choosing a particular constructive system of logic for practical use, an attempt at this stage to put our theorems into constructive form would be putting the cart before the horse.

Those computable functions which take only the values 0 and 1 are of particular importance, since they determine and are determined by computable properties, as may be seen by replacing "0" and "1" by "true" and "false". But, besides this type of property, we may have to consider a different type, which is, roughly speaking, less constructive than the computable properties, but more so than the general predicates of classical mathematics. Suppose that we have a computable function of the natural numbers taking natural numbers as values, then corresponding to this function there is the property of being a value of the function. Such a property we shall describe as "axiomatic"; the reason for using this term is that it is possible to define such a property by giving a set of axioms, the property to hold for a given argument if and only if it is possible to deduce that it holds from the axioms.

Axiomatic properties may also be characterized in this way. A property ψ of positive integers is axiomatic if and only if there is a computable property ϕ of two positive integers, such that $\psi(x)$ is true if and only if there is a positive integer y such that $\phi(x, y)$ is true. Or again ψ is axiomatic if and only if there is a W.F.F. **F** such that $\psi(n)$ is true if and only if **F**(**n**) conv 2.

3. *Number-theoretic theorems.*

By a *number-theoretic theorem*† we shall mean a theorem of the form "$\theta(x)$ vanishes for infinitely many natural numbers x", where $\theta(x)$ is a primitive recursive‡ function.

We shall say that a problem is number-theoretic if it has been shown that any solution of the problem may be put in the form of a proof of one or more number-theoretic theorems. More accurately we may say that a class of problems is number-theoretic if the solution of any one of them can be transformed (by a uniform process) into the form of proofs of number-theoretic theorems.

I shall now draw a few consequences from the definition of "number theoretic theorems", and in section 5 I shall try to justify confining our consideration to this type of problem.

† I believe that there is no generally accepted meaning for this term, but it should be noticed that we are using it in a rather restricted sense. The most generally accepted meaning is probably this: suppose that we take an arbitrary formula of the functional calculus of the first order and replace the function variables by primitive recursive relations. The resulting formula represents a typical number-theoretic theorem in this (more general) sense.

‡ Primitive recursive functions of natural numbers are defined inductively as follows. Suppose that $f(x_1, \ldots, x_{n-1})$, $g(x_1, \ldots, x_n)$, $h(x_1, \ldots, x_{n+1})$ are primitive recursive, then $\phi(x_1, \ldots, x_n)$ is primitive recursive if it is defined by one of the sets of equations (a) to (e).

$$(a)\ \phi(x_1, \ldots, x_n) = h\Big(x_1, \ldots, x_{m-1}, g(x_1, \ldots, x_n), x_{m+1}, \ldots, x_{n-1}, x_m\Big) \quad (1 \leqslant m \leqslant n);$$

$$(b)\ \phi(x_1, \ldots, x_n) = f(x_2, \ldots, x_n);$$

$$(c)\ \phi(x_1) = a, \text{ where } n = 1 \text{ and } a \text{ is some particular natural number};$$

$$(d)\ \phi(x_1) = x_1 + 1 \quad (n = 1);$$

$$(e)\ \phi(x_1, \ldots, x_{n-1}, 0) = f(x_1, \ldots, x_{n-1});$$

$$\phi(x_1, \ldots, x_{n-1}, x_n + 1) = h\Big(x_1, \ldots, x_n, \phi(x_1, \ldots, x_n)\Big).$$

The class of primitive recursive functions is more restricted than the class of computable functions, but it has the advantage that there is a process whereby it can be said of a set of equations whether it defines a primitive recursive function in the manner described above.

If $\phi(x_1, \ldots, x_n)$ is primitive recursive, then $\phi(x_1, \ldots, x_n) = 0$ is described as a primitive recursive relation between $x_1, \ldots, x_n$.

An alternative form for number-theoretic theorems is "for each natural number x there exists a natural number y such that $\phi(x, y)$ vanishes", where $\phi(x, y)$ is primitive recursive. In other words, there is a rule whereby, given the function $\theta(x)$, we can find a function $\phi(x, y)$, or given $\phi(x, y)$, we can find a function $\theta(x)$, such that "$\theta(x)$ vanishes infinitely often" is a necessary and sufficient condition for "for each x there is a y such that $\phi(x, y)=0$". In fact, given $\theta(x)$, we define

$$\phi(x, y)=\theta(x)+\alpha(x, y),$$

where $\alpha(x, y)$ is the (primitive recursive) function with the properties

$$\begin{aligned} \alpha(x, y) &= 1 \quad (y \leqslant x), \\ &= 0 \quad (y > x). \end{aligned}$$

If on the other hand we are given $\phi(x, y)$ we define $\theta(x)$ by the equations

$$\theta_1(0)=3,$$

$$\theta_1(x+1)=2^{(1+\varpi_2(\theta_1(x)))\sigma(\phi(\varpi_3(\theta_1(x))-1,\ \varpi_2(\theta_1(x))))}\,3^{\varpi_3(\theta_1(x))+1-\sigma(\phi(\varpi_3(\theta_1(x))-1,\ \varpi_2(\theta_1(x))))},$$

$$\theta(x)=\phi\Big(\varpi_3\big(\theta_1(x)\big)-1,\ \varpi_2\big(\theta_1(x)\big)\Big),$$

where $\varpi_r(x)$ is defined so as to mean "the largest s for which r^s divides x". The function $\sigma(x)$ is defined by the equations $\sigma(0)=0$, $\sigma(x+1)=1$. It is easily verified that the functions so defined have the desired properties.

We shall now show that questions about the truth of the statements of the form "does $f(x)$ vanish identically", where $f(x)$ is a computable function, can be reduced to questions about the truth of number-theoretic theorems. It is understood that in each case the rule for the calculation of $f(x)$ is given and that we are satisfied that this rule is valid, *i.e.* that the machine which should calculate $f(x)$ is circle free (Turing [**1**], 233).[i] The function $f(x)$, being computable, is general recursive in the Herbrand-Gödel sense, and therefore, by a general theorem due to Kleene†, is expressible in the form

$$\psi\big(\epsilon y[\phi(x, y)=0]\big), \tag{3.2}$$

where $\epsilon y[\mathfrak{A}(y)]$ means "the least y for which $\mathfrak{A}(y)$ is true" and $\psi(y)$ and $\phi(x, y)$ are primitive recursive functions. Without loss of generality, we may suppose that the functions ϕ, ψ take only the values 0, 1. Then, if

† Kleene [2] 727.[ii] This result is really superfluous for our purpose, since the proof that every computable function is general recursive proceeds by showing that these functions are of the form (3 . 2). (Turing [2], 161).

i 119   ii 237

we define $\rho(x)$ by the equations (3.1) and

$$\rho(0) = \psi(0)\big(1-\theta(0)\big),$$

$$\rho(x+1) = 1-\big(1-\rho(x)\big)\,\sigma\Big[1+\theta(x)-\psi\Big\{\varpi_2\big(\theta_1(x)\big)\Big\}\Big]$$

it will be seen that $f(x)$ vanishes identically if and only if $\rho(x)$ vanishes for infinitely many values of x.

The converse of this result is not quite true. We cannot say that the question about the truth of any number-theoretic theorem is reducible to a question about whether a corresponding computable function vanishes identically; we should have rather to say that it is reducible to the problem of whether a certain machine is circle free and calculates an identically vanishing function. But more is true: every number-theoretic theorem is equivalent to the statement that a corresponding machine is circle free. The behaviour of the machine may be described roughly as follows: the machine is one for the calculation of the primitive recursive function $\theta(x)$ of the number-theoretic problem, except that the results of the calculation are first arranged in a form in which the figures 0 and 1 do not occur, and the machine is then modified so that, whenever it has been found that the function vanishes for some value of the argument, then 0 is printed. The machine is circle free if and only if an infinity of these figures are printed, *i.e.* if and only if $\theta(x)$ vanishes for infinitely many values of the argument. That, on the other hand, questions of circle freedom may be reduced to questions of the truth of number-theoretic theorems follows from the fact that $\theta(x)$ is primitive recursive when it is defined to have the value 0 if a certain machine $\mathcal{M}$ prints 0 or 1 in its $(x+1)$-th complete configuration, and to have the value 1 otherwise.

The conversion calculus provides another normal form for the number-theoretic theorems, and the one which we shall find the most convenient to use. Every number-theoretic theorem is equivalent to a statement of the form "$\mathbf{A}(\mathbf{n})$ is convertible to 2 for every W.F.F. $\mathbf{n}$ representing a positive integer", $\mathbf{A}$ being a W.F.F. determined by the theorem; the property of $\mathbf{A}$ here asserted will be described briefly as "$\mathbf{A}$ is dual". Conversely such statements are reducible to number theoretic theorems. The first half of this assertion follows from our results for computable functions, or directly in this way. Since $\theta(x-1)+2$ is primitive recursive, it is formally definable, say, by means of a formula $\mathbf{G}$. Now there is (Kleene [**1**], 232) a W.F.F. $\mathfrak{P}$ with the property that, if $\mathbf{T}(\mathbf{r})$ is convertible to a formula representing a positive integer for each positive integer r, then $\mathfrak{P}(\mathbf{T}, \mathbf{n})$ is convertible to s, where s is the n-th positive integer t (if there is one) for which

$\mathbf{T}(\mathbf{t})$ conv 2; if $\mathbf{T}(\mathbf{t})$ conv 2 for less than n values of t then $\mathcal{P}(\mathbf{T}, \mathbf{n})$ has no normal form. The formula $\mathbf{G}\big(\mathcal{P}(\mathbf{G}, \mathbf{n})\big)$ is therefore convertible to 2 if and only if $\theta(x)$ vanishes for at least n values of x, and is convertible to 2 for every positive integer x if and only if $\theta(x)$ vanishes infinitely often. To prove the second half of the assertion, we take Gödel representations for the formulae of the conversion calculus. Let $c(x)$ be 0 if x is the G.R. of 2 (*i.e.* if x is $2^3 . 3^{10} . 5 . 7^3 . 11^{28} . 13 . 17 . 19^{10} . 23^2 . 29 . 31 . 37^{10} . 41^2 . 43 . 47^{28} . 53^2 . 59^2 . 61^2 . 67^2$) and let $c(x)$ be 1 otherwise. Take an enumeration of the G.R. of the formulae into which $\mathbf{A}(\mathbf{m})$ is convertible: let $a(m, n)$ be the n-th number in the enumeration. We can arrange the enumeration so that $a(m, n)$ is primitive recursive. Now the statement that $\mathbf{A}(\mathbf{m})$ is convertible to 2 for every positive integer m is equivalent to the statement that, corresponding to each positive integer m, there is a positive integer n such that $c\big(a(m, n)\big) = 0$; and this is number-theoretic.

It is easy to show that a number of unsolved problems, such as the problem of the truth of Fermat's last theorem, are number-theoretic. There are, however, also problems of analysis which are number-theoretic. The Riemann hypothesis gives us an example of this. We denote by $\zeta(s)$ the function defined for $\mathfrak{R}s = \sigma > 1$ by the series $\sum_{n=1}^{\infty} n^{-s}$ and over the rest of the complex plane with the exception of the point $s = 1$ by analytic continuation. The Riemann hypothesis asserts that this function does not vanish in the domain $\sigma > \frac{1}{2}$. It is easily shown that this is equivalent to saying that it does not vanish for $2 > \sigma > \frac{1}{2}$, $\mathfrak{I}s = t > 2$, *i.e.* that it does not vanish inside any rectangle $2 > \sigma > \frac{1}{2} + 1/T$, $T > t > 2$, where T is an integer greater than 2. Now the function satisfies the inequalities

$$\left.\begin{aligned} \left|\zeta(s) - \sum_{1}^{N} n^{-s} - \frac{N^{1-s}}{s-1}\right| < 2t(N-2)^{-\frac{1}{2}}, \quad 2 < \sigma < \tfrac{1}{2}, \quad t \geqslant 2, \\ |\zeta(s) - \zeta(s')| < 60t|s-s'|, \quad 2 < \sigma' < \tfrac{1}{2}, \quad t' \geqslant 2, \end{aligned}\right\}$$

and we can define a primitive recursive function $\xi(l, l', m, m', N, M)$ such that

$$\left|\xi(l, l', m, m', N, M) - M\left|\sum_{1}^{N} n^{-s} + \frac{N^{1-s}}{s-1}\right|\right| < 2, \quad \left(s = \frac{l}{l'} + i\frac{m}{m'}\right),$$

and therefore, if we put

$$\xi(l, M, m, M, M^2+2, M) = X(l, m, M),$$

we have

$$\left|\zeta\left(\frac{l+\vartheta}{M}+i\,\frac{m+\vartheta'}{M}\right)\right| \geqslant \frac{X(l,\, m,\, M)-122T}{M},$$

provided that

$$\tfrac{1}{2}+\frac{1}{T} \leqslant \frac{l-1}{M} < \frac{l+1}{M} < 2-\frac{1}{M}, \qquad 2 < \frac{m-1}{M} < \frac{m+1}{M} < T$$

$$(-1<\vartheta<1, \quad -1<\vartheta'<1).$$

If we define $B(M,\, T)$ to be the smallest value of $X(l,\, m,\, M)$ for which

$$\tfrac{1}{2}+\frac{1}{T}+\frac{1}{M} \leqslant \frac{l}{M} < 2-\frac{1}{M}, \qquad 2+\frac{1}{M} < \frac{m}{M} < T-\frac{1}{M},$$

then the Riemann hypothesis is true if for each T there is an M satisfying

$$B(M,\, T) > 122T.$$

If on the other hand there is a T such that, for all M, $B(M,\, T) \leqslant 122T$, the Riemann hypothesis is false; for let l_M, m_M be such that

$$X(l_M,\, m_M,\, M) \leqslant 122T,$$

then

$$\left|\zeta\left(\frac{l_M+im_M}{M}\right)\right| \leqslant \frac{244T}{M}.$$

Now if a is a condensation point of the sequence $(l_M+im_M)/M$ then since $\zeta(s)$ is continuous except at $s=1$ we must have $\zeta(a)=0$ implying the falsity of the Riemann hypothesis. Thus we have reduced the problem to the question whether for each T there is an M for which

$$B(M,\, T) > 122T.$$

$B(M,\, T)$ is primitive recursive, and the problem is therefore number-theoretic.

4. *A type of problem which is not number-theoretic*†.

Let us suppose that we are supplied with some unspecified means of solving number-theoretic problems; a kind of oracle as it were. We shall

† Compare Rosser [1].

not go any further into the nature of this oracle apart from saying that it cannot be a machine. With the help of the oracle we could form a new kind of machine (call them o-machines), having as one of its fundamental processes that of solving a given number-theoretic problem. More definitely these machines are to behave in this way. The moves of the machine are determined as usual by a table except in the case of moves from a certain internal configuration $\mathfrak{o}$. If the machine is in the internal configuration $\mathfrak{o}$ and if the sequence of symbols marked with l is then the well-formed† formula **A**, then the machine goes into the internal configuration $\mathfrak{p}$ or $\mathfrak{t}$ according as it is or is not true that **A** is dual. The decision as to which is the case is referred to the oracle.

These machines may be described by tables of the same kind as those used for the description of a-machines, there being no entries, however, for the internal configuration $\mathfrak{o}$. We obtain description numbers from these tables in the same way as before. If we make the convention that, in assigning numbers to internal configurations, $\mathfrak{o}$, $\mathfrak{p}$, $\mathfrak{t}$ are always to be q_2, q_3, q_4, then the description numbers determine the behaviour of the machines uniquely.

Given any one of these machines we may ask ourselves the question whether or not it prints an infinity of figures 0 or 1; I assert that this class of problem is not number-theoretic. In view of the definition of "number theoretic problem" this means that it is not possible to construct an o-machine which, when supplied‡ with the description of any other o-machine, will determine whether that machine is o-circle free. The argument may be taken over directly from Turing [**1**],[a] § 8. We say that a number is o-satisfactory if it is the description number of an o-circle free machine. Then, if there is an o-machine which will determine of any integer whether it is o-satisfactory, there is also an o-machine to calculate the values of the function $1-\phi_n(n)$. Let $r(n)$ be the n-th o-satisfactory number and let $\phi_n(m)$ be the m-th figure printed by the o-machine whose description number is $r(n)$. This o-machine is circle free and there is therefore an o-satisfactory number K such that $\phi_K(n) = 1-\phi_n(n)$ for all n. Putting $n = K$ yields a contradiction. This completes the proof that problems of circle freedom of o-machines are not number-theoretic.

Propositions of the form that an o-machine is o-circle free can always be put in the form of propositions obtained from formulae of the functional calculus of the first order by replacing *some* of the functional variables by primitive recursive relations. Compare foot-note † on page 162.

† Without real loss of generality we may suppose that **A** is always well formed.

‡ Compare Turing [1],[a] §6, 7.

5. *Syntactical theorems as number-theoretic theorems.*

I now mention a property of number-theoretic theorems which suggests that there is reason for regarding them as of particular importance.

Suppose that we have some axiomatic system of a purely formal nature. We do not concern ourselves at all in interpretations for the formulae of this system; they are to be regarded as of interest for themselves. An example of what is in mind is afforded by the conversion calculus (§ 1). Every sequence of symbols "**A** conv **B**", where **A** and **B** are well formed formulae, is a formula of the axiomatic system and is provable if the W.F.F. **A** is convertible to **B**. The rules of conversion give us the rules of procedure in this axiomatic system.

Now consider a new rule of procedure which is reputed to yield only formulae provable in the original sense. We may ask ourselves whether such a rule is valid. The statement that such a rule is valid would be number-theoretic. To prove this, let us take Gödel representations for the formulae, and an enumeration of the provable formulae; let $\phi(r)$ be the G.R. of the r-th formula in the enumeration. We may suppose $\phi(r)$ to be primitive recursive if we are prepared to allow repetitions in the enumeration. Let $\psi(r)$ be the G.R. of the r-th formula obtained by the new rule, then the statement that this new rule is valid is equivalent to the assertion of

$$(r)(\exists s)[\psi(r) = \phi(s)]$$

(the domain of individuals being the natural numbers). It has been shown in § 3 that such statements are number-theoretic.

It might plausibly be argued that all those theorems of mathematics which have any significance when taken alone are in effect syntactical theorems of this kind, stating the validity of certain "derived rules" of procedure. Without going so far as this, I should assert that theorems of this kind have an importance which makes it worth while to give them special consideration.

6. *Logic formulae.*

We shall call a formula **L** a *logic formula* (or, if it is clear that we are speaking of a W.F.F., simply a *logic*) if it has the property that, if **A** is a formula such that **L**(**A**) conv 2, then **A** is dual.

A logic formula gives us a means of satisfying ourselves of the truth of number-theoretic theorems. For to each number-theoretic proposition there corresponds a W.F.F. **A** which is dual if and only if the proposition is true. Now, if **L** is a logic and **L**(**A**) conv 2, then **A** is dual and we know that the corresponding number-theoretic proposition is true. It does not follow that, if **L** is a logic, we can use **L** to satisfy ourselves of the truth of *any* number-theoretic theorem.

If **L** is a logic, the set of formulae **A** for which **L**(**A**) conv 2 will be called the *extent* of **L**.

It may be proved by the use of (D), (E), p. 160, that there is a formula X such that, if **M** has a normal form, has no free variables and is not convertible to 2, then X(**M**) conv 1, but, if **M** conv 2, then X(**M**) conv 2. If **L** is a logic, then $\lambda x . X\big(\mathbf{L}(x)\big)$ is also a logic whose extent is the same as that of **L**, and which has the property that, if **A** has no free variables, then

$$\big\{\lambda x . X\big(\mathbf{L}(x)\big)\big\}(\mathbf{A})$$

either is always convertible to 1 or to 2 or else has no normal form. A logic with this property will be said to be *standardized*.

We shall say that a logic **L′** is *at least as complete as* a logic **L** if the extent of **L** is a subset of the extent of **L′**. The logic **L′** is *more complete than* **L** if the extent of **L** is a proper subset of the extent of **L′**.

Suppose that we have an effective set of rules by which we can prove formulae to be dual; *i.e.* we have a system of symbolic logic in which the propositions proved are of the form that certain formulae are dual. Then we can find a logic formula whose extent consists of just those formulae which can be proved to be dual by the rules; that is to say, there is a rule for obtaining the logic formula from the system of symbolic logic. In fact the system of symbolic logic enables us to obtain† a computable function of positive integers whose values run through the Gödel representations of the formulae provable by means of the given rules. By the theorem of equivalence of computable and λ-definable functions, there is a formula **J** such that **J**(1), **J**(2), ... are the G.R. of these formulae. Now let

$$W \rightarrow \lambda jv . \mathscr{P}\Big(\lambda u . \delta\big(j(u), v\big), 1, I, 2\Big).$$

Then I assert that W(**J**) is a logic with the required properties. The properties of $\mathscr{P}$ imply that $\mathscr{P}$(**C**, 1) is convertible to the least positive integer **n** for which **C**(**n**) conv 2, and has no normal form if there is no such integer. Consequently $\mathscr{P}$(**C**, 1, I, 2) is convertible to 2 if **C**(**n**) conv 2 for some positive integer n, and it has no normal form otherwise. That is to say that W(**J**, **A**) conv 2 if and only if $\delta\big($**J**(**n**), **A**$\big)$ conv 2, some n, *i.e.* if **J**(**n**) conv **A** some n.

† Compare Turing [1], 252,[i] second footnote, [2], 156.

[i] 138

There is conversely a formula W' such that, if **L** is a logic, then $W'(\mathbf{L})$ enumerates the extent of **L**. For there is a formula Q such that $Q(\mathbf{L}, \mathbf{A}, \mathbf{n})$ conv 2 if and only if $\mathbf{L}(\mathbf{A})$ is convertible to 2 in less than n steps. We then put

$$W' \rightarrow \lambda ln\,.\,\text{form}\Big(\varpi\Big(2,\ \mathcal{P}\Big(\lambda x\,.\,Q\Big(l,\ \text{form}\big(\varpi(2, x)\big),\ \varpi(3, x)\Big),\ n\Big)\Big)\Big).$$

Of course, $W'\big(W(\mathbf{J})\big)$ normally entirely different from **J** and $W\big(W'(\mathbf{L})\big)$ from **L**.

In the case where we have a symbolic logic whose propositions can be interpreted as number-theoretic theorems, but are not expressed in the form of the duality of formulae, we shall again have a corresponding logic formula, but its relation to the symbolic logic is not so simple. As an example let us take the case where the symbolic logic proves that certain primitive recursive functions vanish infinitely often. As was shown in § 3, we can associate with each such proposition a W.F.F. which is dual if and only if the proposition is true. When we replace the propositions of the symbolic logic by theorems on the duality of formulae in this way, our previous argument applies and we obtain a certain logic formula **L**. However, **L** does not determine uniquely which are the propositions provable in the symbolic logic; for it is possible that "$\theta_1(x)$ vanishes infinitely often" and "$\theta_2(x)$ vanishes infinitely often" are both associated with "**A** is dual", and that the first of these propositions is provable in the system, but the second not. However, if we suppose that the system of symbolic logic is sufficiently powerful to be able to carry out the argument on pp. 164–165 then this difficulty cannot arise. There is also the possibility that there may be formulae in the extent of **L** with no propositions of the form "$\theta(x)$ vanishes infinitely often" corresponding to them. But to each such formula we can assign (by a different argument) a proposition p of the symbolic logic which is a necessary and sufficient condition for **A** to be dual. With p is associated (in the first way) a formula $\mathbf{A}'$. Now **L** can always be modified so that its extent contains $\mathbf{A}'$ whenever it contains **A**.

We shall be interested principally in questions of completeness. Let us suppose that we have a class of systems of symbolic logic, the propositions of these systems being expressed in a uniform notation and interpretable as number-theoretic theorems; suppose also that there is a rule by which we can assign to each proposition p of the notation a W.F.F. $\mathbf{A}_p$ which is dual if and only if p is true, and that to each W.F.F. **A** we can assign a proposition $p_{\mathbf{A}}$ which is a necessary and sufficient condition for **A** to be dual. $p_{\mathbf{A}_p}$ is to be expected to differ from p. To each symbolic logic C we can assign two logic formulae $\mathbf{L}_C$ and $\mathbf{L}_C'$. A formula **A** belongs to the extent of $\mathbf{L}_C$ if $p_{\mathbf{A}}$ is provable in C, while the extent of $\mathbf{L}_C'$ consists of all $\mathbf{A}_p$, where p is provable in C. Let us say that the class of symbolic logics is complete if each true proposition is provable in one of them: let us also say that a class of logic formulae is complete if the set-theoretic sum of the extents of these logics includes all dual formulae. I assert that a necessary condition for a class of symbolic logics C to be complete is that the class of logics $\mathbf{L}_C$ is complete, while a sufficient condition is that the class of logics $\mathbf{L}_C'$ is complete. Let us suppose that the class of symbolic logics is complete; consider $p_{\mathbf{A}}$, where **A** is arbitrary but dual. It must be provable in one of the systems, C say. **A** therefore belongs to the extent of $\mathbf{L}_C$, *i.e.* the class of logics $\mathbf{L}_C$ is complete. Now suppose the class of logics $\mathbf{L}_C'$ to be complete. Let p be an arbitrary true proposition of the notation; $\mathbf{A}_p$ must belong to the extent of some $\mathbf{L}_C'$, and this means that p is provable in C.

We shall say that a single logic formula **L** is complete if its extent includes all dual formulae; that is to say, it is complete if it enables us to prove every true number-theoretic theorem. It is a consequence of the theorem of Gödel (if suitably extended) that no logic formula is complete, and this also follows from (C), p. 159, or from the results of Turing [1],[a] § 8, when taken in conjunction with § 3 of the present paper. The idea of completeness of a logic formula is not therefore very important, although it is useful to have a term for it.

Suppose **Y** to be a W.F.F. such that $\mathbf{Y}(\mathbf{n})$ is a logic for each positive integer n. The formulae of the extent of $\mathbf{Y}(\mathbf{n})$ are enumerated by $W\big(\mathbf{Y}(\mathbf{n})\big)$, and the combined extents of these logics by

$$\lambda r\,.\,W\Big(\mathbf{Y}\big(\varpi(2,r),\ \varpi(3,r)\big)\Big).$$

If we put

$$\Gamma \to \lambda y\,.\,W'\bigg(\lambda r\,.\,W\Big(y\big(\varpi(2,r),\ \varpi(3,r)\big)\Big)\bigg),$$

then $\Gamma(\mathbf{Y})$ is a logic whose extent is the combined extent of

$$\mathbf{Y}(1),\quad \mathbf{Y}(2),\quad \mathbf{Y}(3),\quad \ldots.$$

To each W.F.F. **L** we can assign a W.F.F. $V(\mathbf{L})$ such that a necessary and sufficient condition for **L** to be a logic formula is that $V(\mathbf{L})$ is dual. Let Nm be a W.F.F. which enumerates all formulae with normal forms and no free variables. Then the condition for $\mathbf{L}$ to be a logic is that $\mathbf{L}\big(\mathrm{Nm}(\mathbf{r}), \mathbf{s}\big)$ conv 2 for all positive integers r, s, *i.e.* that

$$\lambda a \,.\, \mathbf{L}\Big(\mathrm{Nm}\big(\varpi(2, a)\big),\ \varpi(3, a)\Big)$$

is dual. We may therefore put

$$V \rightarrow \lambda l a \,.\, l\Big(\mathrm{Nm}\big(\varpi(2, a)\big),\ \varpi(3, a)\Big).$$

7. *Ordinals.*

We begin our treatment of ordinals with some brief definitions from the Cantor theory of ordinals, but for the understanding of some of the proofs a greater amount of the Cantor theory is necessary than is set out here.

Suppose that we have a class determined by the propositional function $D(x)$ and a relation $G(x, y)$ ordering its members, *i.e.* satisfying

$$\left.\begin{array}{ll} G(x, y) \,\&\, G(y, z) \supset G(x, z), & \text{(i)} \\ D(x) \,\&\, D(y) \supset G(x, y) \vee G(y, x) \vee x = y, & \text{(ii)} \\ G(x, y) \supset D(x) \,\&\, D(y), & \text{(iii)} \\ \sim G(x, x). & \text{(iv)} \end{array}\right\} \quad (7.1)$$

The class defined by $D(x)$ is then called a *series* with the ordering relation $G(x, y)$. The series is said to be *well ordered* and the ordering relation is called an *ordinal* if every sub-series which is not void has a first term, *i.e.* if

$$(D')\Big\{(\exists x)\big(D'(x)\big) \,\&\, (x)\big(D'(x) \supset D(x)\big)$$
$$\supset (\exists z)(y)\Big[D'(z) \,\&\, \big(D'(y) \supset G(z, y) \vee z = y\big)\Big]\Big\}. \quad (7.2)$$

The condition (7.2) is equivalent to another, more suitable for our purposes, namely the condition that every descending subsequence must terminate; formally

$$(x)\Big\{D'(x) \supset D(x) \,\&\, (\exists y)\big(D'(y) \,\&\, G(y, x)\big)\Big\} \supset (x)\big(\sim D'(x)\big). \quad (7.3)$$

The ordering relation $G(x, y)$ is said to be similar to $G'(x, y)$ if there is a one-one correspondence between the series transforming the one relation into the other. This is best expressed formally, thus

$$(\exists M)\Big[(x)\{D(x)\supset(\exists x')\,M(x,x')\}\,\&\,(x')\{D'(x')\supset(\exists x)\,M(x,x')\}$$
$$\&\Big\{\big(M(x,x')\,\&\,M(x,x'')\big)\vee\big(M(x',x)\,\&\,M(x'',x)\big)\supset x'=x''\Big\}$$
$$\&\Big\{M(x,x')\,\&\,M(y,y')\supset\big(G(x,y)\equiv G(x',y')\big)\Big\}\Big]. \qquad (7.4)$$

Ordering relations are regarded as belonging to the same ordinal if and only if they are similar.

We wish to give names to all the ordinals, but this will not be possible until they have been restricted in some way; the class of ordinals, as at present defined, is more than enumerable. The restrictions that we actually impose are these: $D(x)$ is to imply that x is a positive integer; $D(x)$ and $G(x, y)$ are to be computable properties. Both of the propositional functions $D(x)$, $G(x, y)$ can then be described by means of a single W.F.F. $\mathbf{\Omega}$ with the properties:

$\mathbf{\Omega}(\mathbf{m}, \mathbf{n})$ conv 4 unless both $D(m)$ and $D(n)$ are true,

$\mathbf{\Omega}(\mathbf{m}, \mathbf{m})$ conv 3 if $D(m)$ is true,

$\mathbf{\Omega}(\mathbf{m}, \mathbf{n})$ conv 2 if $D(m)$, $D(n)$, $G(m, n)$, $\sim(m = n)$ are true,

$\mathbf{\Omega}(\mathbf{m}, \mathbf{n})$ conv 1 if $D(m)$, $D(n)$, $\sim G(m, n)$, $\sim(m = n)$ are true.

In consequence of the conditions to which $D(x)$, $G(x, y)$ are subjected, $\mathbf{\Omega}$ must further satisfy:

(*a*) if $\mathbf{\Omega}(\mathbf{m}, \mathbf{n})$ is convertible to 1 or 2, then $\mathbf{\Omega}(\mathbf{m}, \mathbf{m})$ and $\mathbf{\Omega}(\mathbf{n}, \mathbf{n})$ are convertible to 3,

(*b*) if $\mathbf{\Omega}(\mathbf{m}, \mathbf{m})$ and $\mathbf{\Omega}(\mathbf{n}, \mathbf{n})$ are convertible to 3, then $\mathbf{\Omega}(\mathbf{m}, \mathbf{n})$ is convertible to 1, 2, or 3,

(*c*) if $\mathbf{\Omega}(\mathbf{m}, \mathbf{n})$ is convertible to 1, then $\mathbf{\Omega}(\mathbf{n}, \mathbf{m})$ is convertible to 2 and conversely,

(*d*) if $\mathbf{\Omega}(\mathbf{m}, \mathbf{n})$ and $\mathbf{\Omega}(\mathbf{n}, \mathbf{p})$ are convertible to 1, then $\mathbf{\Omega}(\mathbf{m}, \mathbf{p})$ is also,

(*e*) there is no sequence m_1, m_2, ... such that $\mathbf{\Omega}(\mathbf{m}_{i+1}, \mathbf{m}_i)$ conv 2 for each positive integer i,

(*f*) $\mathbf{\Omega}(\mathbf{m}, \mathbf{n})$ is always convertible to 1, 2, 3, or 4.

If a formula $\mathbf{\Omega}$ satisfies these conditions then there are corresponding propositional functions $D(x)$, $G(x, y)$. We shall therefore say that $\mathbf{\Omega}$ is an *ordinal formula* if it satisfies the conditions (a)–(f). It will be seen that a consequence of this definition is that Dt is an ordinal formula; it represents the ordinal ω. The definition that we have given does not pretend to have virtues such as elegance or convenience. It has been introduced rather to fix our ideas and to show how it is possible in principle to describe ordinals by means of well formed formulae. The definitions could be modified in a number of ways. Some such modifications are quite trivial; they are typified by modifications such as changing the numbers 1, 2, 3, 4, used in the definition, to others. Two such definitions will be said to be equivalent; in general, we shall say that two definitions are equivalent if there are W.F.F. **T**, **T′** such that, if **A** is an ordinal formula under one definition and represents the ordinal α, then **T′**(**A**) is an ordinal formula under the second definition and represents the same ordinal; and, conversely, if **A′** is an ordinal formula under the second definition representing α, then **T**(**A′**) represents α under the first definition. Besides definitions equivalent in this sense to our original definition, there are a number of other possibilities open. Suppose for instance that we do not require $D(x)$ and $G(x, y)$ to be computable, but that we require only that $D(x)$ and $G(x, y) \,\&\, x < y$ are axiomatic†. This leads to a definition of an ordinal formula which is (presumably) not equivalent to the definition that we are using‡. There are numerous possibilities, and little to guide us in choosing one definition rather than another. No one of them could well be described as "wrong"; some of them may be found more valuable in applications than others, and the particular choice that we have made has been determined partly by the applications that we have in view. In the case of theorems of a negative character, one would wish to prove them for each one of the possible definitions of "ordinal formula". This programme could, I think, be carried through for the negative results of § 9, 10.

Before leaving the subject of possible ways of defining ordinal formulae, I must mention another definition due to Church and Kleene (Church and Kleene [**1**]). We can make use of this definition in constructing ordinal logics, but it is more convenient to use a slightly different definition which is equivalent (in the sense just described) to the Church-Kleene definition as modified in Church [**4**].

† To require $G(x, y)$ to be axiomatic amounts to requiring $G(x, y)$ to be computable on account of (7. 1) (ii).

‡ On the other hand, if $D(x)$ is axiomatic and $G(x, y)$ is computable in the modified sense that there is a rule for determining whether $G(x, y)$ is true which leads to a definite result in all cases where $D(x)$ and $D(y)$ are true, the corresponding definition of ordinal formula is equivalent to our definition. To give the proof would be too much of a digression. Probably other equivalences of this kind hold.

Introduce the abbreviations

$$U \rightarrow \lambda ufx\,.\,u\Big(\lambda y\,.f\big(y(I,\,x)\big)\Big),$$

$$\text{Suc} \rightarrow \lambda aufx\,.f\big(a(u,\,f,\,x)\big).$$

We define first a partial ordering relation "$<$" which holds between certain pairs of W.F.F. [conditions (1)–(5)].

(1) If $\mathbf{A}$ conv $\mathbf{B}$, then $\mathbf{A} < \mathbf{C}$ implies $\mathbf{B} < \mathbf{C}$ and $\mathbf{C} < \mathbf{A}$ implies $\mathbf{C} < \mathbf{B}$.

(2) $\mathbf{A} < \text{Suc}\,(\mathbf{A})$.

(3) For any positive integers m and n, $\lambda ufx\,.\,\mathbf{R}(\mathbf{n}) < \lambda ufx\,.\,\mathbf{R}(\mathbf{m})$ implies $\lambda ufx\,.\,\mathbf{R}(\mathbf{n}) < \lambda ufx\,.\,u(\mathbf{R})$.

(4) If $\mathbf{A} < \mathbf{B}$ and $\mathbf{B} < \mathbf{C}$, then $\mathbf{A} < \mathbf{C}$. (1)–(4) are required for any W.F.F. $\mathbf{A}$, $\mathbf{B}$, $\mathbf{C}$, $\lambda ufx\,.\,\mathbf{R}$.

(5) The relation $\mathbf{A} < \mathbf{B}$ holds only when compelled to do so by (1)–(4).

We define C-K ordinal formulae by the conditions (6)–(10).

(6) If $\mathbf{A}$ conv $\mathbf{B}$ and $\mathbf{A}$ is a C-K ordinal formula, then $\mathbf{B}$ is a C-K ordinal formula.

(7) U is a C-K ordinal formula.

(8) If $\mathbf{A}$ is a C-K ordinal formula, then Suc ($\mathbf{A}$) is a C-K ordinal formula.

(9) If $\lambda ufx\,.\,\mathbf{R}(\mathbf{n})$ is a C-K ordinal formula and

$$\lambda ufx\,.\,\mathbf{R}(\mathbf{n}) < \lambda ufx\,.\,\mathbf{R}\big(S(\mathbf{n})\big)$$

for each positive integer n, then $\lambda ufx\,.\,u(\mathbf{R})$ is a C-K ordinal formula†.

(10) A formula is a C-K ordinal formula only if compelled to be so by (6)–(9).

† If we also allow $\lambda ufx\,.\,u(\mathbf{R})$ to be a C-K ordinal formula when

$$\lambda ufx\,.\,\mathbf{n}(\mathbf{R}) \text{ conv } \lambda ufx\,.\,S(\mathbf{n},\,\mathbf{R})$$

for all n, then the formulae for sum, product and exponentiation of C-K ordinal formulae can be much simplified. For instance, if $\mathbf{A}$ and $\mathbf{B}$ represent α and β, then

$$\lambda ufx\,.\,\mathbf{B}(u,\,f,\,\mathbf{A}(u,\,f,\,x))$$

represents $\alpha+\beta$. Property (6) remains true.

The representation of ordinals by formulae is described by (11)–(15).

(11) If $\mathbf{A} \operatorname{conv} \mathbf{B}$ and $\mathbf{A}$ represents α, then $\mathbf{B}$ represents α.

(12) U represents 1.

(13) If $\mathbf{A}$ represents α, then Suc $(\mathbf{A})$ represents $\alpha+1$.

(14) If $\lambda ufx . \mathbf{R}(\mathbf{n})$ represents α_n for each positive integer n, then $\lambda ufx . u(\mathbf{R})$ represents the upper bound of the sequence $\alpha_1, \alpha_2, \alpha_3, \ldots$.

(15) A formula represents an ordinal only when compelled to do so by (11)–(14).

We denote any ordinal represented by $\mathbf{A}$ by $\Xi_{\mathbf{A}}$ without prejudice to the possibility that more than one ordinal may be represented by $\mathbf{A}$. We shall write $\mathbf{A} \leqslant \mathbf{B}$ to mean $\mathbf{A} < \mathbf{B}$ or $\mathbf{A} \operatorname{conv} \mathbf{B}$.

In proving properties of C-K ordinal formulae we shall often use a kind of analogue of the principle of transfinite induction. If ϕ is some property and we have:

$$\left.\begin{array}{l} (a) \text{ If } \mathbf{A} \operatorname{conv} \mathbf{B} \text{ and } \phi(\mathbf{A}), \text{ then } \phi(\mathbf{B}), \\ (b)\ \phi(U), \\ (c) \text{ If } \phi(\mathbf{A}), \text{ then } \phi\big(\text{Suc}\,(\mathbf{A})\big), \\ (d) \text{ If } \phi\big(\lambda ufx . \mathbf{R}(\mathbf{n})\big) \text{ and } \lambda ufx . \mathbf{R}(\mathbf{n}) < \lambda ufx . \mathbf{R}\big(S(\mathbf{n})\big) \text{ for each} \\ \text{positive integer } n, \text{ then} \\ \qquad\qquad \phi\big(\lambda ufx . u(\mathbf{R})\big); \end{array}\right\} \quad (7.5)$$

then $\phi(\mathbf{A})$ for each C-K ordinal formula $\mathbf{A}$. To prove the validity of this principle we have only to observe that the class of formulae $\mathbf{A}$ satisfying $\phi(\mathbf{A})$ is one of those of which the class of C-K ordinal formulae was defined to be the smallest. We can use this principle to help us to prove:—

(i) Every C-K ordinal formula is convertible to the form $\lambda ufx . \mathbf{B}$, where $\mathbf{B}$ is in normal form.

(ii) There is a method by which for any C-K ordinal formula, we can determine into which of the forms U, Suc $(\lambda ufx . \mathbf{B})$, $\lambda ufx . u(\mathbf{R})$ (where u is free in $\mathbf{R}$) it is convertible, and by which we can determine $\mathbf{B}$, $\mathbf{R}$. In each case $\mathbf{B}$, $\mathbf{R}$ are unique apart from conversions.

(iii) If $\mathbf{A}$ represents any ordinal, $\Xi_{\mathbf{A}}$ is unique. If $\Xi_{\mathbf{A}}$, $\Xi_{\mathbf{B}}$ exist and $\mathbf{A} < \mathbf{B}$, then $\Xi_{\mathbf{A}} < \Xi_{\mathbf{B}}$.

(iv) If $\mathbf{A}, \mathbf{B}, \mathbf{C}$ are C-K ordinal formulae and $\mathbf{B} < \mathbf{A}$, $\mathbf{C} < \mathbf{A}$, then either $\mathbf{B} < \mathbf{C}$, $\mathbf{C} < \mathbf{B}$, or $\mathbf{B}$ conv $\mathbf{C}$.

(v) A formula $\mathbf{A}$ is a C-K ordinal formula if:

(A) $U \leqslant \mathbf{A}$,

(B) If $\lambda ufx . u(\mathbf{R}) \leqslant \mathbf{A}$ and n is a positive integer, then

$$\lambda ufx . \mathbf{R}(\mathbf{n}) < \lambda ufx . \mathbf{R}\big(S(\mathbf{n})\big),$$

(C) For any two W.F.F. $\mathbf{B}$, $\mathbf{C}$ with $\mathbf{B} < \mathbf{A}$, $\mathbf{C} < \mathbf{A}$ we have $\mathbf{B} < \mathbf{C}$, $\mathbf{C} < \mathbf{B}$, or $\mathbf{B}$ conv $\mathbf{C}$, but never $\mathbf{B} < \mathbf{B}$,

(D) There is no infinite sequence $\mathbf{B}_1, \mathbf{B}_2, \ldots$ for which

$$\mathbf{B}_r < \mathbf{B}_{r-1} < \mathbf{A}$$

for each r.

(vi) There is a formula H such that, if $\mathbf{A}$ is a C-K ordinal formula, then $H(\mathbf{A})$ is an ordinal formula representing the same ordinal. $H(\mathbf{A})$ is not an ordinal formula unless $\mathbf{A}$ is a C-K ordinal formula.

Proof of (i). Take $\phi(\mathbf{A})$ to be "$\mathbf{A}$ is convertible to the form $\lambda ufx . \mathbf{B}$, where $\mathbf{B}$ is in normal form". The conditions (*a*) and (*b*) are trivial. For (*c*), suppose that $\mathbf{A}$ conv $\lambda ufx . \mathbf{B}$, where $\mathbf{B}$ is in normal form; then

$$\text{Suc}\,(\mathbf{A}) \text{ conv } \lambda ufx . f(\mathbf{B})$$

and $f(\mathbf{B})$ is in normal form. For (*d*) we have only to show that $u(\mathbf{R})$ has a normal form, *i.e.* that $\mathbf{R}$ has a normal form; and this is true since $\mathbf{R}(1)$ has a normal form.

Proof of (ii). Since, by hypothesis, the formula is a C-K ordinal formula we have only to perform conversions on it until it is in one of the forms described. It is not possible to convert it into two of these three forms. For suppose that $\lambda ufx . f\big(\mathbf{A}(u, f, x)\big)$ conv $\lambda ufx . u(\mathbf{R})$ and is a C-K ordinal formula; it is then convertible to the form $\lambda ufx . \mathbf{B}$, where $\mathbf{B}$ is in normal form. But the normal form of $\lambda ufx . u(\mathbf{R})$ can be obtained by conversions on $\mathbf{R}$, and that of $\lambda ufx . f\big(\mathbf{A}(u, f, x)\big)$ by conversions on $\mathbf{A}(u, f, x)$ (as follows from Church and Rosser [1], Theorem 2); this, however, would imply that the formula in question had two normal forms, one of form $\lambda ufx . u(\mathbf{S})$ and one of form $\lambda ufx . f(\mathbf{C})$, which is impossible. Or let U conv $\lambda ufx . u(\mathbf{R})$, where $\mathbf{R}$ is a well formed formula with u as a free variable. We may suppose $\mathbf{R}$ to be in normal form. Now U is $\lambda ufx . u\Big(\lambda y . f\big(y(I, x)\big)\Big)$. By

(A), p. 159, **R** is identical with $\lambda y . f\big(y(I, x)\big)$, which does not have u as a free variable. It now remains to show only that if

$$\text{Suc}\,(\lambda ufx . \mathbf{B}) \text{ conv Suc}\,(\lambda ufx . \mathbf{B}') \quad \text{and} \quad \lambda ufx . u(\mathbf{R}) \text{ conv } \lambda ufx . u(\mathbf{R}'),$$

then **B** conv **B**′ and **R** conv **R**′.

If $$\text{Suc}\,(\lambda ufx . \mathbf{B}) \text{ conv Suc}\,(\lambda ufx . \mathbf{B}'),$$

then $$\lambda ufx . f(\mathbf{B}) \text{ conv } \lambda ufx . f(\mathbf{B}');$$

but both of these formulae can be brought to normal form by conversions on **B**, **B**′ and therefore **B** conv **B**′. The same argument applies in the case in which $\lambda ufx . u(\mathbf{R})$ conv $\lambda ufx . u(\mathbf{R}')$.

Proof of (iii). To prove the first half, take $\phi(\mathbf{A})$ to be "$\Xi_{\mathbf{A}}$ is unique". Then (7.5) (*a*) is trivial, and (*b*) follows from the fact that U is not convertible either to the form Suc (**A**) or to $\lambda ufx . u(\mathbf{R})$, where **R** has u as a free variable. For (*c*): Suc (**A**) is not convertible to the form $\lambda ufx . u(\mathbf{R})$; the possibility that Suc (**A**) represents an ordinal on account of (12) or (14) is therefore eliminated. By (13), Suc (**A**) represents $\alpha'+1$ if **A**′ represents α' and Suc (**A**) conv Suc (**A**′). If we suppose that **A** represents α, then **A**, **A**′, being C-K ordinal formulae, are convertible to the forms $\lambda ufx . \mathbf{B}$, $\lambda ufx . \mathbf{B}'$; but then, by (ii), **B** conv **B**′, *i.e.* **A** conv **A**′, and therefore $\alpha = \alpha'$ by the hypothesis $\phi(\mathbf{A})$. Then $\Xi_{\text{Suc}\,(\mathbf{A})} = \alpha'+1$ is unique. For (*d*): $\lambda ufx . u(\mathbf{R})$ is not convertible to the form Suc (**A**) or to U if **R** has u as a free variable. If $\lambda ufx . u(\mathbf{R})$ represents an ordinal, it is so therefore in virtue of (14), possibly together with (11). Now, if $\lambda ufx . u(\mathbf{R})$ conv $\lambda ufx . u(\mathbf{R}')$, then **R** conv **R**′, so that the sequence $\lambda ufx . \mathbf{R}(1)$, λufx, $\mathbf{R}(2)$, ... in (14) is unique apart from conversions. Then, by the induction hypothesis, the sequence $\alpha_1, \alpha_2, \alpha_3, \ldots$ is unique. The only ordinal that is represented by $\lambda ufx . u(\mathbf{R})$ is the upper bound of this sequence; and this is unique.

For the second half we use a type of argument rather different from our transfinite induction principle. The formulae **B** for which $\mathbf{A} < \mathbf{B}$ form the smallest class for which:

> Suc (**A**) belongs to the class.
> If **C** belongs to the class, then Suc (**C**) belongs to it.
> If $\lambda ufx . \mathbf{R}(\mathbf{n})$ belongs to the class and
>
> $$\lambda ufx . \mathbf{R}(\mathbf{n}) < \lambda ufx . \mathbf{R}(\mathbf{m}), \qquad (7.6)$$
>
> where m, n are some positive integers, then $\lambda ufx . u(\mathbf{R})$ belongs to it.
> If **C** belongs to the class and **C** conv **C**′, then **C**′ belongs to it.

It will be sufficient to prove that the class of formulae **B** for which either $\Xi_{\mathbf{B}}$ does not exist or $\Xi_{\mathbf{A}} < \Xi_{\mathbf{B}}$ satisfies the conditions (7.6). Now

$$\Xi_{\mathrm{Suc}\,(\mathbf{A})} = \Xi_{\mathbf{A}} + 1 > \Xi_{\mathbf{A}},$$

$$\Xi_{\mathrm{Suc}\,(\mathbf{C})} > \Xi_{\mathbf{C}} > \Xi_{\mathbf{A}} \text{ if } \mathbf{C} \text{ is in the class.}$$

If $\Xi_{\lambda ufx\,.\,\mathbf{R}(\mathbf{n})}$ does not exist, then $\Xi_{\lambda ufx\,.\,u(\mathbf{R})}$ does not exist, and therefore $\lambda ufx\,.\,u(\mathbf{R})$ is in the class. If $\Xi_{\lambda ufx\,.\,\mathbf{R}(\mathbf{n})}$ exists and is greater than $\Xi_{\mathbf{A}}$, and $\lambda ufx.\,\mathbf{R}(\mathbf{n}) < \lambda ufx\,.\,\mathbf{R}(\mathbf{m})$, then

$$\Xi_{\lambda ufx\,.\,u(\mathbf{R})} \geqslant \Xi_{\lambda ufx\,.\,\mathbf{R}(\mathbf{n})} > \Xi_{\mathbf{A}},$$

so that $\lambda ufx\,.\,u(\mathbf{R})$ belongs to the class.

Proof of (iv). We prove this by induction with respect to **A**. Take $\phi(\mathbf{A})$ to be "whenever $\mathbf{B} < \mathbf{A}$ and $\mathbf{C} < \mathbf{A}$ then $\mathbf{B} < \mathbf{C}$ or $\mathbf{C} < \mathbf{B}$ or **B** conv **C**". $\phi(U)$ follows from the fact that we never have $\mathbf{B} < U$. If we have $\phi(\mathbf{A})$ and $\mathbf{B} < \mathrm{Suc}\,(\mathbf{A})$, then either $\mathbf{B} < \mathbf{A}$ or **B** conv **A**; for we can find **D** such that $\mathbf{B} \leqslant \mathbf{D}$, and then $\mathbf{D} < \mathrm{Suc}\,(\mathbf{A})$ can be proved without appealing either to (1) or (5); (4) does not apply, so we must have **D** conv **A**. Then, if $\mathbf{B} < \mathrm{Suc}\,(\mathbf{A})$ and $\mathbf{C} < \mathrm{Suc}\,(\mathbf{A})$, we have four possibilities,

B conv **A**,	**C** conv **A**,
B conv **A**,	$\mathbf{C} < \mathbf{A}$,
$\mathbf{B} < \mathbf{A}$,	**C** conv **A**,
$\mathbf{B} < \mathbf{A}$,	$\mathbf{C} < \mathbf{A}$.

In the first case **B** conv **C**, in the second $\mathbf{C} < \mathbf{B}$, in the third $\mathbf{B} < \mathbf{C}$, and in the fourth the induction hypothesis applies.

Now suppose that $\lambda ufx\,.\,\mathbf{R}(\mathbf{n})$ is a C-K ordinal formula, that

$$\lambda ufx\,.\,\mathbf{R}(\mathbf{n}) < \lambda ufx\,.\,\mathbf{R}\big(S(\mathbf{n})\big) \quad \text{and} \quad \phi\big(\mathbf{R}(\mathbf{n})\big),$$

for each positive integer n, and that **A** conv $\lambda ufx\,.\,u(\mathbf{R})$. Then, if $\mathbf{B} < \mathbf{A}$, this means that $\mathbf{B} < \lambda ufx\,.\,\mathbf{R}(\mathbf{n})$ for some n; if we have also $\mathbf{C} < \mathbf{A}$, then $\mathbf{B} < \lambda ufx\,.\,\mathbf{R}(\mathbf{q})$, $\mathbf{C} < \lambda ufx\,.\,\mathbf{R}(\mathbf{q})$ for some q. Thus, for these **B** and **C**, the required result follows from $\phi\big(\lambda ufx\,.\,\mathbf{R}(\mathbf{q})\big)$.

Proof of (v). The conditions (C), (D) imply that the classes of inter-convertible formulae **B**, $\mathbf{B} < \mathbf{A}$ are well-ordered by the relation "$<$". We prove (v) by (ordinary) transfinite induction with respect to the order type α of the series formed by these classes; (α is, in fact, the solution of the equation $1+\alpha=\Xi_{\mathbf{A}}$, but we do not need this). We suppose then that (v) is true for all order types less than α. If $\mathbf{E}<\mathbf{A}$, then $\mathbf{E}$ satisfies the conditions of (v) and the corresponding order type is smaller: $\mathbf{E}$ is therefore a C-K ordinal formula. This expresses all consequences of the induction hypothesis that we need. There are three cases to consider:

(x) $\alpha=0$.

(y) $\alpha=\beta+1$.

(z) α is of neither of the forms (x), (y).

In case (x) we must have $\mathbf{A}$ conv U on account of (A). In case (y) there is a formula $\mathbf{D}$ such that $\mathbf{D}<\mathbf{A}$, and $\mathbf{B}\leqslant\mathbf{D}$ whenever $\mathbf{B}<\mathbf{A}$. The relation $\mathbf{D}<\mathbf{A}$ must hold in virtue of either (1), (2), (3), or (4). It cannot be in virtue of (4); for then there would be $\mathbf{B}$, $\mathbf{B}<\mathbf{A}$, $\mathbf{D}<\mathbf{B}$ contrary to (C), taken in conjunction with the definition of $\mathbf{D}$. If it is in virtue of (3), then α is the upper bound of a sequence $\alpha_1, \alpha_2, \ldots$ of ordinals, which are increasing by reason of (iii) and the conditions $\lambda ufx\,.\,\mathbf{R}(\mathbf{n})<\lambda ufx\,.\,\mathbf{R}\big(S(\mathbf{n})\big)$ in (B). This is inconsistent with $\alpha=\beta+1$. This means that (2) applies [after we have eliminated (1) by suitable conversions on $\mathbf{A}$, $\mathbf{D}$] and we see that $\mathbf{A}$ conv Suc ($\mathbf{D}$); but, since $\mathbf{D}<\mathbf{A}$, $\mathbf{D}$ is a C-K ordinal formula, and $\mathbf{A}$ must therefore be a C-K ordinal formula by (8). Now take case (z). It is impossible for $\mathbf{A}$ to be of the form Suc ($\mathbf{D}$), for then we should have $\mathbf{B}<\mathbf{D}$ whenever $\mathbf{B}<\mathbf{A}$, and this would mean that we had case (y). Since $U<\mathbf{A}$, there must be an $\mathbf{F}$ such that $\mathbf{F}<\mathbf{A}$ is demonstrable either by (2) or by (3) (after a possible conversion on $\mathbf{A}$); it must of course be demonstrable by (3). Then $\mathbf{A}$ is of the form $\lambda ufx\,.\,u(\mathbf{R})$. By (3), (B) we see that $\lambda ufx\,.\,\mathbf{R}(\mathbf{n})<\mathbf{A}$ for each positive integer n; each $\lambda ufx\,.\,\mathbf{R}(\mathbf{n})$ is therefore a C-K ordinal formula. Applying (9), (B) we see that $\mathbf{A}$ is a C-K ordinal formula.

Proof of (vi). To prove the first half, it is sufficient to find a method whereby from a C-K ordinal formula $\mathbf{A}$ we can find the corresponding ordinal formula $\mathbf{\Omega}$. For then there is a formula H_1 such that $H_1(\mathbf{a})$ conv $\mathbf{p}$ if a is the G.R. of $\mathbf{A}$ and p is that of $\mathbf{\Omega}$. H is then to be defined by

$$H \rightarrow \lambda a\,.\,\text{form}\Big(H_1\big(\text{Gr}(a)\big)\Big).$$

The method of finding $\mathbf{\Omega}$ may be replaced by a method of finding $\mathbf{\Omega}(\mathbf{m},\mathbf{n})$, given $\mathbf{A}$ and any two positive integers m, n. We shall arrange the method so that, whenever $\mathbf{A}$ is not an ordinal formula, either the calculation of the values does not terminate or else the values are not consistent with

$\boldsymbol{\Omega}$ being an ordinal formula. In this way we can prove the second half of (vi).

Let Ls be a formula such that Ls(**A**) enumerates the classes of formulae **B**, **B** $<$ **A** [*i.e.* if **B** $<$ **A** there is one and only one positive integer n for which Ls(**A**, **n**) conv **B**]. Then the rule for finding the value of $\boldsymbol{\Omega}$(**m**, **n**) is as follows:—

First determine whether $U \leqslant$ **A** and whether **A** is convertible to the form **r**(Suc, U). This terminates if **A** is a C-K ordinal formula.

If **A** conv **r**(Suc, U) and either $m > r+1$ or $n > r+1$, then the value is 4. If $m < n \leqslant r+1$, the value is 2. If $n < m \leqslant r+1$, the value is 1. If $m = n \leqslant r+1$, the value is 3.

If **A** is not convertible to this form, we determine whether either **A** or Ls(**A**, **m**) is convertible to the form $\lambda ufx.u$(**R**); and if either of them is, we verify that λufx.**R**(**n**) $<$ λufx.**R**$\big(S$(**n**)$\big)$. We shall eventually come to an affirmative answer if **A** is a C-K ordinal formula.

Having checked this, we determine concerning m and n whether Ls(**A**, **m**) $<$ Ls(**A**, **n**), Ls(**A**, **n**) $<$ Ls(**A**, **m**), or $m = n$, and the value is to be accordingly 1, 2, or 3.

If **A** is a C-K ordinal formula, this process certainly terminates. To see that the values so calculated correspond to an ordinal formula, and one representing $\Xi_{\mathbf{A}}$, first observe that this is so when $\Xi_{\mathbf{A}}$ is finite. In the other case (iii) and (iv) show that $\Xi_{\mathbf{B}}$ determines a one-one correspondence between the ordinals β, $1 \leqslant \beta \leqslant \Xi_{\mathbf{A}}$, and the classes of interconvertible formulae **B**, **B** $<$ **A**. If we take $G(m, n)$ to be Ls(**A**, **m**) $<$ Ls(**A**, **n**), we see that $G(m, n)$ is the ordering relation of a series of order type† $\Xi_{\mathbf{A}}$ and on the other hand that the values of $\boldsymbol{\Omega}$(**m**, **n**) are related to $G(m, n)$ as on p. 173.

To prove the second half suppose that **A** is not a C-K ordinal formula. Then one of the conditions (A)–(D) in (v) must not be satisfied. If (A) is not satisfied we shall not obtain a result even in the calculation of $\boldsymbol{\Omega}$(1, 1). If (B) is not satisfied, we shall have for some positive integers p and q,

$$\text{Ls}(\mathbf{A}, \mathbf{p}) \text{ conv } \lambda ufx.u(\mathbf{R})$$

but not λufx.**R**(**q**) $<$ λufx.**R**$\big(S$(**q**)$\big)$. Then the process of calculating $\boldsymbol{\Omega}$(**p**, **q**) will not terminate. In case of failure of (C) or (D) the values of $\boldsymbol{\Omega}$(**m**, **n**) may all be calculable, but if so conditions (a)–(f), p. 173, will be violated. Thus, if **A** is not a C-K ordinal formula, then H(**A**) is not an ordinal formula.

† The order type is β, where $1+\beta = \Xi_{\mathbf{A}}$; but $\beta = \Xi_{\mathbf{A}}$, since $\Xi_{\mathbf{A}}$ is infinite.

I propose now to define three formulae Sum, Lim, Inf of importance in connection with ordinal formulae. Since they are comparatively simple, they will for once be given almost in full. The formula Ug is one with the property that Ug($\mathbf{m}$) is convertible to the formula representing the largest odd integer dividing m: it is not given in full. P is the predecessor function; $P\big(S(\mathbf{m})\big)$ conv $\mathbf{m}$, $P(1)$ conv 1.

$$\mathrm{Al} \rightarrow \lambda pxy\,.\,p\big(\lambda guv\,.\,g(v, u),\ \lambda uv\,.\,u(I, v),\ x,\ y\big),$$

$$\mathrm{Hf} \rightarrow \lambda m\,.\,P\Big(m\Big(\lambda guv\,.\,g\big(v, S(u)\big),\ \lambda uv\,.\,v(I, u),\ 1,\ 2\Big)\Big),$$

$$\mathrm{Bd} \rightarrow \lambda ww'aa'x\,.\,\mathrm{Al}\Big(\lambda f\,.\,w\big(a, a, w'(a', a', f)\big),\ x,\ 4\Big),$$

$$\mathrm{Sum} \rightarrow \lambda ww'pq\,.\,\mathrm{Bd}\Big(w,\ w',\ \mathrm{Hf}(p),\ \mathrm{Hf}(q),$$

$$\mathrm{Al}\Big(p,\ \mathrm{Al}\big(q, w'\big(\mathrm{Hf}(p), \mathrm{Hf}(q)\big),\ 1\big),\ \mathrm{Al}\big(S(q),\ w\big(\mathrm{Hf}(p), \mathrm{Hf}(q)\big),\ 2\big)\Big)\Big),$$

$$\mathrm{Lim} \rightarrow \lambda zpq\,.\,\Big\{\lambda ab\,.\,\mathrm{Bd}\Big(z(a),\ z(b),\ \mathrm{Ug}(p),\ \mathrm{Ug}(q),\ \mathrm{Al}\Big(\mathrm{Dt}(a, b)+\mathrm{Dt}(b, a),$$

$$\mathrm{Dt}(a, b),\ z\big(a,\ \mathrm{Ug}(p),\ \mathrm{Ug}(q)\big)\Big)\Big)\Big\}\big(\varpi(2, p),\ \varpi(2, q)\big),$$

$$\mathrm{Inf} \rightarrow \lambda wapq\,.\,\mathrm{Al}\Big(\lambda f\,.\,w\big(a,\ p,\ w(a, q, f)\big),\ w(p, q),\ 4\Big).$$

The essential properties of these formulae are described by:

Al($2\mathbf{r}-1$, $\mathbf{m}$, $\mathbf{n}$) conv $\mathbf{m}$, Al($2\mathbf{r}$, $\mathbf{m}$, $\mathbf{n}$) conv $\mathbf{n}$,

Hf($2\mathbf{m}$) conv $\mathbf{m}$, Hf($2\mathbf{m}-1$) conv $\mathbf{m}$,

Bd($\mathbf{\Omega}$, $\mathbf{\Omega}'$, $\mathbf{a}$, $\mathbf{a}'$, $\mathbf{x}$) conv 4, unless both

$\mathbf{\Omega}(\mathbf{a}, \mathbf{a})$ conv 3 and $\mathbf{\Omega}'(\mathbf{a}', \mathbf{a}')$ conv 3,

it is then convertible to $\mathbf{x}$.

If $\mathbf{\Omega}$, $\mathbf{\Omega}'$ are ordinal formulae representing α, β respectively, then Sum($\mathbf{\Omega}$, $\mathbf{\Omega}'$) is an ordinal formula representing $\alpha+\beta$. If $\mathbf{Z}$ is a W.F.F. enumerating a sequence of ordinal formulae representing $\alpha_1, \alpha_2, \ldots$, then Lim($\mathbf{Z}$) is an ordinal formula representing the infinite sum $\alpha_1+\alpha_2+\alpha_3\ldots$.

If $\mathbf{\Omega}$ is an ordinal formula representing a, then Inf ($\mathbf{\Omega}$) enumerates a sequence of ordinal formulae representing all the ordinals less than a without repetitions other than repetitions of the ordinal 0.

To prove that there is no general method for determining about a formula whether it is an ordinal formula, we use an argument akin to that leading to the Burali-Forti paradox; but the emphasis and the conclusion are different. Let us suppose that such an algorithm is available. This enables us to obtain a recursive enumeration $\mathbf{\Omega}_1$, $\mathbf{\Omega}_2$, ... of the ordinal formulae in normal form. There is a formula $\mathbf{Z}$ such that $\mathbf{Z}(\mathbf{n})$ conv $\mathbf{\Omega}_\mathbf{n}$. Now Lim ($\mathbf{Z}$) represents an ordinal greater than any represented by an $\mathbf{\Omega}_\mathrm{n}$, and it has therefore been omitted from the enumeration.

This argument proves more than was originally asserted. In fact, it proves that, if we take any class E of ordinal formulae in normal form, such that, if $\mathbf{A}$ is any ordinal formula, then there is a formula in E representing the same ordinal as $\mathbf{A}$, then there is no method whereby one can determine whether a W.F.F. in normal form belongs to E.

8. *Ordinal logics.*

An ordinal logic is a W.F.F. $\mathbf{\Lambda}$ such that $\mathbf{\Lambda}(\mathbf{\Omega})$ is a logic formula whenever $\mathbf{\Omega}$ is an ordinal formula.

This definition is intended to bring under one heading a number of ways of constructing logics which have recently been proposed or which are suggested by recent advances. In this section I propose to show how to obtain some of these ordinal logics.

Suppose that we have a class W of logical systems. The symbols used in each of these systems are the same, and a class of sequences of symbols called "formulae" is defined, independently of the particular system in W. The rules of procedure of a system C define an axiomatic subset of the formulae, which are to be described as the "provable formulae of C." Suppose further that we have a method whereby, from any system C of W, we can obtain a new system C', also in W, and such that the set of provable formulae of C' includes the provable formulae of C (we shall be most interested in the case in which they are included as a proper subset). It is to be understood that this "method" is an effective procedure for obtaining the rules of procedure of C' from those of C.

Suppose that to certain of the formulae of W we make number-theoretic theorems correspond: by modifying the definition of formula, we may suppose that this is donef or all formulae. We shall say that one of the systems C is *valid* if the provability of a formula in C implies the truth of the corresponding number-theoretic theorem. Now let the relation of

C' to C be such that the validity of C implies the validity of C', and let there be a valid system C_0 in W. Finally, suppose that, given any computable sequence $C_1, C_2, \ldots$ of systems in W, the "limit system", in which a formula is provable if and only if it is provable in one of the systems C_j, also belongs to W. These limit systems are to be regarded, not as functions of the sequence given in extension, but as functions of the rules of formation of their terms. A sequence given in extension may be described by various rules of formation, and there will be several corresponding limit systems. Each of these may be described as *a* limit system of the sequence.

In these circumstances we may construct an ordinal logic. Let us associate positive integers with the systems in such a way that to each C there corresponds a positive integer m_C, and that m_C completely describes the rules of procedure of C. Then there is a W.F.F. **K**, such that

$$\mathbf{K}(\mathbf{m}_C) \text{ conv } \mathbf{m}_{C'}$$

for each C in W, and there is a W.F.F. $\mathbf{\Theta}$ such that, if $\mathbf{D}(\mathbf{r}) \text{ conv } \mathbf{m}_{C_r}$ for each positive integer r, then $\mathbf{\Theta}(\mathbf{D}) \text{ conv } \mathbf{m}_C$, where C is a limit system of $C_1, C_2, \ldots$. With each system C of W it is possible to associate a logic formula $\mathbf{L}_C$: the relation between them is that, if G is a formula of W and the number-theoretic theorem corresponding to G (assumed expressed in the conversion calculus form) asserts that **B** is dual, then $\mathbf{L}_C(\mathbf{B}) \text{ conv } 2$ if and only if G is provable in C. There is a W.F.F. **G** such that

$$\mathbf{G}(\mathbf{m}_C) \text{ conv } \mathbf{L}_C$$

for each C of W. Put

$$\mathbf{N} \to \lambda a \,.\, \mathbf{G}\big(a(\mathbf{\Theta}, \mathbf{K}, \mathbf{m}_{C_0})\big).$$

I assert that $\mathbf{N}(\mathbf{A})$ is a logic formula for each C-K ordinal formula **A**, and that, if $\mathbf{A} < \mathbf{B}$, then $\mathbf{N}(\mathbf{B})$ is more complete than $\mathbf{N}(\mathbf{A})$, provided that there are formulae provable in C' but not in C for each valid C of W.

To prove this we shall show that to each C-K ordinal formula **A** there corresponds a unique system $C[\mathbf{A}]$ such that:

(i) $\mathbf{A}(\mathbf{\Theta}, \mathbf{K}, \mathbf{m}_{C_0}) \text{ conv } \mathbf{m}_{C[\mathbf{A}]}$,

and that it further satisfies:

(ii) $C[U]$ is a limit system of $C_0', C_0', \ldots$,

(iii) $C[\mathrm{Suc}\,(\mathbf{A})]$ is $(C[\mathbf{A}])'$,

(iv) $C[\lambda ufx \,.\, u(\mathbf{R})]$ is a limit system of $C[\lambda ufx \,.\, \mathbf{R}(1)]$, $C[\lambda ufx \,.\, \mathbf{R}(2)]$, ...,

A and $\lambda ufx \,.\, u(\mathbf{R})$ being assumed to be C-K ordinal formulae. The uniqueness of the system follows from the fact that m_C determines C completely. Let us try to prove the existence of $C[\mathbf{A}]$ for each C-K ordinal formula **A**. As we have seen (p. 176) it is sufficient to prove

(*a*) $C[U]$ exists,

(*b*) if $C[\mathbf{A}]$ exists, then $C[\mathrm{Suc}\,(\mathbf{A})]$ exists,

(*c*) if $C[\lambda ufx \,.\, \mathbf{R}(1)]$, $C[\lambda ufx \,.\, \mathbf{R}(2)]$, ... exist, then $C[\lambda ufx \,.\, u(\mathbf{R})]$ exists.

Proof of (*a*).

$$\{\lambda y \,.\, \mathbf{K}\big(y(I,\, \mathbf{m}_{C_0})\big)\}\,(\mathbf{n}) \text{ conv } \mathbf{K}(\mathbf{m}_{C_0}) \text{ conv } \mathbf{m}_{C_0'}$$

for all positive integers n, and therefore, by the definition of Θ, there is a system, which we call $C[U]$ and which is a limit system of C_0', C_0', ..., satisfying

$$\Theta\Big(\lambda y \,.\, \mathbf{K}\big(y(I,\, \mathbf{m}_{C_0})\big)\Big) \text{ conv } \mathbf{m}_{C[U]}.$$

But, on the other hand,

$$U\big(\Theta,\, \mathbf{K},\, \mathbf{m}_{C_0}\big) \text{ conv } \Theta\Big(\lambda y \,.\, \mathbf{K}\big(y(I,\, \mathbf{m}_{C_0})\big)\Big).$$

This proves (*a*) and incidentally (ii).

Proof of (*b*).

$$\begin{aligned} \mathrm{Suc}\,(\mathbf{A},\, \Theta,\, \mathbf{K},\, \mathbf{m}_{C_0}) &\text{ conv } \mathbf{K}\big(\mathbf{A}(\Theta,\, \mathbf{K},\, \mathbf{m}_{C_0})\big) \\ &\text{ conv } \mathbf{K}(\mathbf{m}_{C[\mathbf{A}]}) \\ &\text{ conv } \mathbf{m}_{(C[\mathbf{A}])'}. \end{aligned}$$

Hence $C[\mathrm{Suc}\,(\mathbf{A})]$ exists and is given by (iii).

Proof of (*c*).

$$\begin{aligned} \big\{\{\lambda ufx \,.\, \mathbf{R}\}(\Theta,\, \mathbf{K},\, \mathbf{m}_{C_0})\big\}\,(\mathbf{n}) &\text{ conv } \{\lambda ufx \,.\, \mathbf{R}(\mathbf{n})\}(\Theta,\, \mathbf{K},\, \mathbf{m}_{C_0}) \\ &\text{ conv } \mathbf{m}_{C[\lambda ufx \,.\, \mathbf{R}(\mathbf{n})]} \end{aligned}$$

by hypothesis. Consequently, by the definition of Θ, there exists a C which is a limit system of

$$C[\lambda ufx \,.\, \mathbf{R}(1)], \quad C[\lambda ufx \,.\, \mathbf{R}(2)], \; ...,$$

and satisfies

$$\Theta\Big(\{\lambda ufx\,.\,u(\mathbf{R})\}(\Theta,\,\mathbf{K},\,\mathbf{m}_{C_0})\Big)\ \mathrm{conv}\ \mathbf{m}_{C}.$$

We define $C[\lambda ufx\,.\,u(\mathbf{R})]$ to be this C. We then have (iv) and

$$\{\lambda ufx\,.\,u(\mathbf{R})\}(\Theta,\,\mathbf{K},\,\mathbf{m}_{C_0})\ \mathrm{conv}\ \Theta\Big(\{\lambda ufx\,.\,\mathbf{R}\}(\Theta,\,\mathbf{K},\,\mathbf{m}_{C_0})\Big)$$

$$\mathrm{conv}\ \mathbf{m}_{C[\lambda ufx\,.\,u(\mathbf{R})]}.$$

This completes the proof of the properties (i)–(iv). From (ii), (iii), (iv), the fact that C_0 is valid, and that C' is valid when C is valid, we infer that $C[\mathbf{A}]$ is valid for each C-K ordinal formula $\mathbf{A}$: also that there are more formulae provable in $C[\mathbf{B}]$ than in $C[\mathbf{A}]$ when $\mathbf{A} < \mathbf{B}$. The truth of our assertions regarding **N** now follows in view of (i) and the definitions of **N** and **G**.

We cannot conclude that **N** is an ordinal logic, since the formulae **A** are C-K ordinal formulae; but the formula H enables us to obtain an ordinal logic from **N**. By the use of the formula Gr we obtain a formula Tn such that, if **A** has a normal form, then Tn(**A**) enumerates the G.R.'s of the formulae into which **A** is convertible. Also there is a formula Ck such that, if h is the G.R. of a formula $H(\mathbf{B})$, then Ck(**h**) conv **B**, but otherwise Ck(**h**) conv U. Since $H(\mathbf{B})$ is an ordinal formula only if **B** is a C-K ordinal formula, $\mathrm{Ck}\big(\mathrm{Tn}(\mathbf{\Omega},\,\mathbf{n})\big)$ is a C-K ordinal formula for each ordinal formula Ω and each integer n. For many ordinal formulae it will be convertible to U, but, for suitable $\mathbf{\Omega}$, it will be convertible to any given C-K ordinal formula. If we put

$$\Lambda \rightarrow \lambda wa\,.\,\Gamma\Big(\lambda n\,.\,\mathbf{N}\Big(\mathrm{Ck}\Big(\mathrm{Tn}(w,\,n)\Big)\Big),\,a\Big),$$

Λ is the required ordinal logic. In fact, on account of the properties of Γ, $\Lambda(\mathbf{\Omega},\,\mathbf{A})$ will be convertible to 2 if and only if there is a positive integer n such that

$$\mathbf{N}\Big(\mathrm{Ck}\Big(\mathrm{Tn}(\mathbf{\Omega},\,\mathbf{n})\Big),\,\mathbf{A}\Big)\ \mathrm{conv}\ 2.$$

If $\mathbf{\Omega}$ conv $H(\mathbf{B})$, there will be an integer n such that $\mathrm{Ck}\big(\mathrm{Tn}(\mathbf{\Omega},\,\mathbf{n})\big)$ conv **B**, and then

$$\mathbf{N}\Big(\mathrm{Ck}\Big(\mathrm{Tn}(\mathbf{\Omega},\,\mathbf{n})\Big),\,\mathbf{A}\Big)\ \mathrm{conv}\ \mathbf{N}(\mathbf{B},\,\mathbf{A}).$$

For any n, $\mathrm{Ck}\big(\mathrm{Tn}(\boldsymbol{\Omega}, \mathbf{n})\big)$ is convertible to U or to some $\mathbf{B}$, where $\boldsymbol{\Omega}$ conv $H(\mathbf{B})$. Thus $\boldsymbol{\Lambda}(\boldsymbol{\Omega}, \mathbf{A})$ conv 2 if $\boldsymbol{\Omega}$ conv $H(\mathbf{B})$ and $\mathbf{N}(\mathbf{B}, \mathbf{A})$ conv 2 or if $\mathbf{N}(U, \mathbf{A})$ conv 2, but not in any other case.

We may now specialize and consider particular classes W of systems. First let us try to construct the ordinal logic described roughly in the introduction. For W we take the class of systems arising from the system of *Principia Mathematica*† by adjoining to it axiomatic (in the sense described on p. 161) sets of axioms‡. Gödel has shown that primitive recursive relations§ can be expressed by means of formulae in P. In fact, there is a rule whereby, given the recursion equations defining a primitive recursive relation, we can find a formula‖ $\mathfrak{A}[x_0, \ldots, z_0]$ such that

$$\mathfrak{A}[f^{(m_1)}0, \ldots, f^{(m_r)}0]$$

is provable in P if $F(m_1, \ldots, m_r)$ is true, and its negation is provable otherwise. Further, there is a method by which we can determine about a formula $\mathfrak{A}[x_0, \ldots, z_0]$ whether it arises from a primitive recursive relation in this way, and by which we can find the equations which defined the relation. Formulae of this kind will be called *recursion formulae*. We shall make use of a property that they possess, which we cannot prove formally here without giving their definition in full, but which is essentially trivial. $\mathrm{Db}[x_0, y_0]$ is to stand for a certain recursion formula such that $\mathrm{Db}[f^{(m)}0, f^{(n)}0]$ is provable in P if $m = 2n$ and its negation is provable otherwise. Suppose that $\mathfrak{A}[x_0]$, $\mathfrak{B}[x_0]$ are two recursion formulae. Then the theorem which I am assuming is that there is a recursion relation $\mathfrak{C}_{\mathfrak{A},\mathfrak{B}}[x_0]$ such that we can prove

$$\mathfrak{C}_{\mathfrak{A},\mathfrak{B}}[x_0] \equiv (\exists y_0)\big((\mathrm{Db}[x_0, y_0] . \mathfrak{A}[y_0]) \vee (\mathrm{Db}[fx_0, fy_0] . \mathfrak{B}[y_0])\big) \qquad (8.1)$$

in P.

† Whitehead and Russell [1]. The axioms and rules of procedure of a similar system P will be found in a convenient form in Gödel [1],[a] and I follow Gödel. The symbols for the natural numbers in P are $0, f0, ff0, \ldots, f^{(n)}0 \ldots$. Variables with the suffix "0" stand for natural numbers.

‡ It is sometimes regarded as necessary that the set of axioms used should be computable, the intention being that it should be possible to verify of a formula reputed to be an axiom whether it really is so. We can obtain the same effect with axiomatic sets of axioms in this way. In the rules of procedure describing which are the axioms, we incorporate a method of enumerating them, and we also introduce a rule that in the main part of the deduction, whenever we write down an axiom as such, we must also write down its position in the enumeration. It is possible to verify whether this has been done correctly.

§ A relation $F(m_1, \ldots, m_r)$ is primitive recursive if it is a necessary and sufficient condition for the vanishing of a primitive recursive function $\phi(m_1, \ldots, m_r)$.

‖ Capital German letters will be used to stand for variable or undetermined formulae in P. An expression such as $\mathfrak{A}[\mathfrak{B}, \mathfrak{C}]$ stands for the result of substituting $\mathfrak{B}$ and $\mathfrak{C}$ for x_0 and y_0 in $\mathfrak{A}$.

The significant formulae in any of our extensions of P are those of the form

$$(x_0)(\exists y_0)\,\mathfrak{A}[x_0, y_0], \tag{8.2}$$

where $\mathfrak{A}[x_0, y_0]$ is a recursion formula, arising from the relation $R(m, n)$ let us say. The corresponding number-theoretic theorem states that for each natural number m there is a natural number n such that $R(m, n)$ is true.

The systems in W which are not valid are those in which a formula of the form (8.2) is provable, but at the same time there is a natural number, m say, such that, for each natural number n, $R(m, n)$ is false. This means to say that $\sim \mathfrak{A}[f^{(m)}0, f^{(n)}0]$ is provable for each natural number n. Since (8.2) is provable, $(\exists x_0)\,\mathfrak{A}[f^{(m)}0, y_0]$ is provable, so that

$$(\exists y_0)\,\mathfrak{A}[f^{(m)}0, y_0], \quad \sim \mathfrak{A}[f^{(m)}0, 0], \quad \sim \mathfrak{A}[f^{(m)}0, f0], \quad \ldots \tag{8.3}$$

are all provable in the system. We may simplify (8.3). For a given m we may prove a formula of the form $\mathfrak{A}[f^{(m)}0, y_0] \equiv \mathfrak{B}[y_0]$ in P, where $\mathfrak{B}[x_0]$ is a recursion formula. Thus we find that a necessary and sufficient condition for a system of W to be valid is that for no recursion formula $\mathfrak{B}[x_0]$ are all of the formulae

$$(\exists x_0)\,\mathfrak{B}[x_0], \quad \sim \mathfrak{B}[0], \quad \sim \mathfrak{B}[f0], \quad \ldots \tag{8.4}$$

provable. An important consequence of this is that, if

$$\mathfrak{A}_1[x_0], \quad \mathfrak{A}_2[x_0], \quad \ldots, \quad \mathfrak{A}_n[x_0]$$

are recursion formulae, if

$$(\exists x_0)\,\mathfrak{A}_1[x_0] \vee \ldots \vee (\exists x_0)\,\mathfrak{A}_n[x_0] \tag{8.5}$$

is provable in C, and C is valid, then we can prove $\mathfrak{A}_r[f^{(a)}0]$ in C for some natural numbers r, a, where $1 \leqslant r \leqslant n$. Let us define $\mathfrak{D}_r$ to be the formula

$$(\exists x_0)\,\mathfrak{A}_1[x_0] \vee \ldots \vee (\exists x_0)\,\mathfrak{A}_r[x_0]$$

and then define $\mathfrak{E}_r[x_0]$ recursively by the condition that $\mathfrak{E}_1[x_0]$ is $\mathfrak{A}_1[x_0]$ and $\mathfrak{E}_{r+1}[x_0]$ be $\mathfrak{E}_{\mathfrak{E}_r, \mathfrak{A}_{r+1}}[x_0]$. Now I say that

$$\mathfrak{D}_r \supset (\exists x_0)\,\mathfrak{E}_r[x_0] \tag{8.6}$$

is provable for $1 \leqslant r \leqslant n$. It is clearly provable for $r = 1$: suppose it to be provable for a given r. We can prove

$$(y_0)(\exists x_0)\,\mathrm{Db}[x_0, y_0]$$

and
$$(y_0)(\exists x_0)\,\mathrm{Db}[fx_0, fy_0],$$

from which we obtain

$$\mathfrak{E}_r[y_0] \supset (\exists x_0)\Big((\mathrm{Db}[x_0, y_0] \,.\, \mathfrak{E}_r[y_0]) \vee (\mathrm{Db}[fx_0, fy_0] \,.\, \mathfrak{A}_{r+1}[y_0])\Big)$$

and

$$\mathfrak{A}_{r+1}[y_0] \supset (\exists x_0)\Big((\mathrm{Db}[x_0, y_0] \,.\, \mathfrak{E}_r[y_0]) \vee (\mathrm{Db}[fx_0, fy_0] \,.\, \mathfrak{A}_{r+1}[y_0])\Big).$$

These together with (8.1) yield

$$(\exists y_0)\,\mathfrak{E}_r[y_0] \vee (\exists y_0)\,\mathfrak{A}_{r+1}[y_0] \supset (\exists x_0)\,\mathfrak{E}_{\mathfrak{E}_r,\, \mathfrak{A}_{r+1}}[x_0],$$

which is sufficient to prove (8.6) for $r+1$. Now, since (8.5) is provable in C, $(\exists x_0)\,\mathfrak{E}_n[x_0]$ must also be provable, and, since C is valid, this means that $\mathfrak{E}_n[f^{(m)}0]$ must be provable for some natural number m. From (8.1) and the definition of $\mathfrak{E}_n[x_0]$ we see that this implies that $\mathfrak{A}_r[f^{(a)}0]$ is provable for some natural numbers a and r, $1 \leqslant r \leqslant n$.

To any system C of W we can assign a primitive recursive relation $P_C(m, n)$ with the intuitive meaning "m is the G.R. of a proof of the formula whose G.R. is n". We call the corresponding recursion formula $\mathrm{Proof}_C[x_0, y_0]$ (*i.e.* $\mathrm{Proof}_C[f^{(m)}0, f^{(n)}0]$ is provable when $P_C(m, n)$ is true, and its negation is provable otherwise). We can now explain what is the relation of a system C' to its predecessor C. The set of axioms which we adjoin to P to obtain C' consists of those adjoined in obtaining C, together with all formulae of the form

$$(\exists x_0)\,\mathrm{Proof}_C[x_0, f^{(m)}0] \supset \mathfrak{F}, \tag{8.7}$$

where m is the G.R. of $\mathfrak{F}$.

We want to show that a contradiction can be obtained by assuming C' to be invalid but C to be valid. Let us suppose that a set of formulae of the form (8.4) is provable in C'. Let $\mathfrak{A}_1, \mathfrak{A}_2, \ldots, \mathfrak{A}_k$ be those axioms of C' of the form (8.7) which are used in the proof of $(\exists x_0)\,\mathfrak{B}[x_0]$. We may suppose that none of them is provable in C. Then by the deduction theorem we see that

$$(\mathfrak{A}_1 \,.\, \mathfrak{A}_2 \ldots \mathfrak{A}_k) \supset (\exists x_0)\,\mathfrak{B}[x_0] \tag{8.8}$$

is provable in C. Let $\mathfrak{A}_l$ be $(\exists x_0)\,\mathrm{Proof}_C[x_0, f^{(m_l)}0] \supset \mathfrak{F}_l$. Then from (8.8) we find that

$$(\exists x_0)\,\mathrm{Proof}_C[x_0, f^{(m_1)}0] \vee \ldots \vee (\exists x_0)\,\mathrm{Proof}_C[x_0, f^{(m_k)}0] \vee (\exists x_0)\,\mathfrak{B}[x_0]$$

is provable in C. It follows from a result which we have just proved that either $\mathfrak{B}[f^{(c)}0]$ is provable for some natural number c, or else $\mathrm{Proof}_C[f^{(n)}0, f^{(m_l)}0]$ is provable in C for some natural number u and some l, $1 \leqslant l \leqslant k$: but this would mean that $\mathfrak{F}_l$ is provable in C (this is one of the points where we assume the validity of C) and therefore also in C', contrary to hypothesis. Thus $\mathfrak{B}[f^{(c)}0]$ must be provable in C'; but we are also assuming $\sim\mathfrak{B}[f^{(c)}0]$ to be provable in C'. There is therefore a contradiction in C'. Let us suppose that the axioms $\mathfrak{A}_1', \ldots, \mathfrak{A}_{k'}'$, of the form (8.7), when adjoined to C are sufficient to obtain the contradiction and that none of these axioms is that provable in C. Then

$$\sim\mathfrak{A}_1' \vee \sim\mathfrak{A}_2' \vee \ldots \vee \sim\mathfrak{A}_{k'}'$$

is provable in C, and if $\mathfrak{A}_l'$ is $(\exists x_0)\,\mathrm{Proof}_C[x_0, f^{(m_l')}0] \supset \mathfrak{F}_l'$ then

$$(\exists x_0]\,\mathrm{Proof}_C[x_0, f^{(m_1')}0] \vee \ldots \vee (\exists x_0)\,\mathrm{Proof}\,[x_0, f^{(m_{k'}')}0]$$

is provable in C. But, by repetition of a previous argument, this means that $\mathfrak{A}_l'$ is provable for some l, $1 \leqslant l \leqslant k'$, contrary to hypothesis. This is the required contradiction.

We may now construct an ordinal logic in the manner described on pp. 184–187. We shall, however, carry out the construction in rather more detail, and with some modifications appropriate to the particular case. Each system C of our set W may be described by means of a W.F.F. M_C which enumerates the G.R.'s of the axioms of C. There is a W.F.F. E such that, if a is the G.R. of some proposition $\mathfrak{F}$, then $E(M_C, \mathbf{a})$ is convertible to the G.R. of

$$(\exists x_0)\,\mathrm{Proof}_C[x_0, f^{(a)}0] \supset \mathfrak{F}.$$

If $\mathbf{a}$ is not the G.R. of any proposition in P, then $E(M_C, \mathbf{a})$ is to be convertible to the G.R. of $0 = 0$. From E we obtain a W.F.F. K such that $K(M_C, 2\mathbf{n}+1)$ conv $M_C(\mathbf{n})$, $K(M_C, 2\mathbf{n})$ conv $E(M_C, \mathbf{n})$. The successor system C' is defined by $K(M_C)$ conv $M_{C'}$. Let us choose a formula G such that $G(M_C, \mathbf{A})$ conv 2 if and only if the number-theoretic theorem equivalent to "$\mathbf{A}$ is dual" is provable in C. Then we define Λ_P by

$$\Lambda_P \rightarrow \lambda wa\,.\,\Gamma\Big(\lambda y\,.\,G\Big(\mathrm{Ck}\Big(\mathrm{Tn}(w, y),\ \lambda mn\,.\,m\big(\varpi(2, n), \varpi(3, n)\big), K, M_P\Big)\Big), a\Big).$$

This is an ordinal logic provided that P is valid.

Another ordinal logic of this type has in effect been introduced by Church†. Superficially this ordinal logic seems to have no more in common with Λ_P than that they both arise by the method which we have described, which uses C-K ordinal formulae. The initial systems are entirely different. However, in the relation between C and C' there is an interesting analogy. In Church's method the step from C to C' is performed by means of subsidiary axioms of which the most important (Church [**2**], p. 88, 1_m) is almost a direct translation into his symbolism of the rule that we may take any formula of the form (8.4) as an axiom. There are other extra axioms, however, in Church's system, and it is therefore not unlikely that it is in some respects more complete than Λ_P.

There are other types of ordinal logic, apparently quite unrelated to the type that we have so far considered. I have in mind two types of ordinal logic, both of which can be best described directly in terms of ordinal formulae without any reference to C-K ordinal formulae. I shall describe here a specimen Λ_H of one of these types of ordinal logic. Ordinal logics of this kind were first considered by Hilbert (Hilbert [**1**], 183ff), and have also been used by Tarski (Tarski [**1**], 395ff); see also Gödel [**1**][a] foot-note 48^a.

Suppose that we have selected a particular ordinal formula $\mathbf{\Omega}$. We shall construct a modification P_Ω of the system P of Gödel (see foot-note † on p. 187. We shall say that a natural number n is a *type* if it is either even or $2p-1$, where $\mathbf{\Omega}(\mathbf{p}, \mathbf{p})$ conv 3. The definition of a variable in P is to be modified by the condition that the only admissible subscripts are to be the types in our sense. Elementary expressions are then defined as in P: in particular the definition of an elementary expression of type 0 is unchanged. An elementary formula is defined to be a sequence of symbols of the form $\mathfrak{A}_m \mathfrak{A}_n$, where $\mathfrak{A}_m$, $\mathfrak{A}_n$ are elementary expressions of types m, n satisfying one of the conditions (a), (b), (c).

(a) m and n are both even and m exceeds n,

(b) m is odd and n is even,

(c) $m = 2p-1$, $n = 2q-1$, and $\mathbf{\Omega}(\mathbf{p}, \mathbf{q})$ conv 2.

With these modifications the formal development of P_Ω is the same as that of P. We want, however, to have a method of associating number-theoretic theorems with certain of the formulae of P_Ω. We cannot take over directly the association which we used in P. Suppose that G is a

† In outline Church [**1**], 279–280. In greater detail Church [**2**], Chap. X.

formula in P interpretable as a number-theoretic theorem in the way described in the course of constructing Λ_P (p. 187). Then, if every type suffix in G is doubled, we shall obtain a formula in P_Ω which is to be interpreted as the same number-theoretic theorem. By the method of § 6 we can now obtain from P_Ω a formula L_Ω which is a logic formula if P_Ω is valid; in fact, given $\mathbf{\Omega}$ there is a method of obtaining L_Ω, so that there is a formula Λ_H such that $\Lambda_H(\mathbf{\Omega})$ conv L_Ω for each ordinal formula $\mathbf{\Omega}$.

Having now familiarized ourselves with ordinal logics by means of these examples we may begin to consider general questions concerning them.

9. *Completeness questions.*

The purpose of introducing ordinal logics was to avoid as far as possible the effects of Gödel's theorem. It is a consequence of this theorem, suitably modified, that it is impossible to obtain a complete logic formula, or (roughly speaking now) a complete system of logic. We were able, however, from a given system to obtain a more complete one by the adjunction as axioms of formulae, seen intuitively to be correct, but which the Gödel theorem shows are unprovable† in the original system; from this we obtained a yet more complete system by a repetition of the process, and so on. We found that the repetition of the process gave us a new system for each C-K ordinal formula. We should like to know whether this process suffices, or whether the system should be extended in other ways as well. If it were possible to determine about a W.F.F. in normal form whether it was an ordinal formula, we should know for certain that it was necessary to make extensions in other ways. In fact for any ordinal formula $\mathbf{\Lambda}$ it would then be possible to find a single logic formula $\mathbf{L}$ such that, if $\mathbf{\Lambda}(\mathbf{\Omega}, \mathbf{A})$ conv 2 for some ordinal formula $\mathbf{\Omega}$, then $\mathbf{L}(\mathbf{A})$ conv 2. Since $\mathbf{L}$ must be incomplete, there must be formulae $\mathbf{A}$ for which $\mathbf{\Lambda}(\mathbf{\Omega}, \mathbf{A})$ is not convertible to 2 for any ordinal formula $\mathbf{\Omega}$. However, in view of the fact, proved in § 7, that there is no method of determining about a formula in normal form whether it is an ordinal formula, the case does not arise, and there is still a possibility that some ordinal logics may be complete in some sense. There is a quite natural way of defining completeness.

Definition of completeness of an ordinal logic. We say that an ordinal logic $\mathbf{\Lambda}$ is complete if corresponding to each dual formula $\mathbf{A}$ there is an ordinal formula $\mathbf{\Omega}_\mathbf{A}$ such that $\mathbf{\Lambda}(\mathbf{\Omega}_\mathbf{A}, \mathbf{A})$ conv 2.

† In the case of P we adjoined all of the axioms $(\exists x_0)$ Proof$[x_0, f^{(m)}0] \supset \mathfrak{F}$, where m is the G.R. of $\mathfrak{F}$; the Gödel theorem shows that *some* of them are unprovable in P.

As has been explained in § 2, the reference in the definition to the existence of $\Omega_{\mathbf{A}}$ for each **A** is to be understood in the same naïve way as any reference to existence in mathematics.

There is room for modification in this definition: we might require that there is a formula **X** such that $\mathbf{X}(\mathbf{A})$ conv $\Omega_{\mathbf{A}}$, $\mathbf{X}(\mathbf{A})$ being an ordinal formula whenever **A** is dual. There is no need, however, to discuss the relative merits of these two definitions, because in all cases in which we prove an ordinal logic to be complete we shall prove it to be complete even in the modified sense; but in cases in which we prove an ordinal logic to be incomplete, we use the definition as it stands.

In the terminology of § 6, Λ is complete if the class of logics $\Lambda(\Omega)$ is complete when Ω runs through all ordinal formulae.

There is another completeness property which is related to this one. Let us for the moment describe an ordinal logic Λ as *all inclusive* if to each logic formula **L** there corresponds an ordinal formula $\Omega_{(\mathbf{L})}$ such that $\Lambda(\Omega_{(\mathbf{L})})$ is as complete as **L**. Clearly every all inclusive ordinal logic is complete; for, if **A** is dual, then $\delta(\mathbf{A})$ is a logic with **A** in its extent. But, if Λ is complete and

$$\mathrm{Ai} \to \lambda kw \,.\, \Gamma\Big(\lambda ra \,.\, \delta\Big(4,\, \delta\Big(2,\, k\Big(w,\, V\big(\mathrm{Nm}(r)\big)\Big)\Big) + \delta\big(2,\, \mathrm{Nm}(r, a)\big)\Big)\Big),$$

then $\mathrm{Ai}(\Lambda)$ is an all inclusive ordinal logic. For, if **A** is in the extent of $\Lambda(\Omega_{\mathbf{A}})$ for each **A**, and we put $\Omega_{(\mathbf{L})} \to \Omega_{V(\mathbf{L})}$, then I say that, if **B** is in the extent of **L**, it must be in the extent of $\mathrm{Ai}(\Lambda, \Omega_{(\mathbf{L})})$. In fact, we see that $\mathrm{Ai}(\Lambda, \Omega_{V(\mathbf{L})}, \mathbf{B})$ is convertible to

$$\Gamma\Big(\lambda ra \,.\, \delta\Big(4,\, \delta\Big(2,\, \Lambda\Big(\Omega_{V(\mathbf{L})},\, V\big(\mathrm{Nm}(r)\big)\Big)\Big) + \delta\big(2,\, \mathrm{Nm}(r, a)\big)\Big),\, \mathbf{B}\Big).$$

For suitable n, $\mathrm{Nm}(\mathbf{n})$ conv **L** and then

$$\Lambda\Big(\Omega_{V(\mathbf{L})},\, V\big(\mathrm{Nm}(\mathbf{n})\big)\Big) \text{ conv } 2,$$

$$\mathrm{Nm}(\mathbf{n}, \mathbf{B}) \text{ conv } 2,$$

and therefore, by the properties of Γ and δ

$$\mathrm{Ai}(\Lambda, \Omega_{V(\mathbf{L})}, \mathbf{B}) \text{ conv } 2.$$

Conversely $\mathrm{Ai}(\Lambda, \Omega_{V(\mathbf{L})}, \mathbf{B})$ can be convertible to 2 only if both $\mathrm{Nm}(\mathbf{n}, \mathbf{B})$

and $\Lambda\Big(\Omega_{V(\mathbf{L})}, V\big(\mathrm{Nm}(\mathbf{n})\big)\Big)$ are convertible to 2 for some positive integer n; but, if $\Lambda\Big(\Omega_{V(\mathbf{L})}, V\big(\mathrm{Nm}(\mathbf{n})\big)\Big)$ conv 2, then $\mathrm{Nm}(\mathbf{n})$ must be a logic, and, since $\mathrm{Nm}(\mathbf{n}, \mathbf{B})$ conv 2, $\mathbf{B}$ must be dual.

It should be noticed that our definitions of completeness refer only to number-theoretic theorems. Although it would be possible to introduce formulae analogous to ordinal logics which would prove more general theorems than number-theoretic ones, and have a corresponding definition of completeness, yet, if our theorems are too general, we shall find that our (modified) ordinal logics are never complete. This follows from the argument of § 4. If our "oracle" tells us, not whether any given number-theoretic statement is true, but whether a given formula is an ordinal formula, the argument still applies, and we find that there are classes of problem which cannot be solved by a uniform process even with the help of this oracle. This is equivalent to saying that there is no ordinal logic of the proposed modified type which is complete with respect to these problems. This situation becomes more definite if we take formulae satisfying conditions (a)–(e), (f') (as described at the end of § 12) instead of ordinal formulae; it is then not possible for the ordinal logic to be complete with respect to any class of problems more extensive than the number-theoretic problems.

We might hope to obtain some intellectually satisfying system of logical inference (for the proof of number-theoretic theorems) with some ordinal logic. Gödel's theorem shows that such a system cannot be wholly mechanical; but with a complete ordinal logic we should be able to confine the non-mechanical steps entirely to verifications that particular formulae are ordinal formulae.

We might also expect to obtain an interesting classification of number-theoretic theorems according to "depth". A theorem which required an ordinal α to prove it would be deeper than one which could be proved by the use of an ordinal β less than α. However, this presupposes more than is justified. We now define

Invariance of ordinal logics. An ordinal logic $\mathbf{\Lambda}$ is said to be *invariant up to* an ordinal α if, whenever $\mathbf{\Omega}$, $\mathbf{\Omega}'$ are ordinal formulae representing the same ordinal less than α, the extent of $\mathbf{\Lambda}(\mathbf{\Omega})$ is identical with the extent of $\mathbf{\Lambda}(\mathbf{\Omega}')$. An ordinal logic is *invariant* if it is invariant up to each ordinal represented by an ordinal formula.

Clearly the classification into depths presupposes that the ordinal logic used is invariant.

Among the questions that we should now like to ask are

(*a*) Are there any complete ordinal logics?

(*b*) Are there any complete invariant ordinal logics?

To these we might have added "are all ordinal logics complete?"; but this is trivial; in fact, there are ordinal logics which do not suffice to prove any number-theoretic theorems whatever.

We shall now show that (*a*) must be answered affirmatively. In fact, we can write down a complete ordinal logic at once. Put

$$\mathrm{Od} \rightarrow \lambda a . \{\lambda fmn . \mathrm{Dt}\big(f(m), f(n)\big)\}\Big(\lambda s . \mathscr{P}\big(\lambda r . r\big(I, a(s)\big), 1, s\big)\Big)$$

and
$$\mathrm{Comp} \rightarrow \lambda wa . \delta\big(w, \mathrm{Od}(a)\big).$$

I shall show that Comp is a complete ordinal logic.

For if, Comp $(\mathbf{\Omega}, \mathbf{A})$ conv 2, then

$\mathbf{\Omega}$ conv Od $(\mathbf{A})$

$$\mathrm{conv}\ \lambda mn . \mathrm{Dt}\Big(\mathscr{P}\big(\lambda r . r\big(I, \mathbf{A}(m)\big), 1, m\big), \ \mathscr{P}\big(\lambda r . r\big(I, \mathbf{A}(n)\big), 1, n\big)\Big).$$

$\mathbf{\Omega}(\mathbf{m}, \mathbf{n})$ has a normal form if $\mathbf{\Omega}$ is an ordinal formula, so that then

$$\mathscr{P}\big(\lambda r . r\big(I, \mathbf{A}(\mathbf{m})\big), 1\big)$$

has a normal form; this means that $\mathbf{r}\big(I, \mathbf{A}(\mathbf{m})\big)$ conv 2 some r, *i.e.* $\mathbf{A}(\mathbf{m})$ conv 2. Thus, if Comp$(\mathbf{\Omega}, \mathbf{A})$ conv 2 and $\mathbf{\Omega}$ is an ordinal formula, then $\mathbf{A}$ is dual. Comp is therefore an ordinal logic. Now suppose conversely that $\mathbf{A}$ is dual. I shall show that Od$(\mathbf{A})$ is an ordinal formula representing the ordinal ω. For

$$\mathscr{P}\big(\lambda r . r\big(I, \mathbf{A}(\mathbf{m})\big), 1, \mathbf{m}\big) \ \mathrm{conv}\ \mathscr{P}\big(\lambda r . r(I, 2), 1, \mathbf{m}\big)$$

$$\mathrm{conv}\ 1(\mathbf{m})\ \mathrm{conv}\ \mathbf{m},$$

$$\mathrm{Od}(\mathbf{A}, \mathbf{m}, \mathbf{n})\ \mathrm{conv}\ \mathrm{Dt}(\mathbf{m}, \mathbf{n}),$$

i.e. Od$(\mathbf{A})$ is an ordinal formula representing the same ordinal as Dt. But

$$\mathrm{Comp}\big(\mathrm{Od}(\mathbf{A}), \mathbf{A}\big)\ \mathrm{conv}\ \delta\big(\mathrm{Od}(\mathbf{A}), \mathrm{Od}(\mathbf{A})\big)\ \mathrm{conv}\ 2.$$

This proves the completeness of Comp.

Of course Comp is not the kind of complete ordinal logic that we should really wish to use. The use of Comp does not make it any easier to see that **A** is dual. In fact, if we really want to use an ordinal logic a proof, of completeness for that particular ordinal logic will be of little value; the ordinals given by the completeness proof will not be ones which can easily be seen intuitively to be ordinals. The only value in a completeness proof of this kind would be to show that, if any objection is to be raised against an ordinal logic, it must be on account of something more subtle than incompleteness.

The theorem of completeness is also unexpected in that the ordinal formulae used are all formulae representing ω. This is contrary to our intentions in constructing Λ_P for instance; implicitly we had in mind large ordinals expressed in a simple manner. Here we have small ordinals expressed in a very complex and artificial way.

Before trying to solve the problem (*b*), let us see how far Λ_P and Λ_H are invariant. We should certainly not expect Λ_P to be invariant, since the extent of $\Lambda_P(\mathbf{\Omega})$ will depend on whether $\mathbf{\Omega}$ is convertible to a formula of the form $H(\mathbf{A})$: but suppose that we call an ordinal logic Λ "C-K invariant up to α" if the extent of $\Lambda\big(H(\mathbf{A})\big)$ is the same as the extent of $\Lambda\big(H(\mathbf{B})\big)$ whenever **A** and **B** are C-K ordinal formulae representing the same ordinal less than α. How far is Λ_P C-K invariant? It is not difficult to see that it is C-K invariant up to any finite ordinal, that is to say up to ω. It is also C-K invariant up to $\omega+1$, as follows from the fact that the extent of

$$\Lambda_P\Big(H\big(\lambda ufx\,.\,u(\mathbf{R})\big)\Big)$$

is the set-theoretic sum of the extents of

$$\Lambda_P\Big(H\big(\lambda ufx\,.\,\mathbf{R}(1)\big)\Big),\quad \Lambda_P\Big(H\big(\lambda ufx\,.\,\mathbf{R}(2)\big)\Big),\quad \ldots.$$

However, there is no obvious reason for believing that it is C-K invariant up to $\omega+2$, and in fact it is demonstrable that this is not the case (see the end of this section). Let us find out what happens if we try to prove that the extent of

$$\Lambda_P\Big(H\Big(\mathrm{Suc}\big(\lambda ufx\,.\,u(\mathbf{R}_1)\big)\Big)\Big)$$

is the same as the extent of

$$\Lambda_P\Big(H\Big(\mathrm{Suc}\big(\lambda ufx\,.\,u(\mathbf{R}_2)\big)\Big)\Big),$$

where $\lambda ufx.u(\mathbf{R}_1)$ and $\lambda ufx.u(\mathbf{R}_2)$ are two C-K ordinal formulae representing ω. We should have to prove that a formula interpretable as a number-theoretic theorem is provable in $C\left[\mathrm{Suc}\left(\lambda ufx.u(\mathbf{R}_1)\right)\right]$ if, and only if, it is provable in $C\left[\mathrm{Suc}\left(\lambda ufx.u(\mathbf{R}_2)\right)\right]$. Now $C\left[\mathrm{Suc}\left(\lambda ufx.u(\mathbf{R}_1)\right)\right]$ is obtained from $C[\lambda ufx.u(\mathbf{R}_1)]$ by adjoining all axioms of the form

$$(\exists x_0)\ \mathrm{Proof}_{C[\lambda ufx.u(\mathbf{R}_1)]}[x_0, f^{(m)}0) \supset \mathfrak{F}, \tag{9.1}$$

where m is the G.R. of $\mathfrak{F}$, and $C\left[\mathrm{Suc}\left(\lambda ufx.u(\mathbf{R}_2)\right)\right]$ is obtained from $C[\lambda ufx.u(\mathbf{R}_2)]$ by adjoining all axioms of the form

$$(\exists x_0)\ \mathrm{Proof}_{C[\lambda ufx.u(\mathbf{R}_2)]}[x_0, f^{(m)}0] \supset \mathfrak{F}. \tag{9.2}$$

The axioms which must be adjoined to P to obtain $C[\lambda ufx.u(\mathbf{R}_1)]$ are essentially the same as those which must be adjoined to obtain the system $C[\lambda ufx.u(\mathbf{R}_2)]$: however the *rules of procedure which have to be applied before these axioms can be written down are in general quite different in the two cases.* Consequently (9.1) and (9.2) are quite different axioms, and there is no reason to expect their consequences to be the same. A proper understanding of this will make our treatment of question (*b*) much more intelligible. See also footnote ‡ on page 187.

Now let us turn to Λ_H. This ordinal logic is invariant. Suppose that $\mathbf{\Omega}$, $\mathbf{\Omega}'$ represent the same ordinal, and suppose that we have a proof of a number-theoretic theorem G in P_{Ω}. The formula expressing the number-theoretic theorem does not involve any odd types. Now there is a one-one correspondence between the odd types such that if $2m-1$ corresponds to $2m'-1$ and $2n-1$ to $2n'-1$ then $\mathbf{\Omega}(\mathbf{m}, \mathbf{n})$ conv 2 implies $\mathbf{\Omega}'(\mathbf{m}', \mathbf{n}')$ conv 2. Let us modify the odd type-subscripts occurring in the proof of G, replacing each by its mate in the one-one correspondence. There results a proof in $P_{\Omega'}$ with the same end formula G. That is to say that if G is provable in P_{Ω} it is provable in $P_{\Omega'}$. Λ_H is invariant.

The question (*b*) must be answered negatively. Much more can be proved, but we shall first prove an even weaker result which can be established very quickly, in order to illustrate the method.

I shall prove that an ordinal logic $\mathbf{\Lambda}$ cannot be invariant and have the property that the extent of $\mathbf{\Lambda}(\mathbf{\Omega})$ is a strictly increasing function of the ordinal represented by $\mathbf{\Omega}$. Suppose that $\mathbf{\Lambda}$ has these properties; then we shall obtain a contradiction. Let $\mathbf{A}$ be a W.F.F. in normal form and without free variables, and consider the process of carrying out conversions on $\mathbf{A}(1)$ until we have shown it convertible to 2, then converting $\mathbf{A}(2)$ to 2, then $\mathbf{A}$ (3) and so on: suppose that after r steps we are still performing the conversion on $\mathbf{A}(\mathbf{m}_r)$. There is a formula Jh such that Jh($\mathbf{A}$, $\mathbf{r}$) conv $\mathbf{m}_r$ for each positive integer r. Now let Z be a formula such that, for each positive integer n, $Z(\mathbf{n})$ is an ordinal formula representing ω^n, and suppose $\mathbf{B}$ to be a member of the extent of $\Lambda\Big(\mathrm{Suc}\big(\mathrm{Lim}(Z)\big)\Big)$ but not of the extent of $\Lambda\big(\mathrm{Lim}(Z)\big)$. Put

$$\mathbf{K}^* \rightarrow \lambda a \,.\, \Lambda\Bigg(\mathrm{Suc}\Big(\mathrm{Lim}\Big(\lambda r \,.\, Z\big(\mathrm{Jh}(a, r)\big)\Big)\Big), \mathbf{B}\Bigg);$$

then $\mathbf{K}^*$ is a complete logic. For, if $\mathbf{A}$ is dual, then

$$\mathrm{Suc}\Big(\mathrm{Lim}\Big(\lambda r \,.\, Z\big(\mathrm{Jh}(\mathbf{A}, r)\big)\Big)\Big)$$

represents the ordinal $\omega^\omega + 1$, and therefore $\mathbf{K}^*(\mathbf{A})$ conv 2; but, if $\mathbf{A}(\mathbf{c})$ is not convertible to 2, then

$$\mathrm{Suc}\Big(\mathrm{Lim}\Big(\lambda r \,.\, Z\big(\mathrm{Jh}(\mathbf{A}, r)\big)\Big)\Big)$$

represents an ordinal not exceeding $\omega^c + 1$, and $\mathbf{K}^*(\mathbf{A})$ is therefore not convertible to 2. Since there are no complete logic formulae, this proves our assertion.

We may now prove more powerful results.

Incompleteness theorems. (A) If an ordinal logic Λ is invariant up to an ordinal α, then for any ordinal formula $\mathbf{\Omega}$ representing an ordinal β, $\beta < \alpha$, the extent of $\Lambda(\mathbf{\Omega})$ is contained in the (set-theoretic) sum of the extents of the logics $\Lambda(\mathbf{P})$, where $\mathbf{P}$ is finite.

(B) If an ordinal logic Λ is C-K invariant up to an ordinal α, then for any C-K ordinal formula $\mathbf{A}$ representing an ordinal β, $\beta < \alpha$, the extent of $\Lambda\big(H(\mathbf{A})\big)$ is contained in the (set-theoretic) sum of the extents of the logics $\Lambda\big(H(\mathbf{F})\big)$, where $\mathbf{F}$ is a C-K ordinal formula representing an ordinal less than ω^2.

Proof of (A). It is sufficient to prove that, if $\mathbf{\Omega}$ represents an ordinal γ, $\omega \leqslant \gamma < \alpha$, then the extent of $\Lambda(\mathbf{\Omega})$ is contained in the set-theoretic sum of the extents of the logics $\Lambda(\mathbf{\Omega}')$, where $\mathbf{\Omega}'$ represents an ordinal less than γ. The ordinal γ must be of the form $\gamma_0 + \rho$, where ρ is finite and represented by $\mathbf{P}$ say, and γ_0 is not the successor of any ordinal and is not less than ω. There are two cases to consider; $\gamma_0 = \omega$ and $\gamma_0 \geqslant 2\omega$. In each of them we shall obtain a contradiction from the assumption that there is a W.F.F.

B such that $\Lambda(\Omega, \mathbf{B})$ conv 2 whenever Ω represents γ, but is not convertible to 2 if Ω represents a smaller ordinal. Let us take first the case $\gamma_0 \geqslant 2\omega$. Suppose that $\gamma_0 = \omega + \gamma_1$, and that Ω_1 is an ordinal formula representing γ_1. Let **A** be any W.F.F. with a normal form and no free variables, and let Z be the class of those positive integers which are exceeded by all integers n for which $\mathbf{A}(\mathbf{n})$ is not convertible to 2. Let E be the class of integers $2p$ such that $\Omega(\mathbf{p}, \mathbf{n})$ conv 2 for some n belonging to Z. The class E, together with the class Q of all odd integers, is constructively enumerable. It is evident that the class can be enumerated with repetitions, and since it is infinite the required enumeration can be obtained by striking out the repetitions. There is, therefore, a formula En such that $\text{En}(\Omega, \mathbf{A}, \mathbf{r})$ runs through the formulae of the class $E+Q$ without repetitions as r runs through the positive integers. We define

$$\text{Rt} \to \lambda wamn \,.\, \text{Sum}\Big(\text{Dt},\, w,\, \text{En}\,(w, a, m),\, \text{En}(w, a, n)\Big).$$

Then $\text{Rt}(\Omega_1, \mathbf{A})$ is an ordinal formula which represents γ_0 if **A** is dual, but a smaller ordinal otherwise. In fact

$$\text{Rt}(\Omega_1, \mathbf{A}, \mathbf{m}, \mathbf{n}) \text{ conv } \{\text{Sum}(\text{Dt}, \Omega_1)\}\Big(\text{En}(\Omega_1, \mathbf{A}, \mathbf{m}),\ \text{En}(\Omega_1, \mathbf{A}, \mathbf{n})\Big).$$

Now, if **A** is dual, $E+Q$ includes all integers m for which

$$\{\text{Sum}(\text{Dt}, \Omega_1)\}\,(\mathbf{m}, \mathbf{m}) \text{ conv } 3.$$

(This depends on the particular form that we have chosen for the formula Sum.) Putting "$\text{En}(\Omega_1, \mathbf{A}, \mathbf{p})$ conv $\mathbf{q}$" for $M(p, q)$, we see that condition (7.4) is satisfied, so that $\text{Rt}(\Omega_1, \mathbf{A})$ is an ordinal formula representing γ_0. But, if **A** is not dual, the set $E+Q$ consists of all integers m for which

$$\{\text{Sum}(\text{Dt}, \Omega_1)\}\,(\mathbf{m}, \mathbf{r}) \text{ conv } 2,$$

where r depends only on **A**. In this case $\text{Rt}(\Omega_1, \mathbf{A})$ is an ordinal formula representing the same ordinal as $\text{Inf}\Big(\text{Sum}\,(\text{Dt}, \Omega_1), \mathbf{r}\Big)$, and this is smaller than γ_0. Now consider **K**:

$$\mathbf{K} \to \lambda a \,.\, \Lambda\Big(\text{Sum}\,\Big(\text{Rt}(\Omega_1, \mathbf{A}), \mathbf{P}\Big), \mathbf{B}\Big).$$

If **A** is dual, $\mathbf{K}(\mathbf{A})$ is convertible to 2 since $\text{Sum}\Big(\text{Rt}(\Omega_1, \mathbf{A}), \mathbf{P}\Big)$ represents γ. But, if **A** is not dual, it is not convertible to 2, since $\text{Sum}\Big(\text{Rt}(\Omega_1, \mathbf{A}), \mathbf{P}\Big)$ then represents an ordinal smaller than γ. In **K** we therefore have a complete logic formula, which is impossible.

Now we take the case $\gamma_0 = \omega$. We introduce a W.F.F. Mg such that if n is the D.N. of a computing machine $\mathcal{M}$, and if by the m-th complete configuration of $\mathcal{M}$ the figure 0 has been printed, then $\mathrm{Mg}(\mathbf{n}, \mathbf{m})$ is convertible to $\lambda pq . \mathrm{Al}\big(4(P, 2p+2q), 3, 4\big)$ (which is an ordinal formula representing the ordinal 1), but if 0 has not been printed it is convertible to $\lambda pq . p(q, I, 4)$ (which represents 0). Now consider

$$\mathbf{M} \rightarrow \lambda n . \mathbf{\Lambda}\Big(\mathrm{Sum}\Big(\mathrm{Lim}\big(\mathrm{Mg}(n)\big), \mathbf{P}\Big), \mathbf{B}\Big).$$

If the machine never prints 0, then $\mathrm{Lim}\big(\lambda r . \mathrm{Mg}(\mathbf{n}, r)\big)$ represents ω and $\mathrm{Sum}\Big(\mathrm{Lim}\big(\mathrm{Mg}(\mathbf{n})\big), \mathbf{P}\Big)$ represents γ. This means that $\mathbf{M}(\mathbf{n})$ is convertible to 2. If, however, $\mathcal{M}$ never prints 0, $\mathrm{Sum}\Big(\mathrm{Lim}\big(\mathrm{Mg}(\mathbf{n})\big), \mathbf{P}\Big)$ represents a finite ordinal and $\mathbf{M}(\mathbf{n})$ is not convertible to 2. In $\mathbf{M}$ we therefore have means of determining about a machine whether it ever prints 0, which is impossible† (Turing [1][a] § 8). This completes the proof of (A).

Proof of (B). It is sufficient to prove that, if $\mathbf{C}$ represents an ordinal γ, $\omega^2 \leqslant \gamma < \alpha$, then the extent of $\mathbf{\Lambda}\big(H(\mathbf{C})\big)$ is included in the set-theoretic sum of the extents of $\mathbf{\Lambda}\big(H(\mathbf{G})\big)$, where $\mathbf{G}$ represents an ordinal less than γ. We obtain a contradiction from the assumption that there is a formula $\mathbf{B}$ which is in the extent of $\mathbf{\Lambda}\big(H(\mathbf{G})\big)$ if $\mathbf{G}$ represents γ, but not if it represents any smaller ordinal. The ordinal γ is of the form $\delta+\omega^2+\xi$, where $\xi < \omega^2$. Let $\mathbf{D}$ be a C-K ordinal formula representing δ and $\lambda ufx . \mathbf{Q}\big(u, f, \mathbf{A}(u, f, x)\big)$ one representing $\alpha+\xi$ whenever $\mathbf{A}$ represents α.

We now define a formula Hg. Suppose that $\mathbf{A}$ is a W.F.F. in normal form and without free variables; consider the process of carrying out conversions on $\mathbf{A}(1)$ until it is brought into the form 2, then converting $\mathbf{A}(2)$ to 2, then $\mathbf{A}(3)$, and so on. Suppose that at the r-th step of this process we are doing the n_r-th step in the conversion of $\mathbf{A}(\mathbf{m}_r)$. Thus, for instance, if $\mathbf{A}$ is not convertible to 2, m_r can never exceed 3. Then $\mathrm{Hg}(\mathbf{A}, \mathbf{r})$ is to be convertible to $\lambda f . f(\mathbf{m}_r, \mathbf{n}_r)$ for each positive integer r. Put

$$\mathrm{Sq} \rightarrow \lambda dmn . n\bigg(\mathrm{Suc}, m\Big(\lambda aufx . u\Big(\lambda y . y\big(\mathrm{Suc}, a(u, f, x)\big)\Big), d(u, f, x)\Big)\bigg),$$

$$\mathbf{M} \rightarrow \lambda aufx . \mathbf{Q}\Big(u, f, u\Big(\lambda y . \mathrm{Hg}\big(a, y, \mathrm{Sq}(\mathbf{D})\big)\Big)\Big),$$

$$\mathbf{K}_1 \rightarrow \lambda a . \mathbf{\Lambda}\big(\mathbf{M}(a), \mathbf{B}\big),$$

† This part of the argument can equally well be based on the impossibility of determining about two W.F.F. whether they are interconvertible. (Church [3], 363.)[i]

[i] 107

then I say that $\mathbf{K}_1$ is a complete logic formula. Sq ($\mathbf{D}$, $\mathbf{m}$, $\mathbf{n}$) is a C-K ordinal formula representing $\delta+m\omega+n$, and therefore $\mathrm{Hg}\big(\mathbf{A}, \mathbf{r}, \mathrm{Sq}(\mathbf{D})\big)$ represents an ordinal ζ_r which increases steadily with increasing r, and tends to the limit $\delta+\omega^2$ if $\mathbf{A}$ is dual. Further

$$\mathrm{Hg}\big(\mathbf{A}, \mathbf{r}, \mathrm{Sq}(\mathbf{D})\big) < \mathrm{Hg}\big(\mathbf{A}, S(\mathbf{r}), \mathrm{Sq}(\mathbf{D})\big)$$

for each positive integer r. Therefore $\lambda ufx\,.\,u\Big(\lambda y\,.\,\mathrm{Hg}\big(\mathbf{A}, y, \mathrm{Sq}(\mathbf{D})\big)\Big)$ is a C-K ordinal formula and represents the limit of the sequence $\zeta_1, \zeta_2, \zeta_3, \ldots$. This is $\delta+\omega^2$ if $\mathbf{A}$ is dual, but a smaller ordinal otherwise. Likewise $\mathbf{M}(\mathbf{A})$ represents γ if $\mathbf{A}$ is dual, but is a smaller ordinal otherwise. The formula $\mathbf{B}$ therefore belongs to the extent of $\Lambda\Big(H\big(\mathbf{M}(\mathbf{A})\big)\Big)$ if and only if $\mathbf{A}$ is dual, and this implies that $\mathbf{K}_1$ is a complete logic formula, as was asserted. But this is impossible and we have the required contradiction.

As a corollary to (A) we see that Λ_H is incomplete and in fact that the extent of $\Lambda_H(\mathrm{Dt})$ contains the extent of $\Lambda_H(\mathbf{\Omega})$ for any ordinal formula $\mathbf{\Omega}$. This result, suggested to me first by the solution of question (*b*), may also be obtained more directly. In fact, if a number-theoretic theorem can be proved in any particular P_Ω, it can also be proved in $P_{\lambda mn.m(n,I,4)}$. The formulae describing number-theoretic theorems in P do not involve more than a finite number of types, type 3 being the highest necessary. The formulae describing the number-theoretic theorems in any P_Ω will be obtained by doubling the type subscripts. Now suppose that we have a proof of a number-theoretic theorem G in P_Ω and that the types occurring in the proof are among $0, 2, 4, 6, t_1, t_2, t_3, \ldots$. We may suppose that they have been arranged with all the even types preceding all the odd types, the even types in order of magnitude and the type $2m-1$ preceding $2n-1$ if $\mathbf{\Omega}(\mathbf{m}, \mathbf{n})$ conv 2. Now let each t_r be replaced by $10+2r$ throughout the proof of G. We thus obtain a proof of G in $P_{\lambda mn.(n,I,4)}$.

As with problem (*a*), the solution of problem (*b*) does not require the use of high ordinals [*e.g.* if we make the assumption that the extent of $\Lambda(\Omega)$ is a steadily increasing function of the ordinal represented by $\mathbf{\Omega}$ we do not have to consider ordinals higher than $\omega+2$]. However, if we restrict what we are to call ordinal formulae in some way, we shall have corresponding modified problems (*a*) and (*b*); the solutions will presumably be essentially the same, but will involve higher ordinals. Suppose, for example, that Prod is a W.F.F. with the property that $\mathrm{Prod}(\mathbf{\Omega}_1, \mathbf{\Omega}_2)$ is an ordinal formula representing $\alpha_1\alpha_2$ when $\mathbf{\Omega}_1$, $\mathbf{\Omega}_2$ are ordinal formulae representing α_1, α_2 respectively, and suppose that we call a W.F.F. a 1-ordinal formula when it is convertible to the form $\mathrm{Sum}\big(\mathrm{Prod}(\mathbf{\Omega}, \mathrm{Dt}), \mathbf{P}\big)$, where $\mathbf{\Omega}$, $\mathbf{P}$ are ordinal formulae of which $\mathbf{P}$ represents a finite ordinal. We may define 1-ordinal logics, 1-completeness and 1-invariance in an obvious way, and obtain a solution of problem (*b*) which differs from the solution in the ordinary case in that the ordinals less than ω^2 take the place of the finite ordinals. More generally the cases that I have in mind are covered by the following theorem.

Suppose that we have a class V of formulae representing ordinals in some manner which we do not propose to specify definitely, and a subset† U of the class V such that:

(i) There is a formula $\mathbf{\Phi}$ such that if $\mathbf{T}$ enumerates a sequence of members of U representing an increasing sequence of ordinals, then $\mathbf{\Phi}(\mathbf{T})$ is a member of U representing the limit of the sequence.

(ii) There is a formula $\mathbf{E}$ such that $\mathbf{E}(\mathbf{m}, \mathbf{n})$ is a member of U for each pair of positive integers m, n and, if it represents $\epsilon_{m,n}$, then $\epsilon_{m,n} < \epsilon_{m',n'}$ if either $m < m'$ or $m = m'$, $n < n'$.

(iii) There is a formula $\mathbf{G}$ such that, if $\mathbf{A}$ is a member of U, then $\mathbf{G}(\mathbf{A})$ is a member of U representing a larger ordinal than does $\mathbf{A}$, and such that $\mathbf{G}\big(\mathbf{E}(\mathbf{m}, \mathbf{n})\big)$ always represents an ordinal not larger than $\epsilon_{m,n+1}$.

We define a V-ordinal logic to be a W.F.F. Λ such that $\Lambda(\mathbf{A})$ is a logic whenever $\mathbf{A}$ belongs to V. Λ is V-invariant if the extent of $\Lambda(\mathbf{A})$ depends only on the ordinal represented by $\mathbf{A}$. Then it is not possible for a V-ordinal logic Λ to be V-invariant and have the property that, if $\mathbf{C}_1$ represents a greater ordinal than $\mathbf{C}_2$ ($\mathbf{C}_1$ and $\mathbf{C}_2$ both being members of U), then the extent of $\Lambda(\mathbf{C}_1)$ is greater than the extent of $\Lambda(\mathbf{C}_2)$.

We suppose the contrary. Let $\mathbf{B}$ be a formula belonging to the extent of $\Lambda\Big(\quad\big(\mathbf{\Phi}\big(\lambda r\,.\,\mathbf{E}(r, 1)\big)\big)\Big)$ but not to the extent of $\Lambda\Big(\mathbf{\Phi}\big(\lambda r\,.\,\mathbf{E}(r, 1)\big)\Big)$,

and let $$\mathbf{K}' \to \lambda a\,.\,\Lambda\Big(\mathbf{G}\Big(\mathbf{\Phi}\big(\lambda r\,.\,\mathrm{Hg}(a, r, \mathbf{E})\big)\Big), \mathbf{B}\Big).$$

Then $\mathbf{K}'$ is a complete logic. For

$$\mathrm{Hg}(\mathbf{A}, \mathbf{r}, \mathbf{E})\ \mathrm{conv}\ \mathbf{E}(\mathbf{m}_{\mathbf{r}}, \mathbf{n}_{\mathbf{r}}).$$

† The subset U wholly supersedes V in what follows. The introduction of V serves to emphasise the fact that the set of ordinals represented by members of U may have gaps.

$\mathbf{E}(\mathbf{m}_r, \mathbf{n}_r)$ is a sequence of V-ordinal formulae representing an increasing sequence of ordinals. Their limit is represented by $\mathbf{\Phi}\big(\lambda r\,.\,\mathrm{Hg}(\mathbf{A}, r, \mathbf{E})\big)$; let us see what this limit is. First suppose that $\mathbf{A}$ is dual: then m_r tends to infinity as r tends to infinity, and $\mathbf{\Phi}\big(\lambda r\,.\,\mathrm{Hg}(\mathbf{A}, r, \mathbf{E})\big)$ therefore represents the same ordinal as $\mathbf{\Phi}\big(\lambda r\,.\,\mathbf{E}(r, 1)\big)$. In this case we must have

$$\mathbf{K}'(\mathbf{A}) \text{ conv } 2.$$

Now suppose that $\mathbf{A}$ is not dual: m_r is eventually equal to some constant number, a say, and $\mathbf{\Phi}\big(\lambda r\,.\,\mathrm{Hg}(\mathbf{A}, r, \mathbf{E})\big)$ represents the same ordinal as $\mathbf{\Phi}\big(\lambda r\,.\,\mathbf{E}(\mathbf{a}, r)\big)$, which is smaller than that represented by $\mathbf{\Phi}\big(\lambda r\,.\,\mathbf{E}(r, 1)\big)$. $\mathbf{B}$ cannot therefore belong to the extent of $\mathbf{\Lambda}\Big(\mathbf{G}\Big(\mathbf{\Phi}\big(\lambda r\,.\,\mathrm{Hg}(\mathbf{A}, r, \mathbf{E})\big)\Big)\Big)$, and $\mathbf{K}'(\mathbf{A})$ is not convertible to 2. We have proved that $\mathbf{K}'$ is a complete logic, which is impossible.

This theorem can no doubt be improved in many ways. However, it is sufficiently general to show that, with almost any reasonable notation for ordinals, completeness is incompatible with invariance.

We can still give a certain meaning to the classification into depths with highly restricted kinds of ordinals. Suppose that we take a particular ordinal logic $\mathbf{\Lambda}$ and a particular ordinal formula $\mathbf{\Psi}$ representing the ordinal α say (preferably a large one), and that we restrict ourselves to ordinal formulae of the form $\mathrm{Inf}(\mathbf{\Psi}, \mathbf{a})$. We then have a classification into depths, but the extents of all the logics which we so obtain are contained in the extent of a single logic.

We now attempt a problem of a rather different character, that of the completeness of Λ_P. It is to be expected that this ordinal logic is complete. I cannot at present give a proof of this, but I can give a proof that it is complete as regards a simpler type of theorem than the number-theoretic theorems, viz. those of form "$\theta(x)$ vanishes identically", where $\theta(x)$ is primitive recursive. The proof will have to be much abbreviated since we do not wish to go into the formal details of the system P. Also there is a certain lack of definiteness in the problem as at present stated, owing to the fact that the formulae G, E, M_P were not completely defined. Our attitude here is that it is open to the sceptical reader to give detailed definitions for these formulae and then verify that the remaining details of the proof can be filled in, using his definition. It is not asserted that these details can be filled in whatever be the definitions of G, E, M_P consistent with the properties already required of them, only that they can be filled in with the more natural definitions.

I shall prove the completeness theorem in the following form. If $\mathfrak{B}[x_0]$ is a recursion formula and if $\mathfrak{B}[0], \mathfrak{B}[f0], \ldots$ are all provable in P, then there is a C-K ordinal formula **A** such that $(x_0)\mathfrak{B}[x_0]$ is provable in the system $P^{\mathbf{A}}$ of logic obtained from P by adjoining as axioms all formulae whose G.R.'s are of the form

$$\mathbf{A}\Big(\lambda mn\,.\,m\big(\varpi(2, n),\ \varpi(3, n)\big),\ K,\ M_P,\ \mathbf{r}\Big)$$

(provided they represent propositions).

First let us define the formula **A**. Suppose that **D** is a W.F.F. with the property that $\mathbf{D}(\mathbf{n})$ conv 2 if $\mathfrak{B}[f^{(n-1)}0]$ is provable in P, but $\mathbf{D}(\mathbf{n})$ conv 1 if $\sim\mathfrak{B}[f^{(n-1)}0]$ is provable in P (P is being assumed consistent). Let Θ be defined by

$$\Theta \rightarrow \Big\{\lambda vu\,.\,u\big(v(v, u)\big)\Big\}\Big(\lambda vu\,.\,u\big(v(v, u)\big)\Big),$$

and let Vi be a formula with the properties

$$\mathrm{Vi}(2) \text{ conv } \lambda u\,.\,u(\mathrm{Suc},\ U),$$

$$\mathrm{Vi}(1) \text{ conv } \lambda u\,.\,u\big(I,\ \Theta(\mathrm{Suc})\big).$$

The existence of such a formula is established in Kleene [**1**], corollary on p. 220. Now put

$$\mathbf{A}^* \rightarrow \lambda ufx\,.\,u\Big(\lambda y\,.\,\mathrm{Vi}\big(\mathbf{D}(y),\ y,\ u,\ f,\ x\big)\Big),$$

$$\mathbf{A} \rightarrow \mathrm{Suc}(\mathbf{A}^*).$$

I assert that $\mathbf{A}^*$, **A** are C-K ordinal formulae whenever it is true that $\mathfrak{B}[0], \mathfrak{B}[f0], \ldots$ are all provable in P. For in this case $\mathbf{A}^*$ is $\lambda ufx\,.\,u(\mathbf{R})$, where

$$\mathbf{R} \rightarrow \lambda y\,.\,\mathrm{Vi}\big(\mathbf{D}(y),\ y,\ u,\ f,\ x\big),$$

and then

$$\begin{aligned}\lambda ufx\,.\,\mathbf{R}(\mathbf{n}) &\text{ conv } \lambda ufx\,.\,\mathrm{Vi}\big(\mathbf{D}(\mathbf{n}),\ \mathbf{n},\ u,\ f,\ x\big)\\ &\text{ conv } \lambda ufx\,.\,\mathrm{Vi}(2,\ \mathbf{n},\ u,\ f,\ x)\\ &\text{ conv } \lambda ufx\,.\,\{\lambda n\,.\,n(\mathrm{Suc},\ U)\}(\mathbf{n},\ u,\ f,\ x)\\ &\text{ conv } \lambda ufx\,.\,\mathbf{n}(\mathrm{Suc},\ U,\ u\ f\ x), \text{ which is a C-K ordinal formula,}\end{aligned}$$

and

$$\lambda ufx\,.\,S(\mathbf{n},\ \mathrm{Suc},\ U,\ u,\ f,\ x) \text{ conv } \mathrm{Suc}\big(\lambda ufx\,.\,\mathbf{n}(\mathrm{Suc},\ U,\ u,\ f,\ x)\big).$$

These relations hold for an arbitrary positive integer n and therefore $\mathbf{A}^*$ is a C-K ordinal formula [condition (9) p. 175]: it follows immediately that $\mathbf{A}$ is also a C-K ordinal formula. It remains to prove that $(x_0)\mathfrak{B}[x_0]$ is provable in $P^{\mathbf{A}}$. To do this it is necessary to examine the structure of $\mathbf{A}^*$ in the case in which $(x_0)\mathfrak{B}[x_0]$ is false. Let us suppose that $\sim\mathfrak{B}[f^{(a-1)}0]$ is true, so that $\mathbf{D}(\mathbf{a})$ conv 1, and let us consider $\mathbf{B}$ where

$$\mathbf{B} \to \lambda ufx\,.\,\mathrm{Vi}\Big(\mathbf{D}(\mathbf{a}),\ \mathbf{a},\ u,\ f,\ x\Big).$$

If $\mathbf{A}^*$ was a C-K ordinal formula, then $\mathbf{B}$ would be a member of its fundamental sequence; but

$$\mathbf{B}\ \mathrm{conv}\ \lambda ufx\,.\,\mathrm{Vi}(1,\ \mathbf{a},\ u,\ f,\ x)$$

$$\mathrm{conv}\ \lambda ufx\,.\,\Big\{\lambda u\,.\,u\Big(I,\ \Theta(\mathrm{Suc})\Big)\Big\}(\mathbf{a},\ u,\ f,\ x)$$

$$\mathrm{conv}\ \lambda ufx\,.\,\Theta(\mathrm{Suc},\ u,\ f,\ x)$$

$$\mathrm{conv}\ \lambda ufx\,.\,\Big\{\lambda u\,.\,u\Big(\Theta(u)\Big)\Big\}(\mathrm{Suc},\ u,\ f,\ x)$$

$$\mathrm{conv}\ \lambda ufx\,.\,\mathrm{Suc}\Big(\Theta(\mathrm{Suc}),\ u,\ f,\ x\Big)$$

$$\mathrm{conv}\ \mathrm{Suc}\Big(\lambda ufx\,.\,\Theta(\mathrm{Suc},\ u,\ f,\ x)\Big)$$

$$\mathrm{conv}\ \mathrm{Suc}(\mathbf{B}). \qquad (9.3)$$

This, of course, implies that $\mathbf{B} < \mathbf{B}$ and therefore that $\mathbf{B}$ is no C-K ordinal formula. This, although fundamental in the possibility of proving our completeness theorem, does not form an actual step in the argument. Roughly speaking, our argument amounts to this. The relation (9.3) implies that the system $P^{\mathbf{B}}$ is inconsistent and therefore that $P^{\mathbf{A}^*}$ is inconsistent and indeed we can prove in P (and *a fortiori* in $P^{\mathbf{A}}$) that $\sim(x_0)\mathfrak{B}[x_0]$ implies the inconsistency of $P^{\mathbf{A}^*}$. On the other hand in $P^{\mathbf{A}}$ we can prove the consistency of $P^{\mathbf{A}^*}$. The inconsistency of $P^{\mathbf{B}}$ is proved by the Gödel argument. Let us return to the details.

The axioms in $P^{\mathbf{B}}$ are those whose G.R.'s are of the form

$$\mathbf{B}\Big(\lambda mn\,.\,m\Big(\varpi(2,\ n),\ \varpi(3,\ n)\Big),\ K,\ M_P,\ \mathbf{r}\Big).$$

When we replace $\mathbf{B}$, by $\mathrm{Suc}(\mathbf{B})$, this becomes

$$\mathrm{Suc}\Big(\mathbf{B},\ \lambda mn\,.\,m\big(\varpi(2,n),\ \varpi(3,n)\big),\ K,\ M_P,\ \mathbf{r}\Big)$$

$$\operatorname{conv} K\Big(\mathbf{B}\Big(\lambda mn\,.\,m\big(\varpi(2,n),\ \varpi(3,n)\big),\ K,\ M_P,\ \mathbf{r}\Big)\Big)$$

$$\operatorname{conv} \mathbf{B}\Big(\lambda mn\,.\,m\big(\varpi(2,n),\ \varpi(3,n)\big),\ K,\ M_P,\ \mathbf{p}\Big)$$

if $\mathbf{r} \operatorname{conv} 2\mathbf{p}+1$,

$$\operatorname{conv} E\Big(\mathbf{B}\Big(\lambda mn\,.\,m\big(\varpi(2,n),\ \varpi(3,n)\big),\ K,\ M_P\Big),\ \mathbf{p}\Big)$$

if $\mathbf{r} \operatorname{conv} 2\mathbf{p}$.

When we remember the essential property of the formula E, we see that the axioms of $P^{\mathbf{B}}$ include all formulae of the form

$$(\exists x_0)\ \mathrm{Proof}_{P\mathbf{B}}[x_0, f^{(q)}0] \supset \mathfrak{F},$$

where q is the G.R. of the formula $\mathfrak{F}$.

Let b be the G.R. of the formula $\mathfrak{A}$.

$$\sim (\exists x_0)(\exists y_0)\{\mathrm{Proof}_{P\mathbf{B}}[x_0, y_0]\,.\,\mathrm{Sb}[z_0, z_0, y_0]\}. \qquad (\mathfrak{A})$$

$\mathrm{Sb}[x_0, y_0, z_0]$ is a particular recursion formula such that $\mathrm{Sb}[f^{(l)}0, f^{(m)}0, f^{(n)}0]$ holds if and only if n is the G.R. of the result of substituting $f^{(m)}0$ for z_0 in the formula whose G.R. is l at all points where z_0 is free. Let p be the G.R. of the formula $\mathfrak{C}$.

$$\sim (\exists x_0)(\exists y_0)\{\mathrm{Proof}_{P\mathbf{B}}[x_0, y_0]\,.\,\mathrm{Sb}\,[f^{(b)}0,\ f^{(b)}0,\ y_0]\}. \qquad (\mathfrak{C})$$

Then we have as an axiom in P

$$(\exists x_0)\ \mathrm{Proof}_{P\mathbf{B}}[x_0,\ f^{(p)}0] \supset \mathfrak{C},$$

and we can prove in $P^{\mathbf{A}}$

$$(x_0)\{\mathrm{Sb}\,[f^{(b)}0,\ f^{(b)}0,\ x_0] \equiv x_0 = f^{(p)}0\}, \qquad (9.4)$$

since $\mathfrak{C}$ is the result of substituting $f^{(b)}0$ for z_0 in $\mathfrak{A}$; hence

$$\sim (\exists y_0)\ \mathrm{Proof}_{P\mathbf{B}}[y_0,\ f^{(p)}0] \qquad (9.5)$$

is provable in P. Using (9.4) again, we see that $\mathfrak{C}$ can be proved in $P^{\mathbf{B}}$. But, if we can prove $\mathfrak{C}$ in $P^{\mathbf{B}}$, then we can prove its provability in $P^{\mathbf{B}}$, the proof being in P; *i.e.* we can prove

$$(\exists x_0)\ \text{Proof}_{P^{\mathbf{B}}}\ [x_0, f^{(p)}0]$$

in P (since p is the G.R. of $\mathfrak{C}$). But this contradicts (9.5), so that, if

$$\sim \mathfrak{B}\ [f^{(a-1)}0]$$

is true, we can prove a contradiction in $P^{\mathbf{B}}$ or in $P^{\mathbf{A}^*}$. Now I assert that the whole argument up to this point can be carried through formally in the system P, in fact, that, if c is the G.R. of $\sim (0=0)$, then

$$\sim (x_0)\,\mathfrak{B}[x_0] \supset (\exists v_0)\ \text{Proof}_{P^{\mathbf{A}^*}}\ [v_0, f^{(c)}0] \tag{9.6}$$

is provable in P. I shall not attempt to give any more detailed proof of this assertion.

The formula

$$(\exists x_0)\ \text{Proof}_{P^{\mathbf{A}^*}}\ [x_0, f^{(c)}0] \supset \sim (0=0) \tag{9.7}$$

is an axiom in $P^{\mathbf{A}}$. Combining (9.6), (9.7) we obtain $(x_0)\,\mathfrak{B}[x_0]$ in $P^{\mathbf{A}}$.

This completeness theorem as usual is of no value. Although it shows, for instance, that it is possible to prove Fermat's last theorem with Λ_P (if it is true) yet the truth of the theorem would really be assumed by taking a certain formula as an ordinal formula.

That Λ_P is not invariant may be proved easily by our general theorem; alternatively it follows from the fact that, in proving our partial completeness theorem, we never used ordinals higher than $\omega+1$. This fact can also be used to prove that Λ_P is not C-K invariant up to $\omega+2$.

10. *The continuum hypothesis. A digression.*

The methods of § 9 may be applied to problems which are constructive analogues of the continuum hypothesis problem. The continuum hypothesis asserts that $2^{\aleph_0} = \aleph_1$, in other words that, if ω_1 is the smallest ordinal α greater than ω such that a series with order type α cannot be put into one-one correspondence with the positive integers, then the ordinals less than ω_1 can be put into one-one correspondence with the subsets of the positive integers. To obtain a constructive analogue of this proposition we may replace the ordinals less than ω_1 either by the ordinal formulae, or by the ordinals represented by them; we may replace the subsets of the positive integers either by the computable sequences of figures 0, 1, or by the description numbers of the machines which compute these sequences. In the manner in which the correspondence is to be set up there is also more than one possibility. Thus, even when we use only one kind of ordinal formula, there is still great ambiguity concerning what the constructive analogue of the continuum hypothesis should be. I shall prove a single result in this connection†. A number of others may be proved in the same way.

We ask "Is it possible to find a computable function of ordinal formulae determining a one-one correspondence between the ordinals represented by ordinal formulae and the computable sequences of figures 0, 1?" More accurately, "Is there a formula **F** such that if $\mathbf{\Omega}$ is an ordinal formula and n a positive integer then $\mathbf{F}(\mathbf{\Omega}, \mathbf{n})$ is convertible to 1 or to 2, and such that $\mathbf{F}(\mathbf{\Omega}, \mathbf{n})$ conv $\mathbf{F}(\mathbf{\Omega}', \mathbf{n})$ for each positive integer n, if and only if $\mathbf{\Omega}$ and $\mathbf{\Omega}'$ represent the same ordinal?" The answer is "No", as will be seen to be a consequence of the following argument: there is no formula **F** such that $\mathbf{F}(\mathbf{\Omega})$ enumerates one sequence of integers (each being 1 or 2) when $\mathbf{\Omega}$ represents ω and enumerates another sequence when $\mathbf{\Omega}$ represents 0. If there is such an **F**, then there is an a such that $\mathbf{F}(\mathbf{\Omega}, \mathbf{a})$ conv (Dt, **a**) if $\mathbf{\Omega}$ represents ω but $\mathbf{F}(\mathbf{\Omega}, \mathbf{a})$ and $\mathbf{F}(\mathrm{Dt}, \mathbf{a})$ are convertible to different integers (1 or 2) if $\mathbf{\Omega}$ represents 0. To obtain a contradiction from this we introduce a W.F.F. Gm not unlike Mg. If the machine $\mathcal{M}$ whose D.N. is n has printed 0 by the time the m-th complete configuration is reached then

$$\mathrm{Gm}(\mathbf{n}, \mathbf{m}) \text{ conv } \lambda mn \,.\, m(n, I, 4);$$

otherwise $\mathrm{Gm}(\mathbf{n}, \mathbf{m})$ conv $\lambda pq \,.\, \mathrm{Al}\big(4(P, 2p+2q), 3, 4\big)$. Now consider $\mathbf{F}(\mathrm{Dt}, \mathbf{a})$ and $\mathbf{F}\Big(\mathrm{Lim}\big(\mathrm{Gm}(\mathbf{n})\big), \mathbf{a}\Big)$. If $\mathcal{M}$ never prints 0, $\mathrm{Lim}\big(\mathrm{Gm}(\mathbf{n})\big)$ represents the ordinal ω. Otherwise it represents 0. Consequently these two formulae are convertible to one another if and only if $\mathcal{M}$ never prints 0. This gives us a means of determining about any machine whether it ever prints 0, which is impossible.

Results of this kind have of course no real relevance for the classical continuum hypothesis.

11. *The purpose of ordinal logics.*

Mathematical reasoning may be regarded rather schematically as the exercise of a combination of two faculties‡, which we may call *intuition* and *ingenuity*. The activity of the intuition consists in making spontaneous judgments which are not the result of conscious trains

† A suggestion to consider this problem came to me indirectly from F. Bernstein. A related problem was suggested by P. Bernays.

‡ We are leaving out of account that most important faculty which distinguishes topics of interest from others; in fact, we are regarding the function of the mathematician as simply to determine the truth or falsity of propositions.

of reasoning. These judgments are often but by no means invariably correct (leaving aside the question what is meant by "correct"). Often it is possible to find some other way of verifying the correctness of an intuitive judgment. We may, for instance, judge that all positive integers are uniquely factorizable into primes; a detailed mathematical argument leads to the same result. This argument will also involve intuitive judgments, but they will be less open to criticism than the original judgment about factorization. I shall not attempt to explain this idea of "intuition" any more explicitly.

The exercise of ingenuity in mathematics consists in aiding the intuition through suitable arrangements of propositions, and perhaps geometrical figures or drawings. It is intended that when these are really well arranged the validity of the intuitive steps which are required cannot seriously be doubted.

The parts played by these two faculties differ of course from occasion to occasion, and from mathematician to mathematician. This arbitrariness can be removed by the introduction of a formal logic. The necessity for using the intuition is then greatly reduced by setting down formal rules for carrying out inferences which are always intuitively valid. When working with a formal logic, the idea of ingenuity takes a more definite shape. In general a formal logic, will be framed so as to admit a considerable variety of possible steps in any stage in a proof. Ingenuity will then determine which steps are the more profitable for the purpose of proving a particular proposition. In pre-Gödel times it was thought by some that it would probably be possible to carry this programme to such a point that all the intuitive judgments of mathematics could be replaced by a finite number of these rules. The necessity for intuition would then be entirely eliminated.

In our discussions, however, we have gone to the opposite extreme and eliminated not intuition but ingenuity, and this in spite of the fact that our aim has been in much the same direction. We have been trying to see how far it is possible to eliminate intuition, and leave only ingenuity. We do not mind how much ingenuity is required, and therefore assume it to be available in unlimited supply. In our metamathematical discussions we actually express this assumption rather differently. We are always able to obtain from the rules of a formal logic a method of enumerating the propositions proved by its means. We then imagine that all proofs take the form of a search through this enumeration for the theorem for which a proof is desired. In this way ingenuity is replaced by patience. In these heuristic discussions, however, it is better not to make this reduction.

In consequence of the impossibility of finding a formal logic which wholly eliminates the necessity of using intuition, we naturally turn to "non-constructive" systems of logic with which not all the steps in a proof are mechanical, some being intuitive. An example of a non-constructive logic is afforded by any ordinal logic. When we have an ordinal logic, we are in a position to prove number-theoretic theorems by the intuitive steps of recognizing formulae as ordinal formulae, and the mechanical steps of carrying out conversions. What properties do we desire a non-constructive logic to have if we are to make use of it for the expression of mathematical proofs? We want it to show quite clearly when a step makes use of intuition, and when it is purely formal. The strain put on the intuition should be a minimum. Most important of all, it must be beyond all reasonable doubt that the logic leads to correct results whenever the intuitive steps are correct†. It is also desirable that the logic shall be adequate for the expression of number-theoretic theorems, in order that it may be used in metamathematical discussions (cf. § 5).

Of the particular ordinal logics that we have discussed, Λ_H and Λ_P certainly will not satisfy us. In the case of Λ_H we are in no better position than with a constructive logic. In the case of Λ_P (and for that matter also Λ_H) we are by no means certain that we shall never obtain any but true results, because we do not know whether all the number-theoretic theorems provable in the system P are true. To take Λ_P as a fundamental non-constructive logic for metamathematical arguments would be most unsound. There remains the system of Church which is free from these objections. It is probably complete (although this would not necessarily mean much) and it is beyond reasonable doubt that it always leads to correct results‡. In the next section I propose to describe another ordinal logic, of a very different type, which is suggested by the work of Gentzen and which should also be adequate for the formalization of number-theoretic theorems. In particular it should be suitable for proofs of metamathematical theorems (cf. § 5).

† This requirement is very vague. It is not of course intended that the criterion of the correctness of the intuitive steps be the correctness of the final result. The meaning becomes clearer if each intuitive step is regarded as a judgment that a particular proposition is true. In the case of an ordinal logic it is always a judgment that a formula is an ordinal formula, and this is equivalent to judging that a number-theoretic proposition is true. In this case then the requirement is that the reputed ordinal logic *is* an ordinal logic.

‡ This ordinal logic arises from a certain system C_0 in essentially the same way as Λ_P arose from P. By an argument similar to one occurring in § 8 we can show that the ordinal logic leads to correct results if and only if C_0 is valid; the validity of C_0 is proved in Church [**1**], making use of the results of Church and Rosser [**1**].

12. *Gentzen type ordinal logics.*

In proving the consistency of a certain system of formal logic Gentzen (Gentzen [**1**]) has made use of the principle of transfinite induction for ordinals less than ϵ_0, and has suggested that it is to be expected that transfinite induction carried sufficiently far would suffice to solve all problems of consistency. Another suggestion of basing systems of logic on transfinite induction has been made by Zermelo (Zermelo [**1**]). In this section I propose to show how this method of proof may be put into the form of a formal (non-constructive) logic, and afterwards to obtain from it an ordinal logic.

We can express the Gentzen method of proof formally in this way. Let us take the system P and adjoin to it an axiom $\mathfrak{A}_\Omega$ with the intuitive meaning that the W.F.F. $\mathbf{\Omega}$ is an ordinal formula, whenever we feel certain that $\mathbf{\Omega}$ *is* an ordinal formula. This is a non-constructive system of logic which may easily be put into the form of an ordinal logic. By the method of § 6 we make correspond to the system of logic consisting of P with the axiom $\mathfrak{A}_\Omega$ adjoined a logic formula L_Ω: L_Ω is an effectively calculable function of $\mathbf{\Omega}$, and there is therefore a formula $\Lambda_G{}^1$ such that $\Lambda_G{}^1(\mathbf{\Omega})$ conv L_Ω for each formula $\mathbf{\Omega}$. $\Lambda_G{}^1$ is certainly not an ordinal logic unless P is valid, and therefore consistent. This formalization of Gentzen's idea would therefore not be applicable for the problem with which Gentzen himself was concerned, for he was proving the consistency of a system weaker than P. However, there are other ways in which the Gentzen method of proof can be formalized. I shall explain one, beginning by describing a certain logical calculus.

The symbols of the calculus are f, x, ^{1}, $_1$, 0, S, R, Γ, Δ, E, |, $\odot$, !, (,), =, and the comma ",". For clarity we shall use various sizes of brackets (,) in the following. We use capital German letters to stand for variable or undetermined sequences of these symbols.

It is to be understood that the relations that we are about to define hold only when compelled to do so by the conditions that we lay down. The conditions should be taken together as a simultaneous inductive definition of all the relations involved.

Suffixes.

$_1$ is a suffix. If $\mathfrak{S}$ is a suffix then $\mathfrak{S}_1$ is a suffix.

Indices.

^{1} is an index. If $\mathfrak{I}$ is an index then $\mathfrak{I}^1$ is an index.

Numerical variables.

If $\mathfrak{S}$ is a suffix then $x\mathfrak{S}$ is a numerical variable.

Functional variables.

If $\mathfrak{S}$ is a suffix and $\mathfrak{I}$ is an index, then $f\mathfrak{S}\mathfrak{I}$ is a functional variable of index $\mathfrak{I}$.

Arguments.

(,) is an argument of index ^{1}. If $(\mathfrak{A})$ is an argument of index $\mathfrak{I}$ and $\mathfrak{T}$ is a term, then $(\mathfrak{A}\mathfrak{T},)$ is an argument of index $\mathfrak{I}^{1}$.

Numerals.

0 is a numeral.

If $\mathfrak{N}$ is a numeral, then $S(, \mathfrak{N},)$ is a numeral.

In metamathematical statements we shall denote the numeral in which S occurs r times by $S^{(r)}(, 0,)$.

Expressions of a given index.

A functional variable of index $\mathfrak{I}$ is an expression of index $\mathfrak{I}$.

R, S are expressions of index ^{111}, ^{11} respectively.

If $\mathfrak{N}$ is a numeral, then it is also an expression of index ^{1}.

Suppose that $\mathfrak{G}$ is an expression of index $\mathfrak{I}$, $\mathfrak{H}$ one of index $\mathfrak{I}^{1}$ and $\mathfrak{K}$ one of index $\mathfrak{I}^{111}$; then $(\Gamma\mathfrak{G})$ and $(\Delta\mathfrak{G})$ are expressions of index $\mathfrak{I}$, while $(E\mathfrak{G})$ and $(\mathfrak{G}\,|\,\mathfrak{H})$ and $(\mathfrak{G}\odot\mathfrak{K})$ and $(\mathfrak{G}\,!\,\mathfrak{H}\,!\,\mathfrak{K})$ are expressions of index $\mathfrak{I}^{1}$.

Function constants.

An expression of index $\mathfrak{I}$ in which no functional variable occurs is a function constant of index $\mathfrak{I}$. If in addition R does not occur, the expression is called a *primitive function constant.*

Terms.

0 is a term.

Every numerical variable is a term.

If $\mathfrak{G}$ is an expression of index $\mathfrak{I}$ and $(\mathfrak{A})$ is an argument of index $\mathfrak{I}$, then $\mathfrak{G}(\mathfrak{A})$ is a term.

Equations.

If $\mathfrak{T}$ and $\mathfrak{T}'$ are terms, then $\mathfrak{T} = \mathfrak{T}'$ is an equation.

Provable equations.

We define what is meant by the provable equations relative to a given set of equations as axioms.

(a) The provable equations include all the axioms. The axioms are of the form of equations in which the symbols Γ, Δ, E, $|$, $\odot$, ! do not appear.

(b) If $\mathfrak{G}$ is an expression of index $\mathfrak{J}^{11}$ and $(\mathfrak{A})$ is an argument of index $\mathfrak{J}$, then

$$(\Gamma\mathfrak{G})(\mathfrak{A}x_1, x_{11},) = \mathfrak{G}(\mathfrak{A}x_{11}, x_1,)$$

is a provable equation.

(c) If $\mathfrak{G}$ is an expression of index $\mathfrak{J}^1$, and $(\mathfrak{A})$ is an argument of index $\mathfrak{J}$, then

$$(\Delta\mathfrak{G})(\mathfrak{A}x_1,) = \mathfrak{G}(, x_1\,\mathfrak{A})$$

is a provable equation.

(d) If $\mathfrak{G}$ is an expression of index $\mathfrak{J}$, and $(\mathfrak{A})$ is an argument of index $\mathfrak{J}$, then

$$(E\mathfrak{G})(\mathfrak{A}x_1,) = \mathfrak{G}(\mathfrak{A})$$

is a provable equation.

(e) If $\mathfrak{G}$ is an expression of index $\mathfrak{J}$ and $\mathfrak{H}$ is one of index $\mathfrak{J}^1$, and $(\mathfrak{A})$ is an argument of index $\mathfrak{J}$, then

$$(\mathfrak{G}\,|\,\mathfrak{H})(\mathfrak{A}) = \mathfrak{H}\Big(\mathfrak{A}\mathfrak{G}(\mathfrak{A}),\Big)$$

is a provable equation.

(f) If $\mathfrak{N}$ is an expression of index ^{1}, then $\mathfrak{N}(,) = \mathfrak{N}$ is a provable equation.

(g) If $\mathfrak{G}$ is an expression of index $\mathfrak{J}$ and $\mathfrak{K}$ one of index $\mathfrak{J}^{111}$, and $(\mathfrak{A})$ an argument of index $\mathfrak{J}^1$, then

$$(\mathfrak{G} \odot \mathfrak{K})\,(\mathfrak{A}0,) = \mathfrak{G}(\mathfrak{A})$$

and $$(\mathfrak{G} \odot \mathfrak{K})\Big(\mathfrak{A}S(, x_1,),\Big) = \mathfrak{K}\Big(\mathfrak{A}x_1,\ S(, x_1,),\ (\mathfrak{G}\odot\mathfrak{K})(\mathfrak{A}x_1,),\Big)$$

are provable equations. If in addition $\mathfrak{H}$ is an expression of index $\mathfrak{J}^1$ and

$$R\Big(, \mathfrak{G}\Big(\mathfrak{A}S(, x_1,),\Big), x_1,\Big) = 0$$

is provable, then

$$(\mathfrak{G}\,!\,\mathfrak{K}\,!\,\mathfrak{H})\,(\mathfrak{A}0,) = \mathfrak{G}(\mathfrak{A})$$

and

$$(\mathfrak{G}!\,\mathfrak{K}!\,\mathfrak{H})\Big(\mathfrak{A}S(, x_1,),\Big)$$
$$= \mathfrak{K}\Big(\Big(\mathfrak{A}\mathfrak{H}(\mathfrak{A}S(, x_1,),\Big), S(, x_1,), (\mathfrak{G}!\,\mathfrak{K}!\,\mathfrak{H})\Big(\mathfrak{A}\mathfrak{H}\Big(\mathfrak{A}S(, x_1,),\Big),\Big),\Big)$$

are provable.

(*h*) If $\mathfrak{T} = \mathfrak{T}'$ and $\mathfrak{U} = \mathfrak{U}'$ are provable, where $\mathfrak{T}$, $\mathfrak{T}'$, $\mathfrak{U}$ and $\mathfrak{U}'$ are terms, then $\mathfrak{U}' = \mathfrak{U}$ and the result of substituting $\mathfrak{U}'$ for $\mathfrak{U}$ at any particular occurrence in $\mathfrak{T} = \mathfrak{T}'$ are provable equations.

(*i*) The result of substituting any term for a particular numerical variable throughout a provable equation is provable.

(*j*) Suppose that $\mathfrak{G}$, $\mathfrak{G}'$ are expressions of index $\mathfrak{I}^1$, that $(\mathfrak{A})$ is an argument of index $\mathfrak{I}$ not containing the numerical variable $\mathfrak{X}$ and that $\mathfrak{G}(\mathfrak{A}0,) = \mathfrak{G}'(\mathfrak{A}0,)$ is provable. Also suppose that, if we add

$$\mathfrak{G}(\mathfrak{A}\mathfrak{X},) = \mathfrak{G}'(\mathfrak{A}\mathfrak{X},)$$

to the axioms and restrict (*i*) so that it can never be applied to the numerical variable $\mathfrak{X}$, then

$$\mathfrak{G}\Big(\mathfrak{A}S(,\mathfrak{X},),\Big) = \mathfrak{G}'\Big(\mathfrak{A}S(,\mathfrak{X}),\Big)$$

becomes a provable equation; in the hypothetical proof of this equation this rule (*j*) itself may be used provided that a different variable is chosen to take the part of $\mathfrak{X}$.

Under these conditions $\mathfrak{G}(\mathfrak{A}\mathfrak{X},) = \mathfrak{G}'(\mathfrak{A}\mathfrak{X},)$ is a provable equation.

(*k*) Suppose that $\mathfrak{G}$, $\mathfrak{G}'$, $\mathfrak{H}$ are expressions of index $\mathfrak{I}^1$, that $(\mathfrak{A})$ is an argument of index $\mathfrak{I}$ not containing the numerical variable $\mathfrak{X}$ and that

$$\mathfrak{G}(\mathfrak{A}0,) = \mathfrak{G}'(\mathfrak{A}0,) \quad \text{and} \quad R\Big(,\mathfrak{H}\big(\mathfrak{A}S(,\mathfrak{X},),\big),S(,\mathfrak{X},),\Big) = 0$$

are provable equations. Suppose also that, if we add

$$\mathfrak{G}\Big(\mathfrak{A}\mathfrak{H}\big(\mathfrak{A}S(,\mathfrak{X},),\big)\Big) = \mathfrak{G}'\Big(\mathfrak{A}\mathfrak{H}\big(\mathfrak{A}S(,\mathfrak{X},),\big)\Big)$$

to the axioms, and again restrict (*i*) so that it does not apply to $\mathfrak{X}$, then

$$\mathfrak{G}(\mathfrak{A}\mathfrak{X},) = \mathfrak{G}'(\mathfrak{A}\mathfrak{X},) \tag{12.1}$$

becomes a provable equation; in the hypothetical proof of (12.1) the rule (*k*) may be used if a different variable takes the part of $\mathfrak{X}$.

Under these conditions (12.1) is a provable equation.

We have now completed the definition of a provable equation relative to a given set of axioms. Next we shall show how to obtain an ordinal logic from this calculus. The first step is to set up a correspondence between some of the equations and number-theoretic theorems, in other words to show how they can be interpreted as number-theoretic theorems.

Let $\mathfrak{G}$ be a primitive function constant of index [111]. $\mathfrak{G}$ describes a certain primitive recursive function $\phi(m, n)$, determined by the condition that, for all natural numbers m, n, the equation

$$\mathfrak{G}\Big(, S^{(m)}(,0,), S^{(n)}(,0,),\Big) = S^{(\phi(m,n))}(,0,)$$

is provable without using the axioms (a). Suppose also that $\mathfrak{H}$ is an expression of index $\mathfrak{I}$. Then to the equation

$$\mathfrak{G}\Big(, x_1, \mathfrak{H}(, x_1,),\Big) = 0$$

we make correspond the number-theoretic theorem which asserts that for each natural number m there is a natural number n such that $\phi(m, n) = 0$. (The circumstance that there is more than one equation to represent each number-theoretic theorem could be avoided by a trivial but inconvenient modification of the calculus.)

Now let us suppose that some definite method is chosen for describing the sets of axioms by means of positive integers, the null set of axioms being described by the integer 1. By an argument used in § 6 there is a W.F.F. Σ such that, if r is the integer describing a set A of axioms, then $\Sigma(\mathbf{r})$ is a logic formula enabling us to prove just those number-theoretic theorems which are associated with equations provable with the above described calculus, the axioms being those described by the number r.

I explain two ways in which the construction of the ordinal logic may be completed.

In the first method we make use of the theory of general recursive functions (Kleene [**2**])[a]. Let us consider all equations of the form

$$R\Big(, S^{(m)}(,0,), S^{(n)}(,0,),\Big) = S^{(p)}(,0,) \qquad (12.2)$$

which are obtainable from the axioms by the use of rules (h), (i). It is a consequence of the theorem of equivalence of λ-definable and general recursive functions (Kleene [**3**]) that, if $r(m, n)$ is any λ-definable function of two variables, then we can choose the axioms so that (12.2) with $p = r(m, n)$ is obtainable in this way for each pair of natural numbers m, n, and no equation of the form

$$S^{(m)}(,0,) = S^{(n)}(,0,) \quad (m \neq n) \qquad (12.3)$$

is obtainable. In particular, this is the case if $r(m, n)$ is defined by the condition that

$$\mathbf{\Omega}(\mathbf{m}, \mathbf{n}) \text{ conv } S(\mathbf{p}) \quad \text{implies} \quad p = r(m, n),$$

$$r(0, n) = 1, \quad \text{all} \quad n > 0, \quad r(0, 0) = 2,$$

where $\mathbf{\Omega}$ is an ordinal formula. There is a method for obtaining the axioms given the ordinal formula, and consequently a formula Rec such that, for any ordinal formula $\mathbf{\Omega}$, Rec ($\mathbf{\Omega}$) conv $\mathbf{m}$, where m is the integer describing the set of axioms corresponding to $\mathbf{\Omega}$. Then the formula

$$\Lambda_G{}^2 \to \lambda w . \Sigma\Big(\text{Rec}\,(w)\Big)$$

is an ordinal logic. Let us leave the proof of this aside for the present.

Our second ordinal logic is to be constructed by a method not unlike the one which we used in constructing Λ_P. We begin by assigning ordinal formulae to all sets of axioms satisfying certain conditions. For this purpose we again consider that part of the calculus which is obtained by restricting "expressions" to be functional variables or R or S and restricting the meaning of "term" accordingly; the new provable equations are given by conditions (a), (h), (i), together with an extra condition (l).

(l) The equation

$$R\Big(, 0,\ S(, x_1,),\Big) = 0$$

is provable.

We could design a machine which would obtain all equations of the form (12 . 2), with $m \neq n$, provable in this sense, and all of the form (12 . 3), except that it would cease to obtain any more equations when it had once obtained one of the latter "contradictory" equations. From the description of the machine we obtain a formula $\mathbf{\Omega}$ such that

$$\mathbf{\Omega}(\mathbf{m},\ \mathbf{n}) \text{ conv } 2 \quad \text{if} \quad R\Big(,\ S^{(m-1)}(,\ 0,),\ S^{(n-1)}(, 0,),\Big) = 0$$

is obtained by the machine,

$$\mathbf{\Omega}(\mathbf{m},\ \mathbf{n}) \text{ conv } 1 \quad \text{if} \quad R\Big(,\ S^{(n-1)}(, 0,),\ S^{(m-1)}(, 0,),\Big) = 0$$

is obtained by the machine, and

$$\mathbf{\Omega}(\mathbf{m},\ \mathbf{m}) \text{ conv } 3 \text{ always.}$$

The formula $\mathbf{\Omega}$ is an effectively calculable function of the set of axioms, and therefore also of m: consequently there is a formula M such that $M(\mathbf{m})$ conv $\mathbf{\Omega}$ when m describes the set of axioms. Now let Cm be a formula such that, if b is the G.R. of a formula $M(\mathbf{m})$, then Cm(b) conv $\mathbf{m}$, but otherwise Cm(b) conv 1. Let

$$\Lambda_G{}^3 \to \lambda wa . \Gamma\Big(\lambda n . \Sigma\Big(\text{Cm}(\text{Tn}(w,\ n))\Big),\ a\Big).$$

Then Λ_G^3 $(\mathbf{\Omega}, \mathbf{A})$ conv 2 if and only if $\mathbf{\Omega}$ conv $M(\mathbf{m})$, where m describes a set of axioms which, taken with our calculus, suffices to prove the equation which is, roughly speaking, equivalent to "**A** is dual". To prove that Λ_G^3 is an ordinal logic, it is sufficient to prove that the calculus with the axioms described by m proves only true number-theoretic theorems when $\mathbf{\Omega}$ is an ordinal formula. This condition on m may also be expressed in this way. Let us put $m \leqslant n$ if we can prove $R\Big(, S^{(m)}(,0,), \; S^{(n)}(,0,),\Big) = 0$ with (a), (h), (i), (l): the condition is that $m \leqslant n$ is a well-ordering of the natural numbers and that no contradictory equation (12.3) is provable with the same rules (a), (h), (i), (l). Let us say that such a set of axioms is *admissible*. Λ_G^3 is an ordinal logic if the calculus leads to none but true number-theoretic theorems when an admissible set of axioms is used.

In the case of Λ_G^2, Rec $(\mathbf{\Omega})$ describes an admissible set of axioms whenever Ω is an ordinal formula. Λ_G^2 therefore is an ordinal logic if the calculus leads to correct results when admissible axioms are used.

To prove that admissible axioms have the required property, I do not attempt to do more than show how interpretations can be given to the equations of the calculus so that the rules of inference (a)–(k) become intuitively valid methods of deduction, and so that the interpretation agrees with our convention regarding number-theoretic theorems.

Each expression is the name of a function, which may be only partially defined. The expression S corresponds simply to the successor function. If $\mathfrak{G}$ is either R or a functional variable and has $p+1$ symbols in its index, then it corresponds to a function g of p natural numbers defined as follows. If

$$\mathfrak{G}\Big(, S^{(r_1)}(,0,), \; S^{(r_2)}(,0,), \; \ldots, \; S^{(r_p)}(,0,),\Big) = S^{(l)}(,0,)$$

is provable by the use of (a), (h), (i), (l) only, then $g(r_1, r_2, \ldots, r_p)$ has the value p. It may not be defined for all arguments, but its value is always unique, for otherwise we could prove a "contradictory" equation and $M(\mathbf{m})$ would then not be an ordinal formula. The functions corresponding to the other expressions are essentially defined by (b)–(f). For example, if g is the function corresponding to $\mathfrak{G}$ and g' that corresponding to $(\Gamma\mathfrak{G})$, then

$$g'(r_1, r_2, \ldots, r_p, l, m) = g(r_1, r_2, \ldots, r_p, m, l).$$

The values of the functions are clearly unique (when defined at all) if given by one of (b)–(e). The case (f) is less obvious since the function defined appears also in the definiens. I do not treat the case of $(\mathfrak{G} \odot \mathfrak{K})$, since this is the well-known definition by primitive recursion, but I shall show that the values of the function corresponding to $(\mathfrak{G}!\;\mathfrak{K}!\;\mathfrak{H})$ are unique. Without loss of generality we may suppose that $(\mathfrak{A})$ in (f) is of index ^{1}. We have then to show that, if $h(m)$ is the function corresponding to $\mathfrak{H}$ and $r(m, n)$ that corresponding to R, and $k(u, v, w)$ is a given function and a a given natural number, then the equations

$$l(0) = a, \tag{α}$$

$$l(m+1) = k\Big(h(m+1),\ m+1,\ l\big(h(m+1)\big)\Big) \tag{β}$$

do not ever assign two different values for the function $l(m)$. Consider those values of r for which we obtain more than one value of $l(r)$, and suppose that there is at least one such. Clearly 0 is not one, for $l(0)$ can be defined only by (α). Since the relation $\leqslant$ is a well ordering, there is an integer r_0 such that $r_0 > 0$, $l(r_0)$ is not unique, and if $s \neq r_0$ and $l(s)$ is not unique then $r_0 \leqslant s$. We may put $s = h(r_0)$, for, if $l\big(h(r_0)\big)$ were unique, then $l(r_0)$, defined by (β), would be unique. But $r\big(h(r_0), r_0\big) = 0$ *i.e.* $s \leqslant r_0$. There is, therefore, no integer r for which we obtain more than one value for the function $l(r)$.

Our interpretation of expressions as functions gives us an immediate interpretation for equations with no numerical variables. In general we interpret an equation with numerical variables as the (infinite) conjunction of all equations obtainable by replacing the variables by numerals. With this interpretation (*h*), (*i*) are seen to be valid methods of proof. In (*j*) the provability of

$$\mathfrak{G}\big(\mathfrak{A}S(,x_1,),\big) = \mathfrak{G}'\big(\mathfrak{A}S(,x_1,),\big)$$

when $\mathfrak{G}(\mathfrak{A}x_1,) = \mathfrak{G}'(\mathfrak{A}x_1,)$ is assumed to be interpreted as meaning that the implication between these equations holds for all substitutions of numerals for x_1. To justify this, one should satisfy oneself that these implications always hold when the hypothetical proof can be carried out. The rule of procedure (*j*) is now seen to be simply mathematical induction. The rule (*k*) is a form of transfinite induction. In proving the validity of (*k*) we may again suppose ($\mathfrak{A}$) is of index 1. Let $r(m, n)$, $g(m)$, $g_1(m)$, $h(n)$ be the functions corresponding respectively to R, $\mathfrak{G}$, $\mathfrak{G}'$, $\mathfrak{H}$. We shall prove that, if $g(0) = g'(0)$ and $r\big(h(n), n\big) = 0$ for each positive integer n and if $g(n+1) = g'(n+1)$ whenever $g\big(h(n+1)\big) = g'\big(h(n+1)\big)$, then $g(n) = g'(n)$ for each natural number n. We consider the class of natural numbers for which $g(n) = g'(n)$ is not true. If the class is not void it has a positive member n_0 which precedes all other members in the well ordering $\leqslant$. But $h(n_0)$ is another member of the class, for otherwise we should have

$$g\big(h(n_0)\big) = g'\big(h(n_0)\big)$$

and therefore $g(n_0) = g'(n_0)$, *i.e.* n_0 would not be in the class. This implies $n_0 \leqslant h(n_0)$ contrary to $r\big(h(n_0),\ n_0\big) = 0$. The class is therefore void.

It should be noticed that we do not really need to make use of the fact that $\mathbf{\Omega}$ is an ordinal formula. It suffices that $\mathbf{\Omega}$ should satisfy conditions (*a*)–(*e*) (p. 173) for ordinal formulae, and in place of (*f*) satisfy (*f*′).

(*f*′) There is no formula **T** such that **T**(**n**) is convertible to a formula representing a positive integer for each positive integer n, and such that $\mathbf{\Omega}\big(\mathbf{T}(\mathbf{n}),\ \mathbf{n}\big)$ conv 2, for each positive integer n for which $\mathbf{\Omega}(\mathbf{n},\ \mathbf{n})$ conv 3.

The problem whether a formula satisfies conditions (*a*)–(*e*), (*f*′) is number-theoretic. If we use formulae satisfying these conditions instead of ordinal formulae with $\Lambda_G{}^2$ or $\Lambda_G{}^3$, we have a non-constructive logic with certain advantages over ordinal logics. The intuitive judgments that must be made are all judgments of the truth of number theoretic-theorems. We have seen in § 9 that the connection of ordinal logics with the classical theory of ordinals is quite superficial. There seem to be good reasons, therefore, for giving attention to ordinal formulae in this modified sense.

The ordinal logic $\Lambda_G{}^3$ appears to be adequate for most purposes. It should, for instance, be possible to carry out Gentzen's proof of consistency of number theory, or the proof of the uniqueness of the normal form of a well-formed formula (Church and Rosser [**1**]) with our calculus and a fairly simple set of axioms. How far this is the case can, of course, only be determined by experiment.

One would prefer a non-constructive system of logic based on transfinite induction rather simpler than the system which we have described. In particular, it would seem that it should be possible to eliminate the necessity of stating explicitly the validity of definitions by primitive recursions, since this principle itself can be shown to be valid by transfinite induction. It is possible to make such modifications in the system, even in such a way that the resulting system is still complete, but no real advantage is gained by doing so. The effect is always, so far as I know, to restrict the class of formulae provable with a given set of axioms, so that we obtain no theorems but trivial restatements of the axioms. We have therefore to compromise between simplicity and comprehensiveness.

Index of definitions.

No attempt is being made to list heavy type formulae since their meanings are not always constant throughout the paper. Abbreviations for definite well-formed formulae are listed alphabetically.

(*The following refer to* §§1–10 *only.*)

Miscellaneous (in order of appearance).

Bibliography.

Alonzo Church, [1]. "A proof of freedom from contradiction", *Proc. Nat. Acad. Sci.*, 21 (1935), 275–281.

———, [2]. *Mathematical logic*, Lectures at Princeton University (1935–6), mimeographed, 113 pp.

———, [3]. "An unsolvable problem of elementary number theory", *American J. of Math.*, 58 (1936), 345–363.[a]

———, [4]. "The constructive second number class", *Bull. American Math. Soc.*, 44 (1938), 224–238.

G. Gentzen, [1]. "Die Widerspruchsfreiheit der reinen Zahlentheorie", *Math. Annalen*, 112 (1936), 493–565.

K. Gödel, [1]. "Über formal unentscheidbare Sätze der Principia Mathematica und verwandter Systeme, I", *Monatshefte für Math. und Phys.*, 38 (1931), 173–189.[a]

———, [2]. *On undecidable propositions of formal mathematical systems*, Lectures at the Institute for Advanced Study, Princeton, N.J., 1934, mimeographed, 30 pp.[a]

D. Hilbert, [1]. "Über das Unendliche", *Math. Annalen*, 95 (1926), 161–190.

S. C. Kleene, [1]. "A theory of positive integers in formal logic", *American J. of Math.*, 57 (1935), 153–173 and 219–244.

———, [2]. "General recursive functions of natural numbers", *Math. Annalen*, 112 (1935–6), 727–742.[a]

———, [3]. "λ-definability and recursiveness", *Duke Math. Jour.*, 2 (1936), 340–353.

E. L. Post, [1]. "Finite combinatory processes—formulation 1", *Journal Symbolic Logic*, 1 (1936), 103–105.[a]

J. B. Rosser, [1]. "Gödel theorems for non-constructive logics", *Journal Symbolic Logic*, 2 (1937), 129–137.

A. Tarski, [1]. "Der Wahrheitsbegriff in den formalisierten Sprachen", *Studia Philosophica*, 1 (1936), 261–405 (translation from the original paper in Polish dated 1933).

A. M. Turing, [1]. "On computable numbers, with an application to the Entscheidungsproblem", *Proc. London Math. Soc.* (2), 42 (1937), 230–265.[a] A correction to this paper has appeared in the same periodical, 43 (1937), 544–546.[a]

———, [2]. "Computability and λ-definability", *Journal Symbolic Logic*, 2 (1937), 153–163.

E. Zermelo, [1]. "Grundlagen einer allgemeiner Theorie der mathematischen Satzsysteme, I", *Fund. Math.*, 25 (1935), 136–146.

Alonzo Church and S. C. Kleene, [1]. "Formal definitions in the theory of ordinal numbers", *Fund. Math.*, 28 (1936), 11–21.

Alonzo Church and J. B. Rosser, [1]. "Some properties of conversion", *Trans. American Math. Soc.*, 39 (1936), 472–482.

D. Hilbert and W. Ackermann, [1]. *Grundzüge der theoretischen Logik* (2nd edition revised, Berlin, 1938), 130 pp.

A. N. Whitehead and Bertrand Russell, [1]. *Principia Mathematica* (2nd edition, Cambridge, 1925–1927), 3 vols.

J. B. Rosser

AN INFORMAL EXPOSITION OF PROOFS OF GÖDEL'S THEOREM AND CHURCH'S THEOREM

Reprinted from THE JOURNAL OF SYMBOLIC LOGIC, vol. 4 (1939) pp. 53-60.

This paper is an attempt to explain as non-technically as possible the principles and devices used in the various proofs of Gödel's Theorems and Church's Theorem.

Roman numerals in references shall refer to the papers in the bibliography.

In the statements of Gödel's Theorems and Church's Theorem, we will employ the phrase "for suitable *L*." The hidden assumptions which we denote by this phrase have never been put down explicitly in a form intelligible to the average reader.[1] The necessity for thus formulating them has commonly been avoided by proving the theorems for special logics and then remarking that the proofs can be extended to other logics. Hence the conditions necessary for the proofs of Gödel's Theorems and Church's Theorem are at present very indefinite as far as the average reader is concerned. To partly clarify this situation, we will now mention the more prominent of these assumptions.

I. In any proof of Gödel's Theorems or Church's Theorem, two logics are concerned. One serves as the "logic of ordinary discourse" in which the proof is carried out, and the other is a formal logic, *L*, about which the theorem is proved. The first logic may or may not be formal. However *L* must be formal. Among other things, this implies that the propositions of *L* are formulas built according to certain rules of structure. Each formula is to consist of a finite number (counting repetitions) of symbols chosen out of a set (finite or denumerably infinite) which is given at the start; any symbol of the set may be used more than once in any formula. Moreover the symbols have meanings attached, in terms of which propositions of *L* may be interpreted. The rules of structure of the propositions of *L* are supposed to be such that the interpretations of the propositions of *L* will be declarative sentences (not necessarily true) of "ordinary discourse." If *A* is a proposition of *L*, and a certain sentence is the interpretation of *A*, then *A* is said to be the "expression in *L*" of that sentence or any sentence equivalent to it. In general, not all sentences can be expressed in *L*.[2]

Received January 12, 1939.

[1] It is my understanding from conversations with Gödel that an exact formulation of these assumptions was to constitute part of the second part of the paper of which II is the first part. Due to ill health, Gödel has never written this second half. However, in III, Kleene gives an exact statement of a set of assumptions sufficient for his proof of Gödel's First Theorem. Unfortunately they are phrased in terms of general recursive functions, and are illuminating only to someone who is thoroughly familiar with the theory of general recursive functions.

[2] In the *L*'s which receive general attention, no method is apparent of expressing such sentences as "Plato was mortal," "God is good," etc.

II. Amongst the symbols of L must be one, $\sim$, which is interpreted as "not." That is, if A expresses in L a certain sentence, then $\sim A$ expresses in L the contradictory of that sentence.

III. For each positive integer, there must be a particular formula in L which denotes that integer. Also, amongst the symbols of L must be some, called variables, whose mode of interpretation is as follows. If a formula A of L expresses a sentence S and if A contains symbols called variables, $v_1, v_2, \cdots, v_s$, then S contains variables.[3] Moreover, if B is the formula got from A by replacing various of the v_i's of A by other symbols, then the sentence which B expresses is got from S by making corresponding replacements for the variables of S. In particular, if the formula G of L with the symbol v, called a variable, expresses in L the sentence "x has the property Q," with the variable x corresponding to v, and if F is got from G by replacing all the v's of G by the formula denoting the number n, then F expresses in L the sentence "n has the property Q."

IV. Also there must be a process whereby certain of the propositions of L are specified as "provable." The definition of "provable" is always supposed to be made without referring to the meanings of the formulas. However it was always hoped that the set of provable propositions of L would coincide with the set of propositions of L which express true sentences. Gödel's Theorems tell us that such cannot be the case. For Gödel's First Theorem states:

For suitable L, there are undecidable propositions in L; that is, propositions F such that neither F nor $\sim F$ is provable.

As F and $\sim F$ express contradictory sentences, one of them must express a true sentence. So there will be a proposition of L which expresses a true sentence, but nevertheless is not provable. This still leaves open the possibility that all provable propositions of L may express true sentences. As the notion of "truth of a sentence" is vague, it is usual to deal with weaker but more precise notions. For instance, L is said to be "simply consistent" if there is no proposition F such that both F and $\sim F$ are provable. Clearly, if L is not simply consistent, then some provable proposition of L must express a false sentence. However, some provable propositions of L may express false sentences even if L is simply consistent. Tarski[4] showed this by constructing a logic L which was simply consistent but in which one could prove the propositions expressing each sentence of the following infinite set (with Q properly chosen):

Not all positive integers have property Q.
1 has property Q.
2 has property Q.
3 has property Q.
. . . .

[3] I am purposely overlooking the complications due to the use of "apparent variables" as being irrelevant to the present discussion.

[4] Alfred Tarski, *Einige Betrachtungen über die Begriffe der ω-Widerspruchsfreiheit und der ω-Vollständigkeit*, ***Monatshefte für Mathematik und Physik,*** vol. 40 (1933), pp. 97–112.

A logic L in which this latter situation does not occur for any property Q is said to be ω-consistent.

V. If F and $\sim F$ are both provable in L, then all propositions of L are provable. So if L is not simply consistent, it is not ω-consistent. So ω-consistency implies simple consistency. In fact, the non-provability of any formula whatever of L implies the simple consistency of L.

VI. There is a symbol, $\supset$, of L such that if the formula A expresses the sentence S and the formula B expresses the sentence T, then $A \supset B$ expresses the sentence "If S, then T." Also the definition of "provable" shall be such that if A and $A \supset B$ are provable then so is B.

This completes our list. The list was compiled for expository purposes only. Hence the list suffers the double defect of not containing absolutely all necessary assumptions, and of containing some assumptions which may not be necessary. Also the assumptions are not always stated with strict accuracy, on the ground that readers who know of cases not covered by our simplified versions of the assumptions will know the corrections that need to be made, and that readers who do not know of such cases will not fall into error thereby.

Three proofs of Gödel's First Theorem (see above) will be considered in this paper, namely Gödel's proof (II, Satz VI), Rosser's proof (IV, Thm. II), and Kleene's proof (III, Thm. XIII). These proofs will be referred to as GG_1, RG_1, and KG_1 respectively. All three use the general assumptions listed above. In addition, GG_1 assumes that L is ω-consistent, RG_1 assumes that L is simply consistent, and KG_1 assumes a more complicated type of consistency, roughly equivalent to ω-consistency.

Gödel's Second Theorem states:

For suitable L, the simple consistency of L cannot be proved in L.

Gödel proves this statement (II Satz XI) with the special assumption of simple consistency. His proof will be referred to as GG_2.

Church's Theorem states:

For suitable L, there exists no effective method of deciding which propositions of L are provable.

The statement is proved by Church (I, last paragraph) with the special assumption of ω-consistency, and by Rosser (IV, Thm. III) with the special assumption of simple consistency. These proofs will be referred to as CC and RC respectively.

Clearly the existence of CC or RC presupposes a precise definition of "effective." "Effective method" is here used in the rather special sense of a method each step of which is precisely predetermined and which is certain to produce the answer in a finite number of steps. With this special meaning, three different precise definitions have been given to date.[5] The simplest of these to state

---

[5] One definition is given by Church in I. Another definition is due to Jacques Herbrand and Kurt Gödel. It is stated in I, footnote 3, p. 346.[i] The third definition was given independently in two slightly different forms by E. L. Post, *Finite combinatory processes—*

[i] 90

(due to Post and Turing) says essentially that an effective method of solving a certain set of problems exists if one can build a machine which will then solve any problem of the set with no human intervention beyond inserting the question and (later) reading the answer. All three definitions are equivalent, so it does not matter which one is used. Moreover, the fact that all three are equivalent is a very strong argument for the correctness of any one.

All the proofs GG_1, KG_1, RG_1, GG_2, CC, and RC use Gödel's device, which we now describe, for numbering formulas. First assign numbers to the symbols of L in any way that seems suitable. For instance Gödel discusses a logic involving the symbols

$$\sim, \mathbf{v}, \Pi, 0, f, (,),$$

and an infinite set of variables in each of an infinite set of types. He assigns numbers to these symbols as follows: 1 to 0, 3 to f, 5 to $\sim$, 7 to $\mathbf{v}$, 9 to Π, 11 to (, 13 to), and p_i^n (where the p_i's are primes greater than 13) to variables of type n.

Having assigned numbers to symbols, we next assign numbers to formulas as follows. Let $n_1, n_2, \cdots, n_s$ be the numbers of the symbols of a formula F in the order in which they occur in F. Let $p_1, p_2, \cdots, p_s$ be the first s primes in order of increasing magnitude (counting 2 as the first prime). Then the number assigned to F will be $p_1^{n_1} \cdot p_2^{n_2} \cdot \cdots \cdot p_s^{n_s}$. For example, one of the provable formulas of the logic which Gödel used is

$$\sim(x\Pi((\sim(x(fy)))\mathbf{v}(x(0)))),$$

(x and y being variables of types 2 and 1 respectively). The numbers of the symbols of this formula are successively 5, 11, 289, 9, 11, 11, 5, 11, 289, 11, 3, 17, 13, 13, 13, 7, 11, 289, 11, 1, 13, 13, 13, 13. So the number of the formula itself is $2^5 \cdot 3^{11} \cdot 5^{289} \cdot 7^9 \cdot 11^{11} \cdot 13^{11} \cdot 17^5 \cdot 19^{11} \cdot 23^{289} \cdot 29^{11} \cdot 31^3 \cdot 37^{17} \cdot 41^{13} \cdot 43^{13} \cdot 47^{13} \cdot 53^7 \cdot 59^{11} \cdot 61^{289} \cdot 67^{11} \cdot 71 \cdot 73^{13} \cdot 79^{13} \cdot 83^{13} \cdot 89^{13}$.

We see that for every formula, a number is assigned. However, not all numbers are assigned to formulas.[6] If a number is assigned to a formula, the formula can always be found as follows. Factor the number into its prime factors. Then the number of 2's occurring in the factorization is the number of the first symbol of the formula, the number of 3's occurring in the factorization is the number of the second symbol of the formula, the number of 5's occurring in the factorization is the number of the third symbol of the formula, etc.

When numbers have been assigned to formulas, statements about formulas can be replaced by statements about numbers. That is, if P is a property of formulas, we can find a property of numbers, Q, such that the formula A has the

---

formulation 1, this JOURNAL, vol. 1 (1936), pp. 103–105,[i] and A. M. Turing, *On computable numbers, with an application to the Entscheidungsproblem*, ***Proceedings of the London Mathematical Society***, ser. 2 vol. 42 (1937), pp. 230–265[ii] (see also the correction to the above, in the same journal, vol. 43 (1937), pp. 544–546).[iii] The first two definitions are proved equivalent in I. The third is proved equivalent to the first two by A. M. Turing, *Computability and λ-definability*, this JOURNAL, vol. 2 (1937), pp. 153–163.

[6] The number 4 is not assigned to any formula; for $4=2^2$, and so the first and only symbol of the formula must have the number 2 assigned to it, and no symbol has 2 assigned to it.

i 289–291 ii 115–151 iii 152–154

property P if and only if the number of A has the property Q. Throughout the rest of the paper, P will signify a property of formulas, and Q will signify the corresponding property of numbers. That is, Q will be the property of numbers such that we can use the statements "A has property P" and "the number of A has property Q" interchangeably.

Many statements about numbers can be expressed in L, even though all cannot. In particular, if P is properly chosen, we can often express "x has the property Q" in L. If x is taken to be the number of a formula of L, we are then expressing in L a statement about a formula of L. This element of circularity is capitalized in the following basic lemma:[7]

LEMMA 1. *Let "x has the property Q" be expressible in L. Then for suitable L, there can be found a formula F of L, with a number n, such that F expresses "n has the property Q." That is, F expresses "F has the property P."*

We now call attention to an extra assumption implicit in the "for suitable L" of Lemma 1,[8] namely that "$z=\phi(x, x)$" be expressible in L, where $\phi(x, y)$ is the function described below.

DEFINITION. $\phi(x, y)$ is the number of the formula got by taking the formula with the number x and replacing all occurrences of v in it by the formula of L which denotes the number of y.

We now give the proof of Lemma 1. Assume "x has the property Q" and "$z=\phi(x, x)$" are expressible in L. Then "$\phi(x, x)$ has the property Q" is expressible in L.[9] Let G be the formula of L which expresses "$\phi(x, x)$ has the property Q." G has a number, n. Now get F from G by replacing all v's of G by the formula of L which denotes n. Then F denotes "$\phi(n, n)$ has the property Q" (cf. Assumption III). However (cf. the definition of $\phi(x, y)$), $\phi(n, n)$ is the number of F, because F was got by taking the formula with the number n and replacing all occurrences of v in it by the formula of L which denotes n. So F expresses "the number of F has the property Q," that is "F has the property P."

To use Lemma 1, one must know that "$z=\phi(x, x)$" is expressible in L. Gödel proves this for a large class of L's by proving:

(a) $\phi(x, y)$ is "rekursiv" (II, pp. 179–188).[i]

(b) If $\psi(x_1, x_2, \cdots, x_s)$ is "rekursiv," then "$z=\psi(x_1, x_2, \cdots, x_s)$" is expressible in L (II, Satz V).

The proofs of both (a) and (b) are very complicated and technical, and will not even be sketched here.

Lemma 1 is basic in GG_1, GG_2, RG_1, and RC, but is not used in CC or KG_1.

We now outline GG_1, GG_2, RG_1, and RC. Each proof depends on the choice of a suitable property P to be used in Lemma 1. Gödel chooses for P the property of not being provable in L. So if we denote (as Gödel does) "the

---

7 This lemma is due to Gödel. On pp. 187–188[ii] of II, he proves it for a particular Q. However he does not state the lemma explicitly.

8 This assumption was not included in the original list because at that point the idea of the number of a formula had not yet been explained.

9 Note that the statement in question is equivalent to "there is a z such that $z=\phi(x,x)$ and z has the property Q."

[i] 14–25 [ii] 24–25

formula with the number *x* is provable in *L*” by “Bew(*x*),” then “*x* has property *Q*” is equivalent to “not-Bew(*x*).”

By an extensive argument involving “rekursiv” functions, Gödel shows that for a large class of *L*’s:

(c) “Bew(*x*)” (and hence “not-Bew(*x*)”) is expressible in *L*.

(d) If *L* is ω-consistent and if the formula expressing “Bew(*x*)” is provable, then “Bew(*x*)” is true.

(e) If “Bew(*x*)” is true, then the formula expressing “Bew(*x*)” is provable.

Now (Lemma 1), let us find a formula *F* with the number *n*, such that *F* expresses “not-Bew(*n*).”[10]

LEMMA 2. *If L is simply consistent, then F is not provable in L.*

For suppose *F* to be provable. That is, the formula with the number *n* is provable. That is, Bew(*n*). So by (e), the formula which expresses “Bew(*n*)” is provable. However, *F* expresses “not-Bew(*n*),” and so $\sim$F expresses “Bew(*n*)” (Assumption II). So $\sim$F is provable. However we assumed *F* provable, so that *L* is not simply consistent. So if *L* had been simply consistent, *F* would not have been provable.

LEMMA 3. *If L is ω-consistent, then $\sim F$ is not provable in L.*

For suppose *L* to be ω-consistent and pretend that $\sim F$ is provable. $\sim F$ expresses “Bew(*n*).” So by (d), Bew(*n*). That is, *F* is provable. So *L* is not simply consistent. However ω-consistency implies simple consistency (Assumption V), so our pretense that $\sim$F could be provable has to be false.

As ω-consistency implies simple consistency, Lemma 2 and Lemma 3 together give GG_1 (which assumed ω-consistency).

GG_2 runs as follows. Let *A* be a provable proposition of *L*, and let *m* be the number of $\sim$A. If Bew(*m*), then both *A* and $\sim A$ are provable, and *L* is not simply consistent. On the other hand, if *L* is not simply consistent, all propositions of *L* are provable, including $\sim A$, so that Bew(*m*). Hence “not-Bew(*m*)” and “*L* is simply consistent” are equivalent. So Lemma 2 is equivalent to

“If not-Bew(*m*), then not-Bew(*n*),”

since *n* is the number of *F*. Let Wid be the formula of *L* which expresses “not-Bew(*m*).” *F* is the formula of *L* which expresses “not-Bew(*n*).” So

$$\mathrm{Wid} \supset F$$

expresses Lemma 2 in *L* (Assumption VI). Now the proof of Lemma 2 can be carried out in a great many logics, so that in those logics

$$\mathrm{Wid} \supset F$$

is provable. Then if Wid were provable, *F* would be provable (Assumption VI). So Wid is not provable if *L* is simply consistent (by Lemma 2), which is what Gödel’s Second Theorem states.

---

[10] That is, *F* expresses “*F* is not provable.” Naturally one would expect *F* to have certain peculiarities.

For RG_1, Rosser chooses a property, Prov(x), which differs very slightly from Bew(x).[11] By an argument involving a generalization of "rekursiv" functions, Rosser proved that for a large class of logics:

(f) Prov(x) is expressible in L.

(g) If L is simply consistent, then:

(1) If the formula expressing "Prov(x)" is provable, then "not-Bew(Neg(x))" is true.[12]

(2) If $\sim$(the formula expressing "Prov(x)") is provable, then "Bew(x)" is false.

Now (Lemma 1) let us find a formula F with the number n, such that F expresses "not-Prov(n)." Assume that L is simply consistent. Now $\sim F$ expresses "Prov(n)." So if $\sim F$ is provable, then, by (g)(1), $\sim F$ is not provable. Likewise, as F expresses "not-Prov(n)," F is $\sim$(the formula expressing "Prov(x)"). So if F is provable, then, by (g)(2), F is not provable. This completes RG_1.

For RC, we start out by assuming that L is simply consistent and that there is an effective method of deciding which propositions of L are provable. As the Herbrand-Gödel definition of "effective method" involves a generalization of "rekursiv," Rosser was able to prove for a large class of logics, by use of this generalization of "rekursiv," that there must be a property of numbers, Prov(x), such that:

(h) "Prov(x)" is expressible in L.

(i) The formula expressing "Prov(x)" in L is provable in L if and only if the formula with the number x is provable in L.

(j) Either the formula expressing "Prov(x)" or the formula expressing "not-Prov(x)" is provable in L.

Now (Lemma 1) let us choose F with the number n, so that F expresses "not-Prov(n)." By (j), either F or $\sim F$ is provable. If F is provable, then, by (i), $\sim F$ is provable, contradicting our assumption of simple consistency. If $\sim F$ is provable, then, by (i), F is provable.

The proofs KG_1 and CC do not involve Lemma 1.

We now outline KG_1. In III, Kleene shows how general recursive functions (generalizations of "rekursiv" functions) can be defined by positive integers. He further shows that in a large class of logics, "y defines a general recursive function" can be expressed. Let L be one of these logics. Then one can find a general recursive function $f(x)$ such that:

(k) As x runs over the positive integers $f(x)$ runs over those values of y such that the expression of "y is a general recursive function" is a provable formula of L.

That is, $f(x)$ enumerates a certain class of numbers which define general recursive functions, and therefore enumerates a class of (general recursive) functions. To these the diagonal process is applied to get a new function. Explicitly, Kleene defines $g(x)$ as $1+$(the value, for the argument x, of the general recursive function defined by $f(x)$). Then $g(x)$ is a general recursive

[11] Bew(x) and Prov(x) are equivalent if L is simply consistent and only then.

[12] If x is the number of a formula A then Neg(x) is the number of the formula $\sim A$.

function and is defined by an integer m. Let F express "m defines a general recursive function." Then F expresses a true statement, and $\sim F$ a false one, so that $\sim F$ cannot be provable if suitable consistency assumptions are made. If F were provable in L, then by (k) there would be an integer n such that $f(n)=m$. Then $g(n)=1+$(the value, for the argument n, of the general recursive function defined by $f(n))=1+$(the value, for the argument n, of $g(x))=1+g(n)$. This contradiction shows that F cannot be provable.

KG_1 may be contrasted with GG_1 and RG_1 by saying that GG_1 and RG_1 resemble the Epimenides paradox, whereas KG_1 resembles the Richard paradox.

We now outline CC. In I, Church proves of a certain set of sequences that there is no effective method of solving the problem: Given a sequence of the set, does 2 occur in it or not?

Now for a large class of logics, "2 occurs in the sequence s of the set" can be expressed in L by a formula H and moreover, if L is ω-consistent, then H is provable if and only if what it expresses is true. So an effective method of deciding whether a given formula is provable would allow one to decide effectively whether or not H is provable, and hence to solve the problem of whether 2 occurs in a given sequence.

I wish to express here my gratitude to various members of the Department of Mathematics of Cornell University who obliged me by critically reading various drafts of this paper.

BIBLIOGRAPHY

I. Alonzo Church, *An unsolvable problem of elementary number theory*, ***American journal of mathematics***, vol. 58 (1936), pp. 345–363.[a]

II. Kurt Gödel, *Über formal unentscheidbare Sätze der Principia Mathematica und verwandter Systeme I*, ***Monatshefte für Mathematik und Physik***, vol. 38 (1931), pp. 173–198.[a] (A less difficult exposition of Gödel's work is to be found in Carnap's ***The logical syntax of language***.)

III. S. C. Kleene, *General recursive functions of natural numbers*, ***Mathematische Annalen***, vol. 112 (1936), pp. 727–742.[a]

IV. Barkley Rosser, *Extensions of some theorems of Gödel and Church*, this JOURNAL, vol. 1 (1936), pp. 87–91.[a]

CORNELL UNIVERSITY

EXTENSIONS OF SOME THEOREMS OF GÖDEL AND CHURCH

In this paper Rosser shows that the assumption of ω-consistency in Gödel's theorem (this anthology, pp. 23–26) can be replaced by simple consistency. Clearly this assumption can not be further weakened: if the system of logic being considered is inconsistent, then all its assertions are theorems, and hence there is no undecidability.

J. B. Rosser

EXTENSIONS OF SOME THEOREMS OF GÖDEL AND CHURCH

Reprinted from THE JOURNAL OF SYMBOLIC LOGIC, vol. 1 (1936) pp. 87-91.

Introduction. We shall say that a logic is "simply consistent" if there is no formula A such that both A and $\sim A$ are provable. "ω-consistent" will be used in the sense of Gödel.[2] "General recursive" and "primitive recursive" will be used in the sense of Kleene,[2] so that what Gödel calls "rekursiv" will be called "primitive recursive." By an "*Entscheidungsverfahren*" will be meant a general recursive function $\phi(n)$ such that, if n is the Gödel number of a provable formula, $\phi(n)=0$ and, if n is not the Gödel number of a provable formula, $\phi(n)=1$. In specifying that ϕ must be general recursive we are following Church[3] in identifying "general recursiveness" and "effective calculability."

First, a modification is made in Gödel's proofs of his theorems, Satz VI (Gödel, p. 187[iii]—this is the theorem which states that ω-consistency implies the existence of undecidable propositions) and Satz XI (Gödel, p. 196[iv]—this is the theorem which states that simple consistency implies that the formula which states simple consistency is not provable). The modifications of the proofs make these theorems hold for a much more general class of logics. Then, by sacrificing some generality, it is proved that simple consistency implies the existence of undecidable propositions (a strengthening of Gödel's Satz VI and Kleene's Theorem XIII) and that simple consistency implies the non-existence of an *Entscheidungsverfahren* (a strengthening of the result in the last paragraph of Church). The class of logics for which these two results are proved is more general in some respects and less general in other respects than the class of logics for which Gödel's proof of Satz VI holds or the class of logics for which Kleene's proof of Theorem XIII holds.

1. Preliminary lemmas.

LEMMA I. *Given a general recursive function* $\phi(x, y_1, \cdots, y_n)$ *and a number* k *such that* $(Ey_1, \cdots, y_n)[\phi(k, y_1, \cdots, y_n)=0]$, *there is a primitive recursive function* $\gamma(m)$ *such that* $\gamma(0), \gamma(1), \gamma(2), \cdots$ *is an enumeration (allowing repetitions) of the* x*'s such that* $(Ey_1, \cdots, y_n)[\phi(x, y_1, \cdots, y_n)=0]$.

Received September 8, 1936. Presented to the Association for Symbolic Logic and the American Mathematical Society September 1, 1936.

[1] National Research Fellow.

[2] See p. 187[i] of K. Gödel, *Über formal unentscheidbare Sätze der Principia Mathematica und verwandter Systeme I*, ***Monatshefte für Mathematik und Physik,*** vol. 38 (1931), pp. 173–198. We shall assume familiarity with this paper, to which we shall refer as "Gödel," and with the paper to which we shall refer as "Kleene," namely, S. C. Kleene, *General recursive functions of natural numbers*, ***Mathematische Annalen,*** vol. 112 (1936), pp. 727–742. The notations used throughout will be those used in these two papers.

[3] A. Church, *An unsolvable problem of elementary number theory*, ***American journal of mathematics,*** vol. 58 (1936), pp. 345–363. Cf. p. 356.[ii] We shall refer to this paper as "Church".

i 23, 24   *ii* 100   *iii* 24   *iv* 36

Proof.[4] Let $\psi(y)$ and $R(x, y_1, \cdots, y_n, y)$ be chosen for $\phi(x, y_1, \cdots, y_n)$ as in the proof of IV of Kleene. Then put $\gamma(m)=\epsilon p[p\leqq m+k \;\&\; \{\{R(1 \; Gl \; m, \cdots, [n+2]Gl \; m) \;\&\; \psi([n+2]Gl \; m)=0 \;\&\; p=1 \; Gl \; m\} \vee \{\{\overline{R}(1 \; Gl \; m, \cdots, [n+2]Gl \; m) \vee \psi([n+2]Gl \; m)\neq 0\} \;\&\; p=k\}\}]$. We note that the uniqueness clause of Definition 2b of Kleene allows one to add the clause "and $(\mathfrak{x}, y)[R(\mathfrak{x}, y)\rightarrow \psi(y)=\psi(\epsilon y[R(\mathfrak{x}, y)])]$" to IV of Kleene.

Corollary I. *If a class can be enumerated (allowing repetitions) by a general recursive function, it can be enumerated (allowing repetitions) by a primitive recursive function.*

For let $\phi(m)$ be general recursive and $\phi(0), \phi(1), \phi(2), \cdots$ be an enumeration of some class, then that class is just the class of all x's such that $(Ey)\phi(y)=x$. But $[\phi(y)=x]\backsim[(\phi(y)\dot{-}x)+(x\dot{-}\phi(y))=0]$.

Although general recursive enumerability without repetitions is more general than primitive recursive enumerability without repetitions, the two concepts are equivalent when repetitions are allowed. For this reason "recursively enumerable" shall henceforth be understood as allowing repetitions and referring indifferently to enumeration by general or primitive recursive functions.

Corollary II. *A general recursive class which is not null is recursively enumerable.*

Put $n=0$ in Lemma I.

The converse of this corollary is not true. For (cf. Footnote 16 and Theorem XV of Kleene) $(Ey)T_1(x, x, y)$ is a non-recursive class which is recursively enumerable. Also the class of well-formed formulas with normal forms (see Church) is not general recursive (Church, Theorem XVIII) and yet it can be enumerated, *without* repetitions, by a primitive recursive function. In this connection, Kleene has pointed out that if $\gamma(m)$ is primitive recursive and the class $(Em)[\gamma(m)=x]$ is not general recursive, then a primitive recursive function $\xi(m)$ can be defined such that the class $(Em)[\xi(m)=x]$ is not general recursive and $(m, n)\;[\xi(m)=\xi(n)\rightarrow m=n]$. This is done by putting $\xi(m)=\epsilon z[z\leqq 2\gamma(m)+2m+1 \;\&\; \{\{(n)[n<m\rightarrow\gamma(n)\neq\gamma(m)] \;\&\; z=2\gamma(m)\} \vee \{(En)[n<m \;\&\; \gamma(n)=\gamma(m)] \;\&\; z=2m+1\}\}]$.

Definition. A set of rules of procedure for a logic is said to be general recursive if there is a general recursive function $\phi(n, x, y)$ such that "z is an immediate consequence of x and y" is equivalent to "$(En)[\phi(n, x, y)=z]$".

Lemma II. *Let $C_1(x), \cdots, C_r(x)$ be recursively enumerable classes of numbers and $\phi_1(n, x, y), \cdots, \phi_s(n, x, y)$ be the determining functions of general recursive sets of rules of procedure. If $C(x)$ is the least class such that $(x)[C_i(x)\rightarrow C(x)]$ $(i=1, \cdots, r)$ and $(n, x, y)[C(x) \;\&\; C(y)\rightarrow C(\phi_i(n, x, y))]$ $(i=1, \cdots, s)$, then $C(x)$ is recursively enumerable.*

[4] This lemma is a generalization of three lemmas which appeared in an earlier draft of this paper. Upon reading this earlier draft, S. C. Kleene suggested this lemma and furnished the proof of it which is given here.

Proof. Let $\theta_i(n)$ be the functions which enumerate $C_i(x)$ $(i=1, \cdots, r)$. Then put $\phi_{s+i}(n, x, y)=\theta_i(n)$ $(i=1, \cdots, r)$. Then it is easy enough to define recursively

$\phi(n, x, y)$ so that $\phi(n, x, y)=\phi_{\mathrm{Rem}(n,r+s)+1}\left(\left[\frac{n}{r+s}\right], x, y\right)$ (cf. 21 in Kleene).

Then the class $C(x)$ is clearly the same as the least class $K(x)$ such that $K(\theta_1(0))$ and $(n, x, y)[K(x)\ \&\ K(y) \rightarrow K(\phi(n, x, y))]$ and by Theorem I, Kleene, this class is recursively enumerable.

2. Proofs of the theorems. P shall denote the system given in Gödel, pp. 176–178.[i] The theorems shall be proved for P but the method of proof will be general enough to apply to many other systems.

THEOREM I. *If* P_κ *is got by adding various axioms and rules of procedure to* P, *and if the provable formulas of* P_κ *are recursively enumerable:*

A. *If* P_κ *is* ω*-consistent, then there is a primitive recursive class formula,* r, *such that neither* v Gen r *nor* Neg(v Gen r) *is a provable formula in* P_κ (*where* v *is the free variable of* r).

B. *If* P_κ *is simply consistent, then the formal proposition which says that* P_κ *is simply consistent is not provable in* P_κ.

Proof. Let $\phi(m)$ be a primitive recursive function which enumerates the *provable formulas* of P_κ. On pp. 188, 189,[ii] 196 and 197[iii] of Gödel replace $x\ B_\kappa y$ by $\phi(x)=y$ and $\mathrm{Bew}_\kappa(y)$ by $(Ex)[\phi(x)=y]$. Then the proof of Satz VI becomes a proof of A and the proof of Satz XI becomes a proof of B.

By this proof we gain in generality over Gödel's proofs in the following way. Gödel used the hypothesis that the class of axioms was a primitive recursive class and that the rules of procedure were primitive recursive. In view of Lemma II it is sufficient that the class of axioms be recursively enumerable (and by Lemma I, Corollary II, this is even less restrictive than requiring that the class of axioms be general recursive) and that the rules of procedure be general recursive.

THEOREM II. *If* P_κ *is got by adding various axioms and rules of procedure to* P, *and if the provable formulas of* P_κ *are recursively enumerable, and if* P_κ *is simply consistent, then there is a primitive recursive class formula,* r, *such that neither* v Gen r *nor* Neg(v Gen r) *is a provable formula in* P_κ (*where* v *is the free variable of* r).

Proof. Let $\phi(m)$ be a primitive recursive function which enumerates the *provable formulas* and assume that P_κ is simply consistent. Put $x\ B_\kappa y$ for $\phi(x)=y$, $\mathrm{Bew}_\kappa(y)$ for $(Ex)[x\ B_\kappa y]$, $x\ Pr_\kappa y$ for $x\ B_\kappa y\ \&\ \overline{(Ez)}[z \leqq x\ \&\ z\ B_\kappa\ \mathrm{Neg}(y)]$ and $\mathrm{Prov}_\kappa(y)$ for $(Ex)[x\ Pr_\kappa y]$. Then $\mathrm{Bew}_\kappa(y) \sim \mathrm{Prov}_\kappa(y)$. However this equivalence is not provable formally since it requires the hypothesis of simple consistency, which we know by Thm. I to be formally unprovable. Hence the formalization of $\mathrm{Prov}_\kappa(y)$ may, and does, have properties not possessed by the formalization of $\mathrm{Bew}_\kappa(y)$. Such a property is the following (which will be proved shortly). If b is the number of the formalization of $\mathrm{Prov}_\kappa(a)$, then $\mathrm{Prov}_\kappa(a) \rightarrow \mathrm{Prov}_\kappa(b)$ and

[i] 10–13 [ii] 24–26 [iii] 36, 37

$\text{Prov}_\kappa(\text{Neg}(a)) \rightarrow \text{Prov}_\kappa(\text{Neg}(b))$. By use of this property one can proceed as on p. 188[i] of Gödel, but with $x\ Pr_\kappa y$ in place of Gödel's $x\ B_\kappa y$ and $\text{Prov}_\kappa(y)$ in place of Gödel's $\text{Bew}_\kappa(y)$, to find an *undecidable proposition* of the form v Gen r.

We now prove the aforementioned property of "Prov_κ". $x\ B_\kappa y$ and $(Ez)[z \leqq x \ \&\ z\ B_\kappa \text{Neg}(y)]$ are both primitive recursive relations. Hence by Satz V of Gödel, there are *formulas* r and s, both with the *free variables* u and v such that:

$$x\ B_\kappa y \rightarrow \text{Bew}_\kappa\left[Sb\left(r \begin{matrix} u & v \\ Z(x) & Z(y) \end{matrix}\right)\right], \tag{1}$$

$$\overline{x\ B_\kappa y} \rightarrow \text{Bew}_\kappa\left[\text{Neg}\left(Sb\left(r \begin{matrix} u & v \\ Z(x) & Z(y) \end{matrix}\right)\right)\right], \tag{2}$$

$$(Ez)[z \leqq x \ \&\ z\ B_\kappa \text{Neg}(y)] \rightarrow \text{Bew}_\kappa\left[Sb\left(s \begin{matrix} u & v \\ Z(x) & Z(y) \end{matrix}\right)\right], \tag{3}$$

$$\overline{(Ez)}[z \leqq x \ \&\ z\ B_\kappa \text{Neg}(y)] \rightarrow \text{Bew}_\kappa\left[\text{Neg}\left(Sb\left(s \begin{matrix} u & v \\ Z(x) & Z(y) \end{matrix}\right)\right)\right]. \tag{4}$$

If b is the number of the formalization of $\text{Prov}_\kappa(a)$, then clearly $\text{Bew}_\kappa\left(b \text{ Aeq } u \text{ Ex } \left(Sb\left(r \begin{matrix} v \\ Z(a) \end{matrix}\right) \text{Con Neg}\left(Sb\left(s \begin{matrix} v \\ Z(a) \end{matrix}\right)\right)\right)\right)$. Hence $\text{Prov}_\kappa(a) \rightarrow \text{Prov}_\kappa(b)$ by (1), (4) and $\text{Bew}_\kappa(y) \backsim \text{Prov}_\kappa(y)$. Assume $x\ B_\kappa \text{Neg}(a)$. Then by induction within the system P_κ and by use of (3), $\text{Bew}_\kappa\left(Sb\left(s \begin{matrix} u & v \\ x\ N\ R(u) & Z(a) \end{matrix}\right)\right)$ (since this only requires the proof by induction of certain properties of the function χ given on p. 181[ii] of Gödel). Also, by use of (2),

$$\text{Bew}_\kappa\left(\text{Neg}\left(Sb\left(r \begin{matrix} u & v \\ Z(0) & Z(a) \end{matrix}\right)\right)\right),$$

$$\text{Bew}_\kappa\left(\text{Neg}\left(Sb\left(r \begin{matrix} u & v \\ Z(1) & Z(a) \end{matrix}\right)\right)\right),$$

$$\cdots\cdots\cdots\cdots,$$

$$\text{Bew}_\kappa\left(\text{Neg}\left(Sb\left(r \begin{matrix} u & v \\ Z(x) & Z(a) \end{matrix}\right)\right)\right),$$

because $\overline{0\ B_\kappa a}, \overline{1\ B_\kappa a}, \cdots, \overline{x\ B_\kappa a}$ (since $\text{Bew}_\kappa(\text{Neg}(a))$ and P_κ is simply consistent). But the formal analogue of $(z)[z=0 \vee z=1 \vee \cdots \vee z=x \vee (Ew)[z=x+w]]$ is provable in P and hence in P_κ, and so $\text{Bew}_\kappa\left(u \text{ Gen}\left(\text{Neg}\left(Sb\left(r \begin{matrix} v \\ Z(a) \end{matrix}\right)\right) \text{Dis } Sb\left(s \begin{matrix} v \\ Z(a) \end{matrix}\right)\right)\right)$. Hence $\text{Prov}_\kappa(\text{Neg}(a)) \rightarrow \text{Prov}_\kappa(\text{Neg}(b))$.

As it is very easy to tell whether a formula has free variables or not, the existence of an *Entscheidungsverfahren* would imply the existence of a method for

[i] 24 [ii] 16

telling whether or not a formula with no free variables is provable. Hence it is a corollary of Theorem III (stated below) that there is no *Entscheidungsverfahren* for P_κ.

THEOREM III. *If P_κ is got by adding various axioms and rules of procedure to P, and if P_κ is simply consistent, then there is no generally applicable effective process for determining whether or not a formula with no free variables is provable.*

Proof. Assume that P_κ is simply consistent and that there is a generally applicable effective process for determining whether or not a formula with no free variables is provable. Let us define $\phi(n)$ by the rule: $\phi(n)$ shall be 0 if n is the number of a provable formula with no free variables, and 1 otherwise. Then $\phi(n)$ is effectively calculable and we shall follow Church in assuming that this necessitates that $\phi(n)$ be general recursive. Then the class of numbers of provable formulas with no free variables and the class of numbers which are not numbers of provable formulas with no free variables are both recursively enumerable. Also both are non-null. Let $\beta(m)$ and $\gamma(m)$ be primitive recursive functions which enumerate them respectively. Then there is a primitive recursive formula $\theta(m)$ such that $\theta(2n)=\beta(n)$ and $\theta(2n+1)=\gamma(n)$. Then $\theta(m)$ enumerates all numbers in such a way that the numbers occurring in the even places are numbers of provable formulas with no free variables and the numbers occurring at the odd places are not numbers of provable formulas with no free variables. Now put $x\ B_\kappa y$ for $\theta(x)=y\ \&\ x/2$, $\mathrm{Bew}_\kappa(y)$ for $(Ex)[x\ B_\kappa y]$, $x\ Pr_\kappa y$ for $x\ B_\kappa y\ \&\ \overline{(Ez)}[z \leqq x\ \&\ \theta(z)=y\ \&\ \overline{z/2}]$, and $\mathrm{Prov}_\kappa(y)$ for $(Ex)[x\ Pr_\kappa y]$. Then $\mathrm{Bew}_\kappa(y) \backsim \mathrm{Prov}_\kappa(y)$. Now let b be the number of the formalization of $\mathrm{Prov}_\kappa(a)$. By a proof like that in the proof of Thm. II, it follows that $\mathrm{Prov}_\kappa(a) \rightarrow \mathrm{Prov}_\kappa(b)$ and $(Ez)[\theta(z)=a\ \&\ \overline{z/2}] \rightarrow \mathrm{Prov}_\kappa(\mathrm{Neg}(b))$. But if $\overline{\mathrm{Prov}_\kappa(a)}$, then $(Em)[\gamma(m)=a]$, and therefore $\theta(2(\epsilon m[\gamma(m)=a])+1)=a$. Hence $\overline{\mathrm{Prov}_\kappa(a)} \rightarrow \mathrm{Prov}_\kappa(\mathrm{Neg}(b))$. By use of this and $\mathrm{Prov}_\kappa(a) \rightarrow \mathrm{Prov}_\kappa(b)$, one can derive a contradiction by proceeding as on p. 188[i] of Gödel, but with $x\ Pr_\kappa y$ and $\mathrm{Prov}_\kappa(y)$ in place of $x\ B_\kappa y$ and $\mathrm{Bew}_\kappa(y)$ respectively.

With slight modifications the proof above becomes a proof of:

THEOREM IV. *If P_κ is got by adding various axioms and rules of procedure to P, and if P_κ is simply consistent, then the class of numbers of provable formulas and the class of numbers of unprovable formulas are not both recursively enumerable.*

From this theorem and Theorem III follow:

THEOREM V. *If P is simply consistent, then:*

A. *The class of unprovable formulas is not recursively enumerable.*

B. *The class of undecidable formulas is not recursively enumerable.*

C. *The class of provable formulas is recursively enumerable but not general recursive.*

D. *The class of decidable formulas is recursively enumerable but not general recursive.*

I wish to thank S. C. Kleene for reading an earlier draft of this paper and suggesting improvements.

[i] 24

GENERAL RECURSIVE FUNCTIONS OF NATURAL NUMBERS

In this paper, Gödel's technique of arithmetization is applied to the general definition of recursive function given in Gödel's lectures, this anthology, pp. 69-71. This leads to yet another proof that there are unsolvable problems.

Errata and addenda supplied by the author for this anthology appear on p. 253 following this paper.

Stephen C. Kleene

GENERAL RECURSIVE FUNCTIONS OF NATURAL NUMBERS[1]

Reprinted from MATHEMATISCHE ANNALEN Band 112, Heft 5 (1936) pp. 727-742, with the kind permission of Springer-Verlag.

The substitution

$$(1) \qquad \varphi(x_1, \ldots, x_n) = \theta(\chi_1(x_1, \ldots, x_n), \ldots, \chi_m(x_1, \ldots, x_n)),$$

and the ordinary recursion with respect to one variable

$$(2) \qquad \begin{aligned} \varphi(0, x_2, \ldots, x_n) &= \psi(x_2, \ldots, x_n) \\ \varphi(y+1, x_2, \ldots, x_n) &= \chi(y, \varphi(y, x_2, \ldots, x_n), x_2, \ldots, x_n), \end{aligned}$$

where $\theta, \chi_1, \ldots, \chi_m, \psi, \chi$ are given functions of natural numbers, are examples of the definition of a function φ by equations which provide a step by step process for computing the value $\varphi(k_1, \ldots, k_n)$ for any given set $k_1, \ldots, k_n$ of natural numbers. It is known that there are other definitions of this sort, e. g. certain recursions with respect to two or more variables simultaneously, which cannot be reduced to a succession of substitutions and ordinary recursions[2]). Hence, a characterization of the notion of recursive definition in general, which would include all these cases, is desirable. A definition of general recursive function of natural numbers was suggested by Herbrand to Gödel, and was used by Gödel with an important modification in a series of lectures at Princeton in 1934. In this paper we offer several observations on general recursive functions, using essentially Gödel's form of the definition.

The definition will be stated in § 1. It consists in specifying the form of the equations and the nature of the steps admissible in the computation of the values, and in requiring that for each given set of arguments the computation yield a unique number as value. The operations on symbols which occur in the computation have a similarity to ordinary recursive operations on numbers. This similarity will be utilized, by the Gödel method of representing formulas by numbers, to prove that every (general) recursive function is expressible in the form $\psi(\varepsilon y[\varrho(x_1, \ldots, x_n, y) = 0])$ where ψ and ϱ are ordinary or „primitive"

---

[1]) Presented to the American Mathematical Society, September 1935.

[2]) W. Ackermann, Zum Hilbertschen Aufbau der reellen Zahlen, Math. Annalen 99 (1928), S. 118—133; Rózsa Péter, Konstruktion nichtrekursiver Funktionen, Math. Annalen 111 (1935), S. 42—60.

recursive functions and $(x_1, \ldots, x_n)(Ey)[\varrho(x_1, \ldots, x_n, y) = 0]$[3]. Also, it is seen directly that, for any recursive function $\varrho(x_1, \ldots, x_n, y)$, $\varepsilon y[\varrho(x_1, \ldots, x_n, y) = 0]$ is a recursive function, provided $(x_1, \ldots, x_n)(Ey)[\varrho(x_1, \ldots, x_n, y) = 0]$.

In § 2, the problem is raised, which systems of equations define recursive functions under the general definition. The systems which do cannot be recursively enumerated, if by a recursive enumeration is understood one such that the numbers ordered by the Gödel method to the systems of equations in the enumeration are a recursive sequence (i. e. the successive values of a recursive function of one variable), since from any recursive sequence of such numbers we can obtain the recursive definition of a new function by the familiar process of diagonalizing and adding 1. For the same reason, a recursive process of deciding which systems define recursive functions is unattainable, if by a recursive process is meant one such that there is a recursive function of the corresponding numbers whose value is 0 or 1 according to the result obtained. Since the condition under which a recursive function of n variables is defined can be expressed in the form $(x_1, \ldots, x_n)(Ey)[\varrho(x_1, \ldots, x_n, y) = 0]$, we are afforded an approach (somewhat different than Gödel's[4]) to the existence of undecidable number-theoretic propositions in formal logics satisfying certain general conditions. Roughly speaking, every such formal logic must contain undecidable propositions of the form $(x)(Ey)[\varrho(x, y) = 0]$, where $\varrho(x, y)$ is a primitive recursive function, because otherwise the logic could be used to decide recursively which systems of equations define recursive functions, which we know in advance to be impossible. Every problem of the form, whether or not $(x)(Ey)[\sigma(x, y) = 0]$, where $\sigma(x, y)$ is a recursive function, is included in the problem, which systems of equations define recursive functions of one variable.

Also, there are non-recursive functions definable using only one quantifier, thus: $\tau(x) = 0$ if $(y)[\varrho(x, y) = 0]$, $\tau(x) = 1$ otherwise, where $\varrho(x, y)$ is primitive recursive.

---

[3]) In the "functions" which we consider, the arguments are understood to range over the natural numbers (i. e. non-negative integers) and the values to be natural numbers. Also, for abbreviation, we use propositional functions of natural numbers, calling them „relations“ (alternatively "classes", when there is only one variable) and employing the following notations: $(x)A(x)$ [for all natural numbers, $A(x)$], $(Ex)A(x)$ [there is a natural number x such that $A(x)$], $\varepsilon x[A(x)]$ [the least natural number x such that $A(x)$, or 0 if there is no such number], $\overline{}$ [not], $\vee$ [or], & [and], $\rightarrow$ [implies], $\equiv$ [is equivalent to].

[4]) Kurt Gödel, Über formal unentscheidbare Sätze der Principia Mathematica und verwandter Systeme I, Monatsh. für Math. u. Physik 38 (1931), S. 173—198.[a]

§ 1.

The relation between primitive and general recursive functions.

A recursive function (relation) in the sense of Gödel[4]) (S. 179—180) will now be called a *primitive recursive* function (relation). By using

$$(3)\qquad \begin{aligned} &S(x) = x + 1 && \text{(the successor function),} \\ &C(x) = 0 && \text{(the constant function 0),} \\ &U_i^n(x_1, \ldots, x_n) = x_i && \text{(identity functions)} \end{aligned}$$

as initial functions, the definition of primitive recursive function can be phrased thus:

Definition 1. A function is *primitive recursive* if it can be defined from the functions (3) by (zero or more) successive applications of schemas (1) and (2) ($m, n = 1, 2, \ldots; i = 1, \ldots, n$)[5]).

In the study of general recursive functions, we treat the defining equations formally, as sequences of symbols. For abbreviation, we may omit to distinguish between the functions and numbers, and the symbols or sets of symbols which stand for them.

Now consider *expressions* consisting of finite sequences of the following symbols: 0 (the numeral 0), S (the successor function), $w_0, w_1, \ldots$ (numerical variables), $\varrho_0, \varrho_1, \ldots$ (variables for functions of $r_0, r_1, \ldots$ arguments, where $r_0, r_1, \ldots$ is a sequence of positive integers in which each occurs infinitely many times, say 1, 1, 2, 1, 2, 3, ...), (,), , , = (parentheses, comma, equality symbol). We define *term* thus: $0, w_0, w_1, \ldots$ are terms; if $a_1, a_2, \ldots$ are terms, $S(a_1)$, $\varrho_0(a_1, \ldots, a_{r_0})$, $\varrho_1(a_1, \ldots, a_{r_1})$, ... are terms. By *numeral* is meant one of the expressions 0, $S(0)$, $S(S(0))$, ... If a and b are terms (and if $\sigma_1, \ldots, \sigma_n$ are functional variables[6]) such that a least one of $\sigma_1, \ldots, \sigma_n$ occurs in a or b, but no functional variables other than $\sigma_1, \ldots, \sigma_n$ occur in a or b), $a = b$ will be called an *equation* (*in* $\sigma_1, \ldots, \sigma_n$). By a *system* of equations we mean a finite sequence of equations. $\mathrm{S}^{a_1 \ldots a_n}_{b_1 \ldots b_n} A|$ shall denote the result of substituting b_i for a_i ($i = 1, \ldots, n$) throughout A (A itself, if $a_1, \ldots, a_n$

---

[5]) This form of the definition was introduced by Gödel to avoid the necessity of providing for omissions of arguments on the right in schemas (1) and (2). The operations in the construction of primitive recursive functions can be further restricted. See Rózsa Péter, Über den Zusammenhang der verschiedenen Begriffe der rekursiven Funktionen, Math. Annalen **110** (1934), S. 612—632.

[6]) That is, if $\sigma_1, \ldots, \sigma_n$ stand for $\varrho_{\alpha_1}, \ldots, \varrho_{\alpha_n}$ for some set of distinct numbers $\alpha_1, \ldots, \alpha_n$ (then we use s_i for r_{α_i}). Similarly, in R_1 below it is meant that $x_1, \ldots, x_n$ stand for $w_{\beta_1}, \ldots, w_{\beta_n}$ for some set of distinct numbers $\beta_1, \ldots, \beta_n$.

do not occur in A). $E \vdash_{r_1 r_2 \ldots} F$ shall denote that the expression F is derivable from the expressions E by (zero or more) applications of the operations $R_{r_1}, R_{r_2}, \ldots$.

We list the operations on expressions[7]):

R_1: *to replace* A *by* $S^{x_1 \ldots x_n}_{k_1 \ldots k_n} A$, *where* $x_1, \ldots, x_n$ *are the numerical variables which occur in* A, *and* $k_1, \ldots, k_n$ *are numerals.*

R_2: *to pass from* A *and* $\sigma(k_1, \ldots, k_s) = k$ *to the result of substituting* k *for a particular occurrence of* $\sigma(k_1, \ldots, k_s)$ *in* A, *where* $k_1, \ldots, k_s$, k *are numerals.*

R_3: *to pass from* A *and* $B = C$ *to the result of substituting* C *for a particular occurrence of* B *in* A.

The Herbrand-Gödel definition of general recursive function of natural numbers can be formulated thus[8]):

Definition 2a. Given functional variables $\sigma_1, \ldots, \sigma_n$, let E_j^* denote the set of equations $\sigma_j(k_1, \ldots, k_{s_j}) = k$ where k is the "value" of $\sigma_j(k_1, \ldots, k_{s_j})$ as presently defined. The functions $\sigma_1, \ldots, \sigma_n$ are *defined recursively* by the system of equations $(E_1 \ldots E_n)$ if, for each i $(i=1, \ldots, n)$, E_i is a system of equations in $\sigma_1, \ldots, \sigma_i$, each of the form $\sigma_i(a_1, \ldots, a_{s_i}) = b$ where σ_i does not occur in $a_1, \ldots, a_{s_i}$, such that for each set of numerals $k_1, \ldots, k_{s_i}$ there is exactly one numeral k (called the *value* of $\sigma_i(k_1, \ldots, k_{s_i})$) for which $E_1^*, \ldots, E_{i-1}^*, E_i \vdash_{1,2} \sigma_i(k_1, \ldots, k_{s_i}) = k$. A function σ_n is *recursive* if there is an $(E_1 \ldots E_n)$ of this description.

We understand a function $\varphi(x_1, \ldots, x_m)$ to be recursive under this definition, if it is possible to define it by recursion equations of the type described, whether or not originally the function is so defined. More explicitly, a given function $\varphi(x_1, \ldots, x_m)$ is recursive under Def. 2a, if there exists an $(E_1 \ldots E_n)$ as described in Def. 2a in which σ_n may be regarded as representing φ. σ_n may be regarded as representing φ, if $s_n = m$ and whenever $k_1, \ldots, k_{s_n}$ are the numerals $S(\ldots x_1 \text{ times} \ldots S(0)\ldots)$, $\ldots, S(\ldots x_m \text{ times} \ldots S(0)\ldots)$, resp., the "value of $\sigma_n(k_1, \ldots, k_{s_n})$" under Def. 2a is the numeral $S(\ldots \varphi(x_1, \ldots, x_m) \text{ times} \ldots S(0)\ldots)$. A similar remark applies to Def. 2b below.

---

[7]) In these operations we do not require that A and $B = C$ be equations and that σ be a functional variable, since $R_1 - R_3$ as stated when applied to equations generate equations. Thereby, our proof of IV is simplified.

[8]) In what follows, the word "recursive" (when not qualified by the adjective "primitive") will mean recursive under any one of the definitions 2a, 2b and 2c, except when the definition involved is mentioned explicitly (as is necessary in the course of establishing the theorems VI and IX on their equivalence).

We now show that Def. 2a is not more general than the following (which will later be proved equivalent to it):

Definition 2b. The functions $\sigma_1, \ldots, \sigma_n$ are *defined recursively* by E, if E is a system of equations in $\sigma_1, \ldots, \sigma_n$ such that for each i ($i = 1, \ldots, n$) and each set of numerals $k_1, \ldots, k_{s_i}$ there is exactly one numeral k (called the *value* of $\sigma_i(k_1, \ldots, k_{s_i})$) for which $E \vdash_{1,3} \sigma_i(k_1, \ldots, k_{s_i}) = k$. A function σ_n is *recursive* if there is an E of this description[9]).

For the system of equations $(E_1 \ldots E_n)$ of Def. 2a can be proved to be a system E for Def. 2b thus: Clearly, $(E_1 \ldots E_n)$ is a system of equations in $\sigma_1, \ldots, \sigma_n$, and for each i and set of numerals $k_1, \ldots, k_{s_i}$, $(E_1 \ldots E_n) \vdash_{1,3} \sigma_i(k_1, \ldots, k_{s_i}) = k$ where k is the value of $\sigma_i(k_1, \ldots, k_{s_i})$ under Def. 2a. It remains to be shown that $(E_1 \ldots E_n) \vdash_{1,3} \sigma_i(k_1, \ldots, k_{s_i}) = l$ for l a numeral only when $l = k$. Now each equation of $(E_1 \ldots E_n)$ is verifiable (for each replacement of its numerical variables by numerals) by use of the values under Def. 2a, since, on examination, the supposition of the contrary is found to conflict with the hypothesis that for given i and numerals $k_1, \ldots, k_{s_i}$ there is only one numeral k such that $E_1^*, \ldots, E_{i-1}^*, E_i \vdash_{1,2} \sigma_i(k_1, \ldots, k_{s_i}) = k$[9a]). Moreover, R_1 and R_3 applied to verifiable equations yield verifiable equations. Hence, if $(E_1 \ldots E_n) \vdash_{1,3} \sigma_i(k_1, \ldots, k_{s_i}) = l$ where $k_1, \ldots, k_{s_i}$, l are numerals, the values of $\sigma_i(k_1, \ldots, k_{s_i})$ and l must be the same, i. e. l must be the value k of $\sigma_i(k_1, \ldots, k_{s_i})$ under Def. 2a.

The set of operations R_1, R_3 may be replaced in Def. 2b by a set R'_i ($i = 0, 1, 2, \ldots$) of single-valued binary operations, defined over all pairs of equations as follows:

R'_{3i} : *to pass from A and B to* $\left. \mathrm{S}^{w_i}_{S(w_i)} A \right|$.

R'_{3i+1}: *to pass from A and B to* $\left. \mathrm{S}^{w_i}_{0} A \right|$.

R'_{3i+2}: *to pass from A and $B = C$ to the result of replacing the occurrence of B in A beginning with the $i+1^{st}$ symbol by C, if there is such an occurrence; otherwise, to A itself.*

For, under the conditions of Def. 2b, $E \vdash'_{0,1,2,\ldots} \sigma_i(k_1, \ldots, k_{s_i}) = l$ (l a numeral) when l is the value of $\sigma_i(k_1, \ldots, k_{s_i})$ under Def. 2b and only then (as is easily shown).

---

9) A more general definition would not be obtained by allowing under R_3 also the substitution of B for C, since E may be chosen to include $b = a$ whenever $a = b$ is included.

9a) Similarly, the equations of the system E of Def. 2b are verifiable by use of the values under Def. 2b, if they are of the form $\sigma(a_1, \ldots, a_s) = b$.

We now assign numbers to symbols, expressions, finite sequences of expressions, etc., by the Gödel method [*loc. cit.*[4]) S. 179—182];[i] letting numbers correspond to symbols thus:

$$\text{“}0\text{”} \ldots 1,\ \text{“}S\text{”} \ldots 3,\ \text{“}=\text{”} \ldots 5,\ \text{“},\text{”} \ldots 7,\ \text{“}(\text{”} \ldots 11,\ \text{“})\text{”} \ldots 13,$$

$$\text{“}w_i\text{”} \ldots p_{i+7},\ \text{“}\varrho_i\text{”} \ldots p_{i+7}^2,$$

where p_i denotes the ith prime number. Then if the numbers corresponding to $N_1, \ldots, N_k$ are $n_1, \ldots, n_k$, resp., the number corresponding to the sequence $N_1, \ldots, N_k$ is $p_1^{n_1} \ldots p_k^{n_k}$. Employing Gödel's notations (including the use of italics to indicate the correspondent for numbers of a given notion relating to expressions) and his methods of exhibiting the primitive recursiveness of functions and relations[9b]), we adopt 1—10 of his list, modifying 6, and define further primitive recursive functions and relations, as follows:

6. $\quad n\,Gl\,x = \varepsilon y\,[y \leqq x \,\&\, x \,|\, (Pr(n))^y \,\&\, \overline{x \,|\, (Pr(n))^{y+1}}]$.

The finite sequence $n_1, \ldots, n_k$ of positive integers is represented by $p_1^{n_1} \ldots p_k^{n_k}$. Also, we may use the positive integer x to represent any sequence $n_1, \ldots, n_k$ of natural numbers such that $x = p_1^{n_1} \ldots p_k^{n_k}$. The modification in the definition of $n\,Gl\,x$ secures that $n\,Gl\,x$ always be the nth member. The significances ascribed to $l(x)$, $x * y$, etc., refer only to the case $n_1, \ldots, n_k > 0$[10]).

11. $\quad x \dot{-} y = \varepsilon z\,[z \leqq x \,\&\, x = y + z]$.

If $x \geqq y$, $x \dot{-} y = x - y$; if $x \leqq y$, $x \dot{-} y = 0$.

12. $\quad \left[\dfrac{x}{y}\right] = \varepsilon z\,[z \leqq x \,\&\, (z+1)\,y > x]$.

13. $\quad \text{Rem}\,(x, y) = x \dot{-} \left(\left[\dfrac{x}{y}\right]\right) y$.

14. $\quad \text{Dy}\,(0) = 1,$

$$\text{Dy}\,(k+1) = \varepsilon z \Big[z \leqq 3^k \,\&\, \{[1\,Gl\,\text{Dy}\,(k) < 2\,Gl\,\text{Dy}\,(k) \,\&\, z = 2^{[1\,Gl\,\text{Dy}\,(k)]+1}\, 3^{2\,Gl\,\text{Dy}\,(k)}]$$
$$\vee\ [1\,Gl\,\text{Dy}\,(k) \geqq 2\,Gl\,\text{Dy}\,(k) > 0 \,\&\, z = 2^{1\,Gl\,\text{Dy}\,(k)}\, 3^{[2\,Gl\,\text{Dy}\,(k)] \dot{-} 1}]$$
$$\vee\ [2\,Gl\,\text{Dy}\,(k) = 0 \,\&\, z = 3^{[1\,Gl\,\text{Dy}\,(k)]+1}]\} \Big].$$

[9b]) Also see Th. Skolem, Begründung der elementaren Arithmetik durch die rekurrierende Denkweise ohne Anwendung scheinbarer Veränderlichen mit unendlichem Ausdehnungsbereich, Videnskapsselskapets Skrifter 1923. I. Mat.-naturv. Kl., Nr. 6, S. 1—38.

[10]) Note that $l(1) = 0$ and $x * 1 = 1 * x = x\ [l(x) > 0]$.

[i] 13–18

Dy (k) represents the $k+1^{\text{st}}$ pair of numbers in the following order:

$$0\ 0;\ \ 0\ 1,\ \ 1\ 1,\ \ 1\ 0;\ \ 0\ 2,\ \ 1\ 2,\ \ 2\ 2,\ \ 2\ 1,\ \ 2\ 0;\ \ldots.$$

15. $v \operatorname{Occ} x \equiv (En)\,[0 < n \leqq l(x)\ \&\ v = n\, Gl\, x].$

The *symbol* v occurs in the *expression* x.

16. $$Su\, x\binom{n}{y} = \varepsilon z\,\{z \leqq [Pr(l(x)+l(y))]^{x+y}\ \&\ (E u, v)\,[u, v \leqq x\ \&\ x = u * R(n Gl\, x) * v\ \&\ z = u * y * v\ \&\ n = l(u) + 1]\}.$$

„$Su\, x\binom{n}{y}$ entsteht aus x, wenn man an Stelle des n-ten Gliedes von x y einsetzt (vorausgesetzt, daß $0 < n \leqq l(x)$)“ (Gödel, S. 184; Nr. 27).

17. $$\operatorname{Sb}(0, x, v, y) = x,$$

$$\operatorname{Sb}(k+1, x, v, y) = \varepsilon z\Big[z \leqq \operatorname{Sb}(k, x, v, y) + Su\operatorname{Sb}(k, x, v, y)\binom{k+1}{y}\ \&\ \Big\{[k+1\, Gl\, x \neq v\ \&\ z = \operatorname{Sb}(k, x, v, y)] \vee \Big[k+1\, Gl\, x = v\ \&\ z = Su\operatorname{Sb}(k, x, v, y)\binom{k+1}{y}\Big]\Big\}\Big].$$

$\operatorname{Sb}(k, x, v, y)$ is the result of substituting the *expression* y for the *symbol* v throughout the first k *symbols* of the *expression* x (if $k \leqq l(x)$).

18. $S(x, v, y) = \operatorname{Sb}(l(x), x, v, y).$

$S(x, v, y)$ corresponds to the operation $\overset{v}{\underset{y}{\mathrm{S}}}\, x\,|$ (if v is *a symbol* and x and y are *expressions*).

19. $$\operatorname{St}(x, n, a, y) = \varepsilon z\big[z \leqq [Pr(l(x)+l(y))]^{x+y}\ \&\ \{(Ep, q)\,[p, q \leqq x\ \&\ x = p*a*q\ \&\ l(p) = n\ \&\ z = p*y*q] \vee |(p, q)\,[p, q \leqq x\ \&\ x = p*a*q \rightarrow l(p) \neq n]\ \&\ z = x]\}\big].$$

$\operatorname{St}(x, n, a, y)$ is the result of substituting the *expression* y for the occurrence of the *expression* a in the *expression* x beginning with the $n+1^{\text{st}}$ *symbol*, if there is such an occurrence; otherwise, x itself.

20. $R_0''(i, x, y) = S(x, Pr(i+7), R(3) * E(Pr(i+7)))$

$R_1''(i, x, y) = S(x, Pr(i+7), R(1)).$

$$R_2''(i, x, y) = St\,(x, i, \varepsilon p\,[p \leqq y\ \&\ (Eq)\,[q \leqq y\ \&\ y = p * R(5) * q]],\ \varepsilon q\,[q \leqq y\ \&\ (Ep)\,[p \leqq y\ \&\ y = p * R(5) * q]]).$$

i 20

$R_0''(i, x, y)$, $R_1''(i, x, y)$, $R_2''(i, x, y)$ correspond to the operations R'_{3i}, R'_{3i+1}, R'_{3i+2}, resp.

21. $$R'(n, x, y) = \varepsilon z\Big[z \leqq R_0''\Big(\Big[\frac{n}{3}\Big], x, y\Big) + R_1''\Big(\Big[\frac{n \dot{-} 1}{3}\Big], x, y\Big) + R_2''\Big(\Big[\frac{n \dot{-} 2}{3}\Big], x, y\Big) \,\&\, \Big\{\Big[n\,\Big|\,3 \,\&\, z = R_0''\Big(\Big[\frac{n}{3}\Big], x, y\Big)\Big] \vee \Big[n+2\,\Big|\,3 \,\&\, z = R_1''\Big(\Big[\frac{n \dot{-} 1}{3}\Big], x, y\Big)\Big] \vee \Big[n+1\,\Big|\,3 \,\&\, z = R_2''\Big(\Big[\frac{n \dot{-} 2}{3}\Big], x, y\Big)\Big]\Big\}\Big].$$

$R'(n, x, y)$ corresponds to the operation R'_n.

22. $Z(0) = R(1)$,

$Z(n+1) = R(3) * E(Z(n))$.

$Z(n)$ corresponds to the numeral $S(\ldots n \text{ times} \ldots S(0))$.

23. $\mathrm{Eval}_p(n, y, x_1, \ldots, x_p) \equiv (Ex)\{x \leqq y \,\&\, y = R([Pr(n+7)]^2) * E(Z(x_1) * R(7) * \ldots * R(7) * Z(x_p)) * R(5) * Z(x)\}$ (for a fixed number p).

y corresponds to an expression of the form $\varrho_n(x_1, \ldots, x_p) = x$, where x is a numeral.

24. $\mathrm{Val}(y) = \varepsilon x\{x \leqq y \,\&\, (Em)[m \leqq y \,\&\, y = m * Z(x)]\}$.

If y corresponds to an expression of the form $a = x$ where x is a numeral, then $\mathrm{Val}(y) = x$.

Supposing the function $\varphi(n, x, y)$ given, we define a series of functions as follows:

$$\psi(0, x, y) = x,$$
$$\psi(n+1, x, y) = \varphi(n, x, y).$$
$$\lambda(0, z) = l(z),$$
$$\lambda(k+1, z) = [k+1] \cdot \lambda(k, z)^2.$$
$$\tau(0, z) = z,$$
$$\tau(k+1, z) = \prod_{n=0}^{\lambda(k+1, z) \dot{-} 1} [Pr(n+1)] \exp\Big\{\psi\Big(\Big[\frac{n}{\lambda(k, z)^2}\Big], \Big[\big[1\, Gl\, \mathrm{Dy}\big(\mathrm{Rem}(n, \lambda(k, z)^2)\big)\big] + 1\Big] Gl\, \tau(k, z), \Big[\big[2\, Gl\, \mathrm{Dy}\big(\mathrm{Rem}(n, \lambda(k, z)^2)\big)\big] + 1\Big] Gl\, \tau(k, z)\Big)\Big\}.$$

$$\mu(n, z) = \varepsilon t\Big[t \leqq n \,\&\, n < \sum_{i=0}^{t} \lambda(i, z)\Big].$$
$$\nu(n, z) = \Big[\sum_{i=0}^{\mu(n, z)} \lambda(i, z)\Big] \dot{-} n.$$
$$\theta(z, m) = \nu(m, z)\, Gl\, \tau(\mu(m, z), z).$$

Then if z or $\tau(0, z)$ is the Gödel number for the sequence S_0 of the $\lambda(0, z)$ numbers $z_1, \ldots, z_l$ $(z_1, \ldots, z_l > 0)$, $\tau(k+1, z)$ is the Gödel number for the sequence S_{k+1} of the $\lambda(k+1, z)$ numbers $\psi(n, x, y)$, for $n = 0, \ldots, k$ and x and y ranging over S_k, in a certain order. Since $\psi(0, x, y) = x$, S_k includes all numbers in S_j for $0 \leqq j \leqq k$. When $l(z) > 0$, $\mu(n, z)$ and $\nu(n, z)$ as $n = 0, 1, 2, \ldots$ take successively the pairs of values $0\,\lambda(0, z), 0\,\lambda(0, z) - 1, \ldots, 0\,1; 1\,\lambda(1, z), 1\,\lambda(1, z) - 1, \ldots, 1\,1; \ldots$. Hence $\theta(z, m)$ for $m = 0, 1, 2 \ldots$ are the members of $S_k\ (k = 0, 1, 2, \ldots)$. But these are (with repetitions) the numbers obtainable from $z_1, \ldots, z_l$ by zero or more applications of the operations $\varphi(0, x, y), \varphi(1, x, y), \ldots$. Since $\theta(z, m)$ was defined in a manner which shows that it can be obtained from $\varphi(n, x, y)$ and known primitive recursive functions by substitutions and primitive recursions, we have proved:

I. *Given a function $\varphi(n, x, y)$, there is a function $\theta(z, m)$, primitive recursive in $\varphi(n, x, y)$*[11]*, such that, whenever $z = p_1^{z_1} \ldots p_l^{z_l}$ $(z_1, \ldots, z_l > 0)$, then $\theta(z, 0), \theta(z, 1), \ldots$ is an enumeration (with repetitions) of the least class $C(x)$ such that $C(z_1), \ldots, C(z_l)$ and $(n, x, y)\left[C(x) \,\&\, C(y) \rightarrow C(\varphi(n, x, y))\right]$.*

We note here the following two theorems for later use:

II. *Given a class $A(x)$, a relation $x, y\,B\,z$, and a number k which belongs to the least class $C(x)$ such that $(x)\,[A(x) \rightarrow C(x)]$ and $(x, y, z)\,[C(x)$ $\&\, C(y) \,\&\, x, y\,B\,z \rightarrow C(z)]$, there is a function $\eta(m)$, primitive recursive in $A(x)$ and $x, y\,B\,z$, such that $\eta(0), \eta(1), \ldots$ is an enumeration (with repetitions) of $C(x)$.*

$\eta(m)$ is the function $\theta(R(k), m)$ when $\theta(z, m)$ is chosen as in I taking for $\varphi(n, x, y)$ the function

$$\varepsilon z\Big[z \leqq n + k \,\&\, \Big\{\Big\{n \mid 2 \,\&\, \Big[\Big(A\Big[\Big(\frac{n}{2}\Big)\Big] \,\&\, z = \Big[\frac{n}{2}\Big]\Big) \vee \Big(\overline{A\Big(\Big[\frac{n}{2}\Big]\Big)} \,\&\, z = k\Big)\Big]\Big\} \vee \Big\{n + 1 \mid 2 \,\&\, \Big[\Big(x, y\,B\Big[\frac{n \dot{-} 1}{2}\Big] \,\&\, z = \Big[\frac{n \dot{-} 1}{2}\Big]\Big) \vee \Big(\overline{x, y\,B\Big[\frac{n \dot{-} 1}{2}\Big]} \,\&\, z = k\Big)\Big]\Big\}\Big\}\Big]$$ [12].

If a member k of a class $R(x)$ is given, the class is enumerated (allowing repetitions) by the function $\varepsilon y\left[y \leqq m + k \,\&\, \{(R(m) \,\&\, y = m) \vee (\overline{R(m)} \,\&\, y = k)\}\right]$, which is primitive recursive in the class. Similarly:

---

[11]) We call a function φ primitive recursive in other functions ψ_i, if φ becomes primitive recursive under the supposition that ψ_i are primitive recursive. $\prod_{n=0}^{\psi(\mathfrak{x}, \mathfrak{y})} \chi(\mathfrak{x}, \mathfrak{z}, n)$ and $\sum_{n=0}^{\psi(\mathfrak{x}, \mathfrak{y})} \chi(\mathfrak{x}, \mathfrak{z}, n)$ are primitive recursive in $\psi(\mathfrak{x}, \mathfrak{y})$ and $\chi(\mathfrak{x}, \mathfrak{z}, n)$. Here we use $\mathfrak{x}, \mathfrak{y}, \mathfrak{z}$ as abbreviations for $x_1, \ldots, x_n$, $y_1, \ldots, y_m$, $z_1, \ldots, z_l$, resp., and we shall continue to do so when convenient.

[12]) If $k = 0$, replace "$\lambda(0, z) = l(z)$" by "$\lambda(0, z) = 1$" in the definition of $\theta(z, m)$.

III. *Given a relation* $R(x, y)$ *and a number* k *such that* $(E\,y)\,R(k, y)$, *there is a function* $\gamma(m)$, *primitive recursive in* $R(x, y)$, *such that* $\gamma(0), \gamma(1), \ldots$ *is an enumeration (allowing repetitions) of the class* $(E\,y)\,R(x, y)$.

$$\gamma(m) = \varepsilon y\big[y \leqq [1\,Gl\,m] + k \;\&\; \{(R(1\,Gl\,m, 2\,Gl\,m) \;\&\; y = 1\,Gl\,m) \lor (\overline{R(1\,Gl\,m, 2\,Gl\,m)} \;\&\; y = k)\}\big].$$

By applying I, taking for $\varphi(n, x, y)$ the function $R'(n, x, y)$ (21), we obtain a primitive recursive function:

25. $H(z, m)$.

If z corresponds to a system of equations Z, $H(z, 0)$, $H(z, 1)$, ... is an enumeration (with repetitions) of the numbers corresponding to equations Y such that $Z \vdash'_{0,\,1,\,2,\,\ldots}\, Y$.

Now let $\varphi(\mathfrak{x})$ be a recursive function in the sense of Def. 2a or Def. 2b. Then there is a system E of equations defining φ recursively under Def. 2b; suppose that ϱ_a stands for φ in E. The system E has a Gödel number e. Using 23 and 25, if $R(\mathfrak{x}, y) \equiv \mathrm{Eval}_{r_a}(a, H(e, y), \mathfrak{x})$, then, by Def. 2b, $(\mathfrak{x})\,(Ey)\,R(\mathfrak{x}, y)$. Furthermore, using 24, if $\psi(y) = \mathrm{Val}(H(e, y))$, then $\varphi(\mathfrak{x}) = \psi(\varepsilon y[R(\mathfrak{x}, y)])$. We have now proved:

IV. *Every function recursive in the sense of Def.* 2a (*or Def.* 2b) *is expressible in the form* $\psi(\varepsilon y[R(\mathfrak{x}, y)])$, *where* $\psi(y)$ *is a primitive recursive function and* $R(\mathfrak{x}, y)$ *a primitive recursive relation and* $(\mathfrak{x})\,(Ey)\,R(\mathfrak{x}, y)$.

Thus the extension of general over primitive recursive functions consists only in that to substitutions and primitive recursions is added the operation of seeking indefinitely through the series of natural numbers for one satisfying a primitive recursive relation.

By Gödel S. 180[i] IV, $\varepsilon y[R(\mathfrak{x}, y)]$ is primitive recursive in $R(\mathfrak{x}, y)$ and any function $\chi(\mathfrak{x})$ which bounds y. Hence, in a certain sense, the length of the computation algorithm of a recursive function which is not also primitive recursive grows faster with the arguments than the value of any primitive recursive function[13]).

Given a relation $R(\mathfrak{x})$, the function $\varrho(\mathfrak{x})$ which is 0 or 1, according as $R(\mathfrak{x})$ holds or not, may be called the representing function of $R(\mathfrak{x})$. As with primitive recursions, we say that $R(\mathfrak{x})$ is recursive, if its representing function is recursive (under Def. 2a)[14]).

---

[13]) Besides the method, for demonstrating that a function is not primitive recursive (or not definable by given additional means, such as recursions with respect to n variables simultaneously), which consists in finding a lower bound for the values, we have the method, for demonstrating relationships of the opposite kind, which consists in finding an upper bound for the number of steps in the computation algorithm.

[14]) This is equivalent to saying that there is a recursive function $\varrho'(\mathfrak{x})$ such that $R(\mathfrak{x}) \sim [\varrho'(\mathfrak{x}) = 0]$, since then $\varrho(\mathfrak{x}) = 1 \dot{-} (1 \dot{-} \varrho'(\mathfrak{x}))$.

[i] 15, 16

Let $R(\mathfrak{x}, y)$ be a recursive relation such that $(\mathfrak{x})(Ey)\, R(\mathfrak{x}, y)$. Then the function $\pi(\mathfrak{x}, y) = \prod_{i=0}^{y} \varrho(\mathfrak{x}, i)$, where $\varrho(\mathfrak{x}, y)$ is the representing function of $R(\mathfrak{x}, y)$, is recursive (under Def. 2a); and the function $\mu(\mathfrak{x}) = \varepsilon y\,[R(\mathfrak{x}, y)]$ satisfies the following relations in terms of $\pi(\mathfrak{x}, y)$:

$$
\begin{aligned}
&\sigma(0, \mathfrak{x}, y) = y, \\
(4)\qquad &\sigma(S(z), \mathfrak{x}, y) = \sigma\big(\pi(\mathfrak{x}, S(y)), \mathfrak{x}, S(y)\big), \\
&\mu(\mathfrak{x}) = \sigma\big(\pi(\mathfrak{x}, 0), \mathfrak{x}, 0\big).
\end{aligned}
$$

These equations (supplemented by the equations defining $\pi(\mathfrak{x}, y)$ recursively under Def. 2a) form a system E defining $\mu(\mathfrak{x})$ recursively under Def. 2a. Hence:

V. *If $R(\mathfrak{x}, y)$ is a recursive relation, and $(\mathfrak{x})(Ey)\, R(\mathfrak{x}, y)$, then $\varepsilon y\,[R(\mathfrak{x}, y)]$ is recursive (under Def. 2a).*

This shows that the converse of IV is true, and gives us as an operation of recursive definition the formation of $\varepsilon y\,[R(\mathfrak{x}, y)]$ from a recursive relation $R(\mathfrak{x}, y)$ such that $(\mathfrak{x})(Ey)\, R(\mathfrak{x}, y)$[15]). Also, the equivalence of Def. 2a and Def. 2b is now established:

VI. *The class of recursive functions under Def. 2b is identical with that under Def. 2a.*

For, as noted earlier, Def. 2b is not less general than Def. 2a, and now we have by IV and V that any function recursive under Def. 2b is expressible in the form $\psi(\mu(\mathfrak{x}))$ where $\psi(y)$ and $\mu(\mathfrak{x})$ are recursive under Def. 2a.

VII. *Let $R(\mathfrak{x}, y)$ be a relation such that for every $\mathfrak{x}$ $R(\mathfrak{x}, y)$ holds for infinitely many y's, and let $\nu(\mathfrak{x}, n)$ denote the nth y such that $R(\mathfrak{x}, y)$ in order of magnitude. If $R(\mathfrak{x}, y)$ is recursive, then $\nu(\mathfrak{x}, n)$ is recursive.*

For $\nu(\mathfrak{x}, n)$ satisfies the relations $\nu(\mathfrak{x}, 0) = \varepsilon y\,[R(\mathfrak{x}, y)]$ and $\nu(\mathfrak{x}, S(n)) = \xi(\mathfrak{x}, \nu(\mathfrak{x}, n))$ where $\xi(\mathfrak{x}, z) = \varepsilon y\,[R(\mathfrak{x}, y)\ \&\ y > z]$, from which its recursiveness follows by use of V.

The converse of VII holds, since $R(\mathfrak{x}, y) \equiv (En)\,[n \leqq y\ \&\ \nu(\mathfrak{x}, n) = y]$, which is primitive recursive in $\nu(\mathfrak{x}, n)$.

---

[15]) By IV, the use of this operation repeatedly and with $R(\mathfrak{x}, y)$ a general recursive relation gives no extension of the class of functions obtainable by a single application of it with $R(\mathfrak{x}, y)$ primitive recursive.

We had already as an operation of recursive definition the formation of $\varepsilon y\,[R(\mathfrak{x}, y)]$ from a recursive relation $R(\mathfrak{x}, y)$ such that there is a recursive function $\chi(\mathfrak{x})$ for which $R(\mathfrak{x}, y) \rightarrow y \leqq \chi(\mathfrak{x})$ (by Gödel, S. 180[i] IV). This and the present result correspond to different methods of expressing $\varepsilon y\,[R(\mathfrak{x}, y)]$ recursively in terms of $\varrho(\mathfrak{x}, y)$.

[i]15,16

Thus, omitting the parameters $\mathfrak{x}$, an infinite class is recursively enumerable without repetitions in order of magnitude if and only if it is recursive.

VIII. *If the function* $\zeta(x)$ *is recursive and takes infinitely many values, and* $\eta(n)$ *denotes the* n^{th} *in order of first occurrence in* $\zeta(0), \zeta(1), \ldots,$ *then* $\eta(n)$ *is recursive.*

For $\eta(n) = \zeta(\nu(n))$ when $\nu(n)$ is chosen by VII for

$$R(y) \equiv (x)[x < y \rightarrow \zeta(x) \neq \zeta(y)].$$

Thus the recursive enumerability with repetitions of an infinite class implies its recursive enumerability without repetitions [16]).

§ 2.

The undecidability, in general, which systems of equations define recursive functions.

The definition of general recursive function offers no constructive process for determining when a recursive function is defined. This must be the case, if the definition is to be adequate, since otherwise still more general "recursive" functions could be obtained by the diagonal process.

In order to analyze the situation in detail, we utilize the correspondence of systems of equations E to numbers e, under which the problem, which systems E define functions recursively, becomes a number-theoretic one. We introduce for each particular value of n the following primitive recursive relation, where α_n denotes the least i for which $r_i = n$:

26. $$T_n(z, x_1, \ldots, x_n, y) \equiv \mathrm{Eval}_n(\alpha_n, H(z, y), x_1, \ldots, x_n).$$

The relation between the numbers and the recursive functions is simplified under the following definition:

Def. 2c. The number e *defines* (*recursively*) the function $\varphi(x_1, \ldots, x_n) = \mathrm{Val}(H(e, \varepsilon y[T_n(e, x_1, \ldots, x_n, y)]))$ if $(x_1, \ldots, x_n)(Ey)T_n(e, x_1, \ldots, x_n, y)$. A function $\varphi(x_1, \ldots, x_n)$ is *recursive* if there is an e of this description.

IX. *The class of recursive functions under Def.* 2c *is identical with that under Def.* 2a (2b).

For if $\varphi(x_1, \ldots, x_n)$ is recursive under Def. 2a (2b), the system of equations which defines φ recursively under Def. 2a (2b) has, after changing the notation if necessary so that φ is represented in it by ϱ_{α_n}, a Gödel number e which defines φ recursively under Def. 2c (cf. the proof of IV); and conversely, every function recursive under Def. 2c is recursive under Def. 2a (2b) by V (V and VI).

---

[16]) In XV below is given an example $(Ey)\,T_1(x, x, y)$ of a non-recursive class which by III is recursively enumerable.

If a number e defines a function $\varphi(x_1, \ldots, x_n)$ recursively under Def. 2c and is the Gödel number of a system E of equations, E is in general a system determining a multiple-valued function[17]), not necessarily a system defining a function recursively under Def. 2a or 2b.

What follows is stated for $n = 1$, and would hold similarly for any other fixed n[18]).

X. *If* $\theta(x)$ *is a recursive function, and* $(x)(Ey)\,T_1(\theta(x), x, y)$, *there is a number* f *such that* $(x)(Ey)\,T_1(f, x, y)$ *and* $\overline{(Eq)\,[\theta(q) = f]}$.

For then $\eta(x) = \mathrm{Val}\big(H(\theta(x), \varepsilon y\,[T_1(\theta(x), x, y)])\big) + 1$ is a recursive function (by V) such that, for every x for which $\theta(x)$ defines a function φ_x of one variable recursively, $\eta(x) = \varphi_x(x) + 1$ (by Def. 2c). Let f be a number defining $\eta(x)$ recursively. By Def. 2c, $(x)(Ey)\,T_1(f, x, y)$. Also, if there were a q such that $\theta(q)$ is f, we would have $\eta(x) = \varphi_q(x)$, which contradicts the preceding equality when x takes the value q.

In the case that for every x $\theta(x)$ is the Gödel number of a system E_x of equations defining a function φ_x of one variable recursively (φ_x being represented in E_x by ϱ_0), we have that the function $\varphi_x(x) + 1$ is recursive. Thus the diagonal procedure, applied to a sequence of recursive functions which are defined by systems of equations of which the Gödel numbers form a recursive sequence, does not lead outside the class of recursive functions.

XI. *The numbers which define functions* $\varphi(x)$ *recursively are not recursively enumerable, i. e. there is no recursive function* $\theta(m)$ *such that* $(m, x)(Ey)\,T_1(\theta(m), x, y)$ *and* $(z)\{(x)(Ey)\,T_1(z, x, y) \rightarrow (Em)\,[\theta(m) = z]\}$.

For, given any recursive function $\theta(m)$ such that $(m, x)(Ey)\,T_1(\theta(m), x, y)$, then a fortiori $(x)(Ey)\,T_1(\theta(x), x, y)$, and by X there is a number f such that $(x)(Ey)\,T_1(f, x, y)$ but $\overline{(Em)\,[\theta(m) = f]}$.

XII. *The class* $(x)(Ey)\,T_1(z, x, y)$ *of the numbers* z *which define functions* $\varphi(x)$ *recursively is not recursive.*

For if it were recursive, it would be enumerated by a recursive function, contradicting XI.

Indeed, given any recursive class $R(z)$ such that $(z)\{R(z) \rightarrow (x)(Ey)\,T_1(z, x, y)\}$, a number f such that $(x)(Ey)\,T_1(f, x, y)$ but $\overline{R(f)}$ is obtained by X, when $\theta(x) = \varepsilon y\,\big[(R(x)\ \&\ y = x) \vee (\overline{R(x)}\ \&\ y = k)\big]$, where k is any number such that $(x)(Ey)\,T_1(k, x, y)$.

[17]) Then $\varphi(x_1, \ldots, x_n)$ is that one of the values x determined by E for which the Gödel number of $\varrho_{\alpha_n}(x_1, \ldots, x_n) = x$ occurs earliest in the list $H(e, 0), H(e, 1), \ldots$.

[18]) Since the means given for passing from definitions under Def. 2a (2b) to definitions under Def. 2c, and vice versa, are effective, the problem which we now study (which numbers define functions recursively) is equivalent to the one first proposed (which systems of equations define functions recursively).

The definability of a non-recursive class by use of quantifiers applied to a recursive relation gives the existence of undecidable number-theoretic propositions in certain formal logics from the consideration (somewhat different from that employed by Gödel) that otherwise the logics could be used to construct recursive definitions of the class.

XIII. *Given a formal logic S, suppose that the propositions* $(x)(Ey)T_1(z,x,y)$ $(z = 0, 1, 2, \ldots)$ *can be expressed in S by formulas* A_z, *and that numbers can be assigned to the formulas of S, in such a fashion that* (1) *to distinct formulas are assigned distinct numbers,* (2) *the class* $A(x)$ *of the numbers assigned to axioms, and the relation* $x, y\, B\, z$ *between numbers of being assigned to formulas in the relation of immediate consequence, are recursive,* (3) z *is a recursive function* $\beta(a_z)$ *of the number* a_z *of* A_z, (4) *the class* $C(n)$ *of the numbers* a_z *is recursive,* (5) *if* A_z *is provable, then* $(x)(Ey)T_1(z,x,y)$ *is true. Then there are z's for which* A_z *is not provable although* $(x)(Ey)T_1(z,x,y)$ *is true*[19].

For suppose that there is a number k such that A_k is provable. Then, by (2) and II, given a_k, there is a recursive function $H(m)$ which enumerates the numbers assigned to provable formulas of S; and, by (1), (3) and (4), the recursive function $\theta(y) = \beta\left(\varepsilon m\left[\{C(H(y))\ \&\ m = H(y)\} \vee \{\overline{C(H(y))}\ \&\ m = a_k\}\right]\right)$ enumerates the z's for which A_z is provable. By (5), $(x)(Ey)T_1(\theta(x), x, y)$.

Hence, by X, there is a number f such that $\overline{(Eq)[\theta(q) = f]}$ (which implies that A_f is not provable in S) and $(x)(Ey)T_1(f, x, y)$[20].

---

[19]) The relation of "immediate consequence" we suppose to be a given relation between a formula and a pair of formulas, and the class of "provable formulas" to be the least class which contains the given class of "axioms" and has the property that Z is provable whenever X and Y are provable and Z is an immediate consequence of X and Y.

If more details of the structure of S were suitably specified, condition (5) could be given a more metamathematical appearance, such as the following (analogous to Gödel's condition of ω-Widerspruchsfreiheit, S. 187)[i]: for no relation $F(x,y)$ and natural number k are all of the formulas $F(k,0), F(k,1), \ldots, \overline{(Ex)(y)F(x,y)}$ provable. On the further assumption that for no relation $F(x,y)$ and sequence of natural numbers $k_0, k_1, \ldots$ are all of the formulas $F(0,k_0), F(1,k_1), \ldots, \overline{(x)(Ey)F(x,y)}$ provable, the conclusion could be given the form, that there are z's for which A_z is formally undecidable, i. e. for which neither A_z nor $\bar{A}_z$ is provable. (The conditions need to be assumed merely for certain relations $F(x,y)$.)

[20]) The undecidable proposition A_f can be effectively constructed for a given logic, whenever the number a_k, recursive definitions of $A(x)$; $x, y\, B\, z$; $\beta(y)$ and $C(n)$, and effective means of constructing A_f from f, are given.

Whenever the supposition in this proof, that there is a k such that A_k is provable, is not realized, the theorem holds trivially.

i 23,24

XIV. *The function* $\varepsilon y\,[T_1(x, x, y)]$ *is non-recursive*[21]).

For, if $\varrho(x)$ is any recursive function, the function $\eta(x) = \mathrm{Val}\,(H(x, \varrho(x))) + 1$ is recursive, and $\eta(x) = \mathrm{Val}\,(H(f, \varepsilon y\,[T_1(f, x, y)]))$ holds for any number f defining $\eta(x)$ recursively. Now if $\varrho(f) = \varepsilon y\,[T_1(f, f, y)]$, two different values are obtained for $\eta(f)$. Hence $\varrho(f) \neq \varepsilon y\,[T_1(f, f, y)]$. Thus $\varepsilon y\,[T_1(x, x, y)]$ differs from each recursive function for some value of x. Note also that $(E\,y)\,T_1(f, f, y)$.

XV. *The class* $(E\,y)\,T_1(x, x, y)$ *is non-recursive.*

Thus non-recursive functions can be defined by the schema

$$\tau(x) = \begin{cases} 0 \text{ if } (E\,y)\,R(x, y) \\ 1 \text{ if } \overline{(E\,y)\,R(x, y)} \end{cases},$$

where $R(x, y)$ is primitive recursive. This follows from XIV, since, if $(E\,y)\,R(k, y)$ and $\lambda(x) = [1 \dot{-} \tau(x)] \cdot x + \tau(x) \cdot k$, then $\varepsilon y\,[R(x, y)] = [1 \dot{-} \tau(x)] \cdot \varepsilon y\,\big[R(\lambda(x), y)\big]$, which is recursive if $\tau(x)$ is recursive.

To analyze the situation more fully, let $S(x)$ be any recursive class such that $(x)\,\{S(x) \rightarrow (E\,y)\,T_1(x, x, y)\}$, and $\sigma(x)$ the representing function of $S(x)$. If k is any number which defines a function recursively, then $(E\,y)\,T_1(k, k, y)$, and we set $\mu(x) = [1 \dot{-} \sigma(x)] \cdot x + \sigma(x) \cdot k$ and $\varrho(x) = [1 \dot{-} \sigma(x)] \cdot \varepsilon y\,\big[T_1(\mu(x), \mu(x), y)\big]$. $\varrho(x)$ is recursive, and as in the proof of XIV, there is an f such that $\varrho(f) \neq \varepsilon y\,[T_1(f, f, y)]$ and $(E\,y)\,T_1(f, f, y)$. If $S(f)$, then $\sigma(f) = 0$ and $\varrho(f) = \varepsilon y\,[T_1(f, f, y)]$. Hence $\overline{S(f)}$.

XVI. *The class* $\overline{(E\,y)\,T_1(x, x, y)}$ *is not recursively enumerable*[22]).

For by III, the complementary class $(E\,y)\,T_1(x, x, y)$ is enumerated by a recursive function $\gamma(m)$. Now if $\overline{(E\,y)\,T_1(x, x, y)}$ is enumerated by $\varkappa(m)$ and we set $\xi(m) = \varepsilon n \left\{\left[m \,\middle|\, 2 \,\&\, n = \gamma\left(\left[\frac{m}{2}\right]\right)\right] \vee \left[m + 1 \,\middle|\, 2 \,\&\, n = \varkappa\left(\left[\frac{m \dot{-} 1}{2}\right]\right)\right]\right\}$, we have $(E\,y)\,T_1(x, x, y) \equiv \varepsilon m\,[\xi(m) = x]\,|\,2$, which would contradict XV if $\varkappa(m)$ were recursive.

XVII. *Given a recursive relation* $R(\mathfrak{x}, \mathfrak{y})$, *there is a number e such that* $(\mathfrak{x})\,(E\,\mathfrak{y})\,R(\mathfrak{x}, \mathfrak{y}) \equiv (x)\,(E\,y)\,T_1(e, x, y)$. *Given a recursive relation* $R(\mathfrak{y})$, *there is a number e such that* $(E\,\mathfrak{y})\,R(\mathfrak{y}) \equiv (E\,y)\,T_1(e, e, y)$[23]).

---

[21]) We recall that $\varepsilon y\,[R(x, y)] = 0$ when $\overline{(E\,y)\,R(x, y)}$.

[22]) The proof given here is non-constructive. The writer has a constructive proof that for certain recursive relations $R(x, y)$ the class $\overline{(E\,y)\,R(x, y)}$ is not recursively enumerable. From that proof, the existence in certain formal logics of undecidable propositions involving only one quantifier (which can be concluded non-constructively from present results) is obtainable in the same manner as XIII.

[23]) From the great generality of the problems, which e's define recursively functions of one variable, and which e's "determine recursively" the eth value of a function of one variable, as displayed by this theorem, the result, that they are not "effectively" soluble, could have been anticipated.

For, to every proposition of the form $(\mathfrak{x})(E\mathfrak{y})R(\mathfrak{x},\mathfrak{y})$, there is an equivalent proposition of the form $(x)(Ey)R(x,y)$ obtained by utilizing the recursive enumerability of n-tuples of natural numbers, or introducing fictive variables[24]); and the Gödel number e of the system E of equations which defines $\varepsilon y[R(x,y)]$ in the proof of V on the supposition that $(x)(Ey)R(x,y)$ satisfies the present theorem. Similarly, $(E\mathfrak{y})R(\mathfrak{y})$ has an equivalent $(Ey)R(y)$, and for e we may take the Gödel number of the equations defining $\varepsilon y[R(y)\ \&\ x=x]$ on the supposition that $(Ey)R(y)$[25]).

My thanks are due to Prof. Paul Bernays for the suggestion of improvements in the presentation.

24) E. g. $(x_1, x_2, x_3)R(x_1, x_2, x_3) \equiv (x)(Ey)[R(1\,Gl\,x,\ 2\,Gl\,x,\ 3\,Gl\,x)\ \&\ y=y]$.

25) XV, XVI, and XVII are similar, respectively, to results obtained in a different connection by Prof. Alonzo Church (An unsolvable problem of elementary number theory, see Bull. Amer. Math. Soc. Abstract **41**—5—205), Dr. J. B. Rosser (unpublished), and the present writer (A theory of positive integers in formal logic, Part II, Amer. Jour. Math. **57** No. 2, pp. 230 ff.).

Erratum (Journal of Symbolic Logic, vol. 3 (1938) p. 152), correcting an error noticed by Barkley Rosser: On p. 243, in Definition 17, read

$$\text{``}\left(k+1+l(\text{Sb}(\underset{y}{k},\ x,\ v,\ y)) \dot{-} l(x)\right)\text{''} \quad \text{for} \quad \text{``}\binom{k+1}{y}\text{''} \quad \text{twice.}$$

Simplification (Journal of Symbolic Logic, vol. 2 (1937) p. 38 with vol. 4 (1939) p. iv at end), suggested by Rózsa Péter: The theorem I on p. 245 can be proved more simply by defining $\theta(z, m)$ thus

$$\tau(0, z) = z,$$
$$\tau(k+1, z) = \tau(k, z) \cdot Pr(l(z)+k+1) \text{ exp}$$
$$\phi(1\ Gl\ k,\ ((2\ Gl\ k)+1)\ Gl\ \tau(k, z),\ ((3\ Gl\ k)+1)\ Gl\ \tau(k, z)),$$
$$\theta(z, m) = (m+1)\ Gl\ \tau(m, z)$$

(whereupon Definitions 13 and 14 become dispensable).

Addendum to Footnote 22 (circulated privately by the author in 1936), answering a question raised by Paul Bernays:

If $\theta(m)$ is a recursive function, and $(x, y)\ \overline{T_1(\theta(x), \theta(x), y)}$, then there is a number f such that $(y)\ \overline{T_1(f, f, y)}$ and $(q)[\theta(q) \neq f]$.

Proof: Let f be the Gödel number of the system of equations $\boldsymbol{F}$ which would define the function $\epsilon m[\theta(m)=n]$ (represented by ρ_0) as in the proof of V, on the supposition that $(n)(Em)[\theta(m)=n]$. Suppose that $\theta(q)=f$. Then the equations $\boldsymbol{F}$, when f is substituted for the argument of ρ_0, determine a value b $(\leq q)$ of $\epsilon m[\theta(m)=n]$, and hence by the theory of the relation T_1, $(y)\ T_1(\theta(b), \theta(b), y)$. But this contradicts the hypothesis $(x, y)\ \overline{T_1(\theta(x), \theta(x), y)}$. Hence $(q)[\theta(q) \neq f]$ and the equations $\boldsymbol{F}$, when f is substituted for the argument of ρ_0, determine no value of m, and hence, by the theory of the T_1 function, $(y)\ \overline{T_1(f, f, y)}$.

Although this paper does not develop a specific formalism, the arguments really refer to one. The reader is advised to think of the formalism referred to by the author as that of his earlier paper, this anthology, 236-253. In this paper, the results of the earlier paper are reobtained in strengthened form and in a simplified way. But there is also much new ground covered. Recursiveness is extended to relative recursiveness and the classification of arithmetic predicates in the now classical "Kleene hierarchy" according to their quantificational prefix is introduced. Gödel's incompleteness theorem is discussed from a general and revealing point of view. Some of this discussion overlaps Post's paper, this anthology, pp. 304-337. For the reader who compares the two, it may help to note that what Post calls a <u>recursively enumerable set</u> is just a set which can be defined by a predicate of the form $(Ex)R(a,x)$ where $R(a,x)$ is a recursive predicate. It should also be noted that the author uses the term "elementary predicate" for "arithmetic predicate".

The correction below and the addendum on p. 287 were supplied by the author for this anthology:

<u>Correction</u>: Omit §15, because the result claimed from Kleene [5] on which that section depends is false. The error in the supposed proof in [5] was indicated in the bibliographical reference to [5] (=Kleene 1944) on p. 527 of Kleene's <u>Introduction to Metamathematics</u> (1952). A full discussion and corrected result are given in S. C. Kleene, "On the forms of the predicates in the theory of constructive ordinals (second paper)", <u>Amer. J. Math.</u>, vol. 77 (1955), pp. 405-428.

Stephen C. Kleene

RECURSIVE PREDICATES AND QUANTIFIERS [1]

This paper contains a general theorem on the quantification of recursive predicates, with applications to the foundations of mathematics. The theorem (Theorem II) is a slight extension of previous results on Herbrand-Gödel general recursive functions(2), while the applications include theorems of Church (Theorem VII)(3) and Gödel (Theorem VIII)(4) and other incompleteness theorems. It is thought that in this treatment the relationship of the results stands out more clearly than before.

The general theorem asserts that to each of an enumeration of predicate forms, there is a predicate not expressible in that form. The predicates considered belong to elementary number theory.

The possibility that this theorem may apply appears whenever it is proposed to find a necessary and sufficient condition of a certain kind for some given property of natural numbers; in other words, to find a predicate of a given kind equivalent to a given predicate. If the specifications on the predicate which is being sought amount to its having one of the forms listed in the theorem, then for some selection of the given property a necessary and sufficient condition of the desired kind cannot exist.

In particular, it is recognized that to find a complete algorithmic theory for a predicate $P(a)$ amounts to expressing the predicate as a recursive predicate. By one of the cases of the theorem, this is impossible for a certain $P(a)$, which gives us Church's theorem.

Again, when we recognize that to give a complete formal deductive theory (symbolic logic) for a predicate $P(a)$ amounts to finding an equivalent predicate of the form $(Ex)R(a, x)$ where $R(a, x)$ is recursive, we have immediately Gödel's theorem, as another case of the general theorem.

Still another application is made, when we consider the nature of a constructive existence proof. It appears that there is a proposition provable classically for which no constructive proof is possible (Theorem X).

The endeavor has been made to include a fairly complete exposition of definitions and results, including relevant portions of previous theory, so that

(1) A part of the work reported in this paper was supported by the Institute for Advanced Study and the Alumni Research Foundation of the University of Wisconsin.

(2) Gödel [2, §9] (see the bibliography at the end of the paper).

(3) Church [1].

(4) Gödel [1, Theorem VI].

the paper should be self-contained, although some details of proof are omitted.

The general theorem is obtained quickly in Part I from the properties of the μ-operator, or what essentially was called the $\mathfrak{p}$-function in the author's dissertation[5]. Part II contains some variations on the theme of Part I, and may be omitted by the cursory reader. The applications to foundational questions are in Part III, only a few passages of which depend on Part II.

I. The general theorem on recursive predicates and quantifiers

1. **Primitive recursive functions.** The discussion belongs to the context of the informal theory of the natural numbers

$$0, 1, 2, \cdots, x, x', \cdots .$$

The functions which concern us are number-theoretic functions, for which the arguments and values are natural numbers.

We consider the following schemata as operations for the definition of a function ϕ from given functions appearing in the right members of the equations (c is any constant natural number):

$$\text{(I)} \qquad \phi(x) = x',$$

$$\text{(II)} \qquad \phi(x_1, \cdots, x_n) = c,$$

$$\text{(III)} \qquad \phi(x_1, \cdots, x_n) = x_i,$$

$$\text{(IV)} \qquad \phi(x_1, \cdots, x_n) = \theta(\chi_1(x_1, \cdots, x_n), \cdots, \chi_m(x_1, \cdots, x_n)),$$

$$\text{(Va)} \qquad \begin{cases} \phi(0) = c \\ \phi(y') = \chi(y, \phi(y)), \end{cases}$$

$$\text{(Vb)} \qquad \begin{cases} \phi(0, x_1, \cdots, x_n) = \psi(x_1, \cdots, x_n) \\ \phi(y', x_1, \cdots, x_n) = \chi(y, \phi(y, x_1, \cdots, x_n), x_1, \cdots, x_n). \end{cases}$$

Schema (I) introduces the successor function, Schema (II) the constant functions, and Schema (III) the identity functions. Schema (IV) is the schema of definition by substitution, and Schema (V) the schema of primitive recursion. Together we may call them (and more generally, schemata reducible to a series of applications of them) the *primitive recursive* schemata.

A function ϕ which can be defined from given functions $\psi_1, \cdots, \psi_k$ by a series of applications of these schemata we call *primitive recursive* in the given functions; and in particular, a function ϕ definable ab initio by these means, *primitive recursive.*

Now let us consider number-theoretic predicates, that is, propositional functions of natural numbers.

(5) Kleene [1, §18].

In asserting propositions, and in designating predicates, we use a logical symbolism, as follows. Operations of the propositional calculus: & (and), $\vee$ (or), $^{-}$ (not), $\rightarrow$ (implies), $\equiv$ (equivalent). Quantifiers: (x) (for all x), (Ex) (there exists an x such that). These operations may be taken either in the sense of classical mathematics, or in the sense of constructive or intuitionistic mathematics, except where one or the other of the two interpretations is specified.

A predicate $P(x_1, \cdots, x_n)$ is said to be *primitive recursive*, if there is a primitive recursive function $\pi(x_1, \cdots, x_n)$ such that

$$(1) \qquad P(x_1, \cdots, x_n) \equiv \pi(x_1, \cdots, x_n) = 0.$$

We can without loss of generality restrict π to take only 0 and 1 as values, and call it in this case the *representing function* of P.

Under classical interpretations, which give a dichotomy of propositions into true and false, we can assign to any predicate P a *representing function* π which has 0 or 1 as value according as the value of P is true or false; and then say that P is *primitive recursive* if π is.

2. General recursive functions. We shall proceed to the Herbrand-Gödel generalization of the notion of recursive function. We start with a preliminary account, certain features of which we shall then restate carefully.

The way in which the function ϕ is defined from the given functions in an application of one of the primitive recursive schemata amounts to this: the values $\phi(x_1, \cdots, x_n)$ of ϕ for the various sets $x_1, \cdots, x_n$ of arguments are determined unambiguously by the equations and the values of the given functions, using only principles of determination which we can formalize as a substitution rule and a replacement rule.

The formalization presupposes suitable conventions governing the symbolism, which are easily supplied. In particular, we must distinguish between the *variables* for numbers and the *numerals*, that is the expressions for the fixed numbers in terms of the symbols for 0 and the successor operation . The rules are the following.

R1: *to substitute, for the variables* $\mathrm{x}_1, \cdots, \mathrm{x}_n$ *of an equation, numerals* $\mathbf{x}_1, \cdots, \mathbf{x}_n$, *respectively.*

R2: *to replace a part* $\mathrm{f}(\mathbf{x}_1, \cdots, \mathbf{x}_n)$ *of the right member of an equation by* $\mathbf{x}$, *where* f *is a function symbol, where* $\mathbf{x}_1, \cdots, \mathbf{x}_n$, $\mathbf{x}$ *are numerals, and where* $\mathrm{f}(\mathbf{x}_1, \cdots, \mathbf{x}_n) = \mathbf{x}$ *is a given equation.*

By a *given* equation $\mathrm{f}(\mathbf{x}_1, \cdots, \mathbf{x}_n) = \mathbf{x}$ for R2, we mean an equation expressing one of the values of one of the given functions for the schema application, or an equation of this form already derived by R1 and R2 from the equations of the schema application.

Now let us consider any operation or schema, for the definition of a function in terms of given functions, which can be expressed by a system of equations determining the function values in this manner. In general the equations shall be allowed to contain, besides the *principal* function symbol which represents the function defined, and the *given* function symbols which represent the given functions, also *auxiliary* function symbols. The given function symbols shall not appear in the left members of the equations. Such a schema we shall call *general recursive.*

A function ϕ which can be defined from given functions $\psi_1, \cdots, \psi_k$ by a series of applications of general recursive schemata we call *general recursive* in the given functions; and in particular, a function ϕ definable ab initio by these means we call *general recursive.*

Suppose that a function ϕ is defined, either from given functions $\psi_1, \cdots, \psi_k$ or ab initio, by a succession of general recursive operations. Let us combine the successive systems of equations which effect the definition into one system, using different symbols as principal and auxiliary function symbols in each of the successive systems, and in the resulting system considering as auxiliary all of the function symbols but that representing ϕ and those representing $\psi_1, \cdots, \psi_k$. The restriction imposed on a general recursive schema that the given function symbols should not appear on the left will prevent any ambiguity being introduced by the interaction under R1 and R2 of equations in the combined system which were formerly in separate systems. Thus the definition can be considered as effected in a single general recursive operation.

In particular, any general recursive function can be defined ab initio in one operation, so that in the defining equations there are no given function symbols and what we have called the given equations for an application of R2 must all be derivable from the defining equations by previous applications of R1 and R2. For the formal development, it is convenient to adopt the convention that the principal function symbol shall be that one of the function symbols occurring in the equations of the system which comes latest in a preassigned list of function symbols. The function is then completely described by giving the system of defining equations.

We now restate the definition of general recursive function from this point of view.

A function $\phi(x_1, \cdots, x_n)$ is *GENERAL RECURSIVE*, if there is a system E of equations which defines it recursively in the following sense. A system E of equations *defines recursively* a GENERAL RECURSIVE function of n variables if, for each set $x_1, \cdots, x_n$ of natural numbers, an equation of the form $\mathrm{f}(\mathbf{x}_1, \cdots, \mathbf{x}_n) = \mathbf{x}$, where f is the principal function symbol of E, and where $\mathbf{x}_1, \cdots, \mathbf{x}_n$ are the numerals representing the natural numbers $x_1, \cdots, x_n$, is derivable from E by R1 and R2 for EXACTLY one numeral $\mathbf{x}$. The function of n variables which is defined by E in this case is the function ϕ, of which the value $\phi(x_1, \cdots, x_n)$ for $x_1, \cdots, x_n$ as arguments is THE NATURAL NUMBER x REPRESENTED BY THE NUMERAL **x**.

A predicate $P(x_1, \cdots, x_n)$ is *general recursive*, if there is a general recursive function $\pi(x_1, \cdots, x_n)$ taking only 0 and 1 as values such that (1) holds; in this case, π is called the *representing function* of P. (Or, if we introduce the *representing function* π first, P is *general recursive* if π is.)

3. **The μ-operator.** Consider the operator: μy (the least y such that). If this operator is applied to a predicate $R(x_1, \cdots, x_n, y)$ of the $n+1$ variables $x_1, \cdots, x_n, y$, and if this predicate satisfies the condition

$$(2) \qquad (x_1) \cdots (x_n)(Ey)R(x_1, \cdots, x_n, y),$$

we obtain a function $\mu y R(x_1, \cdots, x_n, y)$ of the remaining n free variables $x_1, \cdots, x_n$.

Thence we have a new schema,

$$(\mathrm{VI}_1) \qquad \phi(x_1, \cdots, x_n) = \mu y[\rho(x_1, \cdots, x_n, y) = 0],$$

for the definition of a function ϕ from a given function ρ which satisfies the condition

$$(3) \qquad (x_1) \cdots (x_n)(Ey)[\rho(x_1, \cdots, x_n, y) = 0].$$

We now show that this schema, subject to the condition on ρ, is, like (I)–(V), general recursive. For this purpose, we rewrite it in terms of equations, using an auxiliary function symbol "σ":

$$(\mathrm{VI}_2) \qquad \begin{cases} \sigma(0, x_1, \cdots, x_n, y) = y \\ \sigma(z', x_1, \cdots, x_n, y) = \sigma(\rho(x_1, \cdots, x_n, y'), x_1, \cdots, x_n, y') \\ \qquad \phi(x_1, \cdots, x_n) = \sigma(\rho(x_1, \cdots, x_n, 0), x_1, \cdots, x_n, 0). \end{cases}$$

Assuming the values of ρ, these equations will lead us to the values of ϕ as defined by (VI_1), and to only those values, as follows.

Consider informally any fixed set of values of $x_1, \cdots, x_n$ (formally, this means to substitute the corresponding set of numerals for the variables "x_1", $\cdots$, "x_n"). We seek to obtain the corresponding value of $\phi(x_1, \cdots, x_n)$ by replacements on the third equation, and this is the only possibility we have for obtaining that value under the two principles. First we can replace $\rho(x_1, \cdots, x_n, 0)$ by its value, and this is the only first replacement step possible on that equation. According as that value is 0 or is not 0, we seek the value of σ for the next replacement step from the first or second of the equations, and this is the only possible source for the next replacement value. In the first case, we obtain 0 as that value; in the second, we use the value of $\rho(x_1, \cdots, x_n, 1)$ in the second equation, and then seek another value of σ. We continue thus, with no choice in the procedure at any stage. The first case is first encountered when we come to use the value of $\rho(x_1, \cdots, x_n, y)$ for the first y for which that value is 0, and hence certainly for at most the y given by (3). When this happens, we can complete the pending replacements to obtain that y as the value of $\phi(x_1, \cdots, x_n)$. Thus we get the intended value; and because we had no choice at any stage of the procedure, we can get no other value.

The general recursiveness of the new schema is thus established. Hence, if $R(x_1, \cdots, x_n, y)$ is a general recursive predicate and (2) holds, by taking as ρ the representing function of R, we can conclude that $\mu y R(x_1, \cdots, x_n, y)$ is a general recursive function.

What can we conclude if (2) is not assumed to hold? In this case, $\mu y R(x_1, \cdots, x_n, y)$ may not be completely defined as a function of the variables $x_1, \cdots, x_n$; but for any fixed set of values of $x_1, \cdots, x_n$, the sequence of steps by which we attempt to determine a value for $\phi(x_1, \cdots, x_n)$ from the equations remains as described for the preceding case, only with now the matter of its termination in doubt. If $(Ey)R(x_1, \cdots, x_n, y)$ does hold for that set of values of $x_1, \cdots, x_n$, then it does terminate as described, with $\mu y R(x_1, \cdots, x_n, y)$ as the value; while conversely, if it does terminate, this can only be in consequence of a 0 being encountered among the values of $\rho(x_1, \cdots, x_n, y)$, so that $(Ey)R(x_1, \cdots, x_n, y)$ does hold, and $\mu y R(x_1, \cdots, x_n, y)$ is the value.

Hence, in formal terms, if F is the system of equations obtained by adjoining, to any system E which defines ρ recursively, equations of the form (VI_2), with the notation so arranged that "ϕ" becomes the principal function symbol f, then: an equation of the form $\mathrm{f}(\mathbf{x}_1, \cdots, \mathbf{x}_n) = \mathbf{x}$, where $\mathbf{x}_1, \cdots, \mathbf{x}_n$ are the numerals representing the natural numbers $x_1, \cdots, x_n$, and where $\mathbf{x}$ is a numeral, is derivable from F by R1 and R2 if and only if $(Ey)R(x_1, \cdots, x_n, y)$.

4. **The enumeration theorem.** We introduce a metamathematical predicate $\mathfrak{S}_n$ (for each particular n) as follows.

$\mathfrak{S}_n(\mathrm{Z}, x_1, \cdots, x_n, \mathrm{Y})$: *Z is a system of equations, and* Y *is a formal deduction from* Z *by* R1 *and* R2 *of an equation of the form* $\mathrm{f}(\mathbf{x}_1, \cdots, \mathbf{x}_n) = \mathbf{x}$, *where* f *is the principal function symbol of* Z, *where* $\mathbf{x}_1, \cdots, \mathbf{x}_n$ *are the numerals representing the natural numbers* $x_1, \cdots, x_n$, *and where* $\mathbf{x}$ *is a numeral.*

With this notation, we can state the last result of the preceding section symbolically:

$$(4) \qquad (Ey)R(x_1, \cdots, x_n, y) \equiv (E\mathrm{Y})\mathfrak{S}_n(\mathrm{F}, x_1, \cdots, x_n, \mathrm{Y}).$$

From a like exploration of the possibility that the sequence of steps does not terminate, or simply from (4) by contraposition, we have also:

$$(5) \qquad (y)\overline{R}(x_1, \cdots, x_n, y) \equiv (\mathrm{Y})\overline{\mathfrak{S}}_n(\mathrm{F}, x_1, \cdots, x_n, \mathrm{Y}).$$

Using Gödel's idea of arithmetizing metamathematics[6], suppose that natural numbers have been correlated to the formal objects, distinct numbers to distinct objects. The metamathematical predicate $\mathfrak{S}_n(Z, x_1, \cdots, x_n, Y)$ is carried by the correlation into a number-theoretic predicate $S_n(z, x_1, \cdots, x_n, y)$, the definition of which we complete by taking it as false for values of z, y not both correlated to formal objects.

For a suitably chosen Gödel numbering, we can show, with a little trouble that S_n is primitive recursive.

Now (4) translates under the arithmetization into

$$(6a) \qquad (Ey)R(x_1, \cdots, x_n, y) \equiv (Ey)S_n(f, x_1, \cdots, x_n, y)$$

with f as the Gödel number of the system of equations F. The formula

$$(7a) \qquad (y)R(x_1, \cdots, x_n, y) \equiv (y)\overline{S}_n(g, x_1, \cdots, x_n, y)$$

is obtained likewise from (5), after changing the notation so that R is interchanged with $\overline{R}$.

In stating these results for reference, we shall go over from S_n to a new predicate T_n, which entails no present disadvantage and proves to be of convenience in some further investigations[7]. The predicate T_n is defined from S_n as follows.

$$T_n(z, x_1, \cdots, x_n, y):\ S_n(z, x_1, \cdots, x_n, y)\ \&\ (t)\left[t < y \rightarrow \overline{S}_n(z, x_1, \cdots, x_n, t)\right].$$

By a theorem of Gödel[8], the primitive recursiveness of T_n follows from that of S_n. The formulas (6) and (7) in the theorem follow from (6a) and (7a) by the definition of T_n in terms of S_n.

THEOREM I. *Given a general recursive predicate $R(x_1, \cdots, x_n, y)$, there are numbers f and g such that*

$$(6) \qquad (Ey)R(x_1, \cdots, x_n, y) \equiv (Ey)T_n(f, x_1, \cdots, x_n, y),$$

$$(7) \qquad (y)R(x_1, \cdots, x_n, y) \equiv (y)\overline{T}_n(g, x_1, \cdots, x_n, y).$$

Now $(Ey)T_n(z, x_1, \cdots, x_n, y)$ is a fixed predicate of the form $(Ey)R(z, x_1, \cdots, x_n, y)$ where R is general recursive (in fact, as it happens, primitive recursive). By the theorem, if we take successively $z = 0, 1, 2, \cdots$, we obtain an enumeration (with repetitions) of all predicates of the form $(Ey)R(x_1, \cdots, x_n, y)$ where R is general recursive[9]. Likewise, the theorem gives us a fixed predicate of the form $(y)R(z, x_1, \cdots . x_n, y)$ where R is general recursive which enumerates all predicates of the form $(y)R(x_1, \cdots, x_n, y)$

(6) Gödel [1].

(7) A revision, April 13, 1942.

(8) Gödel [1, IV].

(9) This result entered partly into the last theorem of Kleene [2], but the advantage of using it at an earlier stage was overlooked. In anticipation, we may remark that XI–XVI of that paper are essentially special cases of Theorem II below (with now a constructive proof for XVI).

where R is general recursive. These enumerations form the basis for the application of Cantor's diagonal method in the next section.

5. **The general theorem.** By a familiar rule of classical logic, in each of the following pairs of propositions (with a fixed R for a given pair),

$$\begin{array}{cccc} (Ex)R(x) & (x)(Ey)R(x, y) & (Ex)(y)(Ez)R(x, y, z) & \cdots \\ (x)\overline{R}(x) & (Ex)(y)\overline{R}(x, y) & (x)(Ey)(z)\overline{R}(x, y, z) & \cdots, \end{array}$$

either member is equivalent to the negation of the other. Hence we may assert non-equivalence between the members of the pair. This argument is not good in the intuitionistic logic. However, the non-equivalence for the case of one quantifier,

$$(8) \qquad (Ex)R(x) \not\equiv (x)\overline{R}(x),$$

does hold good intuitionistically.

Consider the predicate form $(x)R(a, x)$ where R is general recursive. This gives a particular predicate of the variable a, whenever we specify the general recursive predicate $R(a, x)$ of two variables. In particular, $(x)\overline{T}_1(a, a, x)$ is a predicate of this form.

We shall show that this predicate is neither general recursive nor expressible in the form $(Ex)R(a, x)$ where R is general recursive.

For this purpose, suppose we have selected any particular general recursive $R(a, x)$, giving a particular predicate of the latter form. By (6), there is for this R a number f such that

$$(9) \qquad (Ex)R(a, x) \equiv (Ex)T_1(f, a, x).$$

Substituting the number f for the variable a,

$$(10) \qquad (Ex)R(f, x) \equiv (Ex)T_1(f, f, x).$$

By (8),

$$(11) \qquad (Ex)T_1(f, f, x) \not\equiv (x)\overline{T}_1(f, f, x).$$

Combining (10) and (11),

$$(12) \qquad (Ex)R(f, x) \not\equiv (x)\overline{T}_1(f, f, x).$$

This refutes, for $a=f$, the equivalence of $(Ex)R(a, x)$ to $(x)\overline{T}_1(a, a, x)$. Since this refutation can be effected, whatever general recursive R we chose, for some f depending on the R, the predicate $(x)\overline{T}_1(a, a, x)$ is not expressible in the form $(Ex)R(a, x)$ where R is general recursive.

A fortiori, $(x)\overline{T}_1(a, a, x)$ is not expressible in the form $R(a)$ where R is general recursive. For were it so expressed, we should then have it in the form $(Ex)R(a, x)$ where R is general recursive, by taking as $R(a, x)$ the predicate $R(a)$ & $x=x$.

This completes the proof of one case of the next theorem.

For another case, consider the predicate form $(Ex)R(a, x)$ where R is general recursive. We can show similarly, using (7) instead of (6), that the predicate $(Ex)T_1(a, a, x)$, which has this form, is neither general recursive nor expressible in the form $(x)R(a, x)$ where R is general recursive.

To illustrate the treatment of a case with more than one quantifier, consider the predicate form $(x)(Ey)(z)R(a, x, y, z)$ where R is general recursive. The predicate $(x)(Ey)(z)\overline{T}_3(a, a, x, y, z)$ has this form. Select any particular general recursive $R(a, x, y, z)$. By (6), for some f depending on this R,

$$(Ez)R(a, x, y, z) \equiv (Ez)T_3(f, a, x, y, z). \tag{13}$$

By corresponding quantifications of these equivalent predicates,

$$(Ex)(y)(Ez)R(a, x, y, z) \equiv (Ex)(y)(Ez)T_3(f, a, x, y, z). \tag{14}$$

Classically, we can complete the argument as before, showing that $(x)(Ey)(z)\overline{T}_3(a, a, x, y, z)$ is not expressible in any of the forms

$$R(a) \quad \begin{array}{ccc} (Ex)R(a, x) & (x)(Ey)R(a, x, y) & (Ex)(y)(Ez)R(a, x, y, z) \\ (x)R(a, x) & (Ex)(y)R(a, x, y) & \end{array}$$

where the R for the form is general recursive.

To obtain an alternative phrasing of the theorem, in which it holds for all cases intuitionistically, we may omit in the classical proof the step which interchanges the two kinds of quantifiers under the operation of negation. We thus show that the predicates $\overline{(Ex)}T_1(a, a, x)$, $\overline{(x)T_1}(a, a, x)$, $\overline{(Ex)}(y)(Ez)T_3(a, a, x, y, z)$, and so on, are neither expressible in the respective forms $(Ex)R(a, x)$, $(x)R(a, x)$, $(Ex)(y)(Ez)R(a, x, y, z)$, and so on, where R is general recursive, nor in any of the forms with fewer quantifiers.

THEOREM II. *Classically, and for the one-quantifier forms intuitionistically: To each of the forms*

$$R(a) \quad \begin{array}{ccc} (Ex)R(a, x) & (x)(Ey)R(a, x, y) & (Ex)(y)(Ez)R(a, x, y, z) \cdots \\ (x)R(a, x) & (Ex)(y)R(a, x, y) & (x)(Ey)(z)R(a, x, y, z) \cdots \end{array}$$

where the R for each is general recursive, after the first, there is a predicate expressible in that form but not in the other form with the same number of quantifiers nor in any of the forms with fewer quantifiers.

Classically, and intuitionistically: To each of the forms, after the first, there is a predicate expressible in the negation of that form but not in that form itself nor in any of the forms with fewer quantifiers.

For simplicity, we have given the theorem for predicates of one variable a, but it holds:

Likewise, replacing the variable a throughout by n variables $a_1, \cdots, a_n$, for any fixed positive integer n.

By an *elementary predicate*, we shall mean one which is expressible in terms of general recursive predicates, the operations &, $\vee$, $^{-}$, $\rightarrow$, $\equiv$ of the propositional calculus, and quantifiers.

Suppose given an expression for a predicate in these terms. By the classical predicate calculus, we can transform the expression so that all quantifiers stand at the front. For each m, let $(x)_1, \cdots, (x)_m$ be a set of m primitive recursive functions of x which as a set ranges, with or without repetitions, over all m-tuples of natural numbers, as x ranges over all natural numbers (such sets of functions are known). The equivalences

$$(15) \qquad (Ex_1) \cdots (Ex_m)A(x_1, \cdots, x_m) \equiv (Ex)A((x)_1, \cdots, (x)_m),$$

$$(16) \qquad (x_1) \cdots (x_m)A(x_1, \cdots, x_m) \equiv (x)A((x)_1, \cdots, (x)_m)$$

enable us to eliminate consecutive occurrences of like quantifiers. These transformations leave as operand of the prefixed quantifiers a general recursive predicate of the free and bound variables. Hence, classically, the predicate forms listed in the theorem for a given n suffice for the expression of every elementary predicate of n variables.

The theorem then says that no finite sublist of the forms would suffice.

Classically, we are led to a classification of the elementary predicates according to the minimum numbers of quantifiers which would suffice for their expression in terms of general recursive predicates and quantifiers.

The analogy between the logical operations of existential and universal quantification and geometrical operations of projection and intersection, respectively, is well known(10). The possibility of a connection between present results and theories of Borel and Baire is suggested(11).

II. Primitive, general, and partial recursive predicates under quantification

6. **Partial recursive functions.** The author's definition of *partial recursive* function extends the Herbrand-Gödel definition of general recursive function to functions ϕ of n variables which need not be defined for all n-tuples of natural numbers as arguments, retaining the characteristic of that definition with respect to each n-tuple for which the function is defined(12). The partial recursive functions include the general recursive functions as those which are defined for all sets of arguments.

For a more complete description, take the definition of general recursive function which is given at the end of §2, and replace the four capitalized phrases by the following, respectively: *PARTIAL RECURSIVE*; PARTIAL RECURSIVE; AT MOST; THE NATURAL NUMBER x REPRE-

(10) In particular, it has been discussed by Tarski.

(11) This suggestion was made to the author by Gödel and by Ulam.

(12) Kleene [4].

SENTED BY THE NUMERAL x IF THAT NUMERAL EXISTS, AND IS OTHERWISE UNDEFINED.

In dealing with functions which may not be completely defined, we interpret the equation $\phi(x_1, \cdots, x_n) = \psi(x_1, \cdots, x_n)$ as the assertion that ϕ and ψ have the same value for $x_1, \cdots, x_n$ as arguments, taking it as undefined (nonsignificant) if either value is undefined. We write $\phi(x_1, \cdots, x_n) \simeq \psi(x_1, \cdots, x_n)$ to express the assertion that, if either of ϕ and ψ is defined for the arguments $x_1, \cdots, x_n$, the other is and the values are the same, and if either of ϕ and ψ is undefined for those arguments, the other is.

Similarly, in dealing with predicates which may not be completely defined, $P(x_1, \cdots, x_n) \equiv Q(x_1, \cdots, x_n)$ expresses equivalence of value, and is undefined if the value of either member is undefined; while $P(x_1, \cdots, x_n) \cong Q(x_1, \cdots, x_n)$ expresses that the definition of either implies mutual definition with equivalence, and the indefinition of either implies mutual indefinition.

A predicate $P(x_1, \cdots, x_n)$ not necessarily defined for all n-tuples of natural numbers as arguments is *partial recursive*, if there is a partial recursive function $\pi(x_1, \cdots, x_n)$ taking only 0 and 1 as values such that

$$P(x_1, \cdots, x_n) \cong \pi(x_1, \cdots, x_n) = 0; \tag{17}$$

in this case, π is called the *representing function* of P. (Or if we first introduce a *representing function* π of P, the value of which is to be 0, 1, or undefined according as the value of P is true, false, or undefined, then P is partial recursive if π is.)

In §§2, 3, we remarked the general recursiveness of Schemata (I)–(VI) with (VI) subjected to the condition (3); and we also considered Schema (VI) for the case that ρ is general recursive but (3) is not required to hold. The method of those sections applies equally well without the restrictions; in explanation of the schemata when the given functions may not be completely defined or (3) not hold for (VI), it will suffice here to remark that the conditions of definition for the functions introduced by the schemata may be inferred a posteriori from the metamathematical results.

THEOREM III. *The class of general recursive functions is closed under applications of Schemata* (I)–(VI) *with* (3) *holding for applications of* (VI).

The class of partial recursive functions is closed under applications of Schemata (I)–(VI).

COROLLARY. *Every function obtainable by applications of Schemata* (I)–(VI) *with* (3) *holding for applications of* (VI) *is general recursive.*

Every function obtainable by applications of Schemata (I)–(VI) *is partial recursive.*

7. **Normal form for recursive functions.** We shall pursue a little further

the method of §4 to obtain the converse of this result. Besides the metamathematical predicate $\mathfrak{S}_n$, we now require a metamathematical function as follows.

$\mathfrak{U}(\mathrm{Y})$: *the natural number x which the numeral* x *represents, in case* Y *is a formal deduction of an equation of the form* t = x, *where* x *is a numeral and* t *is any term; and* 0, *otherwise.*

According to the definition of general recursive function, if ϕ is a general recursive function of n variables, there is a system E of equations such that

$$(18) \qquad (x_1) \cdots (x_n)(E\mathrm{Y})\mathfrak{S}_n(\mathrm{E}, x_1, \cdots, x_n, \mathrm{Y}),$$

$$(19) \qquad (x_1) \cdots (x_n)(\mathrm{Y})[\mathfrak{S}_n(\mathrm{E}, x_1, \cdots, x_n, \mathrm{Y}) \rightarrow \mathfrak{U}(\mathrm{Y}) = \phi(x_1, \cdots, x_n)];$$

and the function $\phi(x_1, \cdots, x_n)$ can be expressed in terms of E thus

$$(20) \qquad \phi(x_1, \cdots, x_n) = \mathfrak{U}(\mu\mathrm{Y}\mathfrak{S}_n(\mathrm{E}, x_1, \cdots, x_n, \mathrm{Y})),$$

if we understand the formal objects to be enumerated in some order, so that the operator μ can be applied with respect to the metamathematical variable Y; we may take the order to be that of the corresponding Gödel numbers.

If ϕ is a partial recursive function of n variables, instead of asserting (18), we can write

$$(E\mathrm{Y})\mathfrak{S}_n(\mathrm{E}, x_1, \cdots, x_n, \mathrm{Y})$$

as the condition on $x_1, \cdots, x_n$ that the function be defined for $x_1, \cdots, x_n$ as arguments; we have (19), taking the implication to be true whenever the first member is false, irrespective of the status of the second member; and our convention calls for rewriting (20) thus,

$$(21) \qquad \phi(x_1, \cdots, x_n) \simeq \mathfrak{U}(\mu\mathrm{Y}\mathfrak{S}_n(\mathrm{E}, x_1, \cdots, x_n, \mathrm{Y})),$$

in order that it be true (and not sometimes undefined) for all values of $x_1, \cdots, x_n$.

By the Gödel numbering already considered, the metamathematical function $\mathfrak{U}(\mathrm{Y})$ is carried into a number-theoretic function $U(y)$, the definition of which we complete by taking the value to be 0 for any y not correlated to a formal object. If the Gödel numbering was suitably chosen, U as well as S_n is primitive recursive.

Now (20), (18) and (19) in terms of $\mathfrak{S}_n$ and $\mathfrak{U}$ are carried into formulas of like form in terms of S_n and U. On passing over from S_n to T_n, we then have the (22), (23) and (24) of the theorem[13]. The part of the theorem which refers to a partial recursive function is obtained similarly.

THEOREM IV. *Given a general recursive function $\phi(x_1, \cdots, x_n)$, there is a*

---

[13] Kleene [2, IV], with some changes in the formulation. The present S_n corresponds to the former T_n, using the Gödel numbering of proofs instead of the enumeration of provable equations.

number e such that

(22) $$\phi(x_1, \cdots, x_n) = U(\mu y T_n(e, x_1, \cdots, x_n, y)),$$

(23) $$(x_1) \cdots (x_n)(Ey)T_n(e, x_1, \cdots, x_n, y),$$

(24) $$(x_1) \cdots (x_n)(y)[T_n(e, x_1, \cdots, x_n, y) \rightarrow U(y) = \phi(x_1, \cdots, x_n)].$$

Given a partial recursive function $\phi(x_1, \cdots, x_n)$, *there is a number e such that*

(25) $$\phi(x_1, \cdots, x_n) \simeq U(\mu y T_n(e, x_1, \cdots, x_n, y)),$$

where

$$(Ey)T_n(e, x_1, \cdots, x_n, y)$$

is the condition of definition of the function, and (24) *holds.*

Thus any general recursive function (any partial recursive function) is expressible in the form $\psi(\mu y R(x_1, \cdots, x_n, y))$ with (2) holding (in the form $\psi(\mu y R(x_1, \cdots, x_n, y)))$ where ψ and R are primitive recursive. Hence:

COROLLARY. *Every general recursive function is obtainable by applications of Schemata* (I)–(VI) *with* (3) *holding for applications of* (VI).

Every partial recursive function is obtainable by applications of Schemata (I)–(VI).

Formula (25) contains the substance of the theorem. For it implies the condition of definition of the function; and, in the case that $\phi(x_1, \cdots, x_n)$ is defined for all sets of arguments, it gives (22) and (23). Moreover by the definition of T_n in terms of S_n, it implies (24).

We say that *e defines* ϕ *recursively*, or *e* is a *Gödel number* of ϕ, if (25) holds(14), in which case *e* has all the properties in relation to ϕ which are specified in the theorem.

It is here that the advantage of using T_n instead of S_n appears. A number *e* which satisfies $\phi(x_1, \cdots, x_n) \simeq U(\mu y S_n(e, x_1, \cdots, x_n, y))$ (which is equivalent to (25)) does not necessarily satisfy $(x_1) \cdots (x_n)(y)[S_n(e, x_1, \cdots, x_n, y) \rightarrow U(y) = \phi(x_1, \cdots, x_n)]$. While we could get around the difficulty by imposing the latter as an additional condition on the Gödel numbers, it is more convenient simply to use T_n instead of S_n. (On the basis of Theorem III and the results which we had in terms of S_n before passing over to T_n, one can set up a primitive recursive function V such that, if *e* satisfies (25), then $V(e)$ has all the properties in terms of S_n.)

The numbers *f* and *g* for Theorem I can be described now as any numbers which define recursively the partial recursive functions $\mu y R(x_1, \cdots, x_n, y)$ and $\mu y \overline{R}(x_1, \cdots, x_n, y)$, respectively.

(14) Kleene [2, Definition 2c, p. 738] and [4, top p. 153]. We have now also the changes in the formulation of Theorem IV.

248

8. **Consistency.** Let us review the arguments used in proof of Theorems I and III. For rigor, these have to be put in metamathematical form. Let E be the system of equations associated with a series of applications of Schemata (I)–(VI). We shall review only the case that no given function symbols occur in E.

In general, we easily establish that, for each of certain sets $x_1, \cdots, x_n$ of natural numbers, an equation of the form $f(\mathbf{x}_1, \cdots, \mathbf{x}_n) = \mathbf{x}$, as described in the definitions of general and partial recursive function, is derivable from E by R1 and R2. In particular, if we are proving that E defines a general recursive function, we must show this for all $x_1, \cdots, x_n$; if we have a prior interpretation of the schemata applications as definition of a (partial or complete) function $\phi(x_1, \cdots, x_n)$, or require that E define a $\phi(x_1, \cdots, x_n)$ already known to us in some other manner, we must show this for all $x_1, \cdots, x_n$ belonging to the range of definition of ϕ, and also show that the $\mathbf{x}$ in the equation is the numeral representing the value of ϕ for $x_1, \cdots, x_n$ as arguments. This property of the equations E and rules R1 and R2, the precise formulation of which depends on the circumstances, we call the "completeness property." (When we wish merely to show that E defines a partial recursive function, the function to be determined a posteriori from E, no completeness property is required.)

The second part of the discussion consists in showing that an equation of the described form $f(\mathbf{x}_1, \cdots, \mathbf{x}_n) = \mathbf{x}$ is derivable from E for at most one numeral $\mathbf{x}$; or if we have already established completeness in one of the above senses, that the equations $f(\mathbf{x}_1, \cdots, \mathbf{x}_n) = \mathbf{x}$ referred to in the discussion of completeness, for various $x_1, \cdots, x_n$, are the only equations of that form which are derivable from E by R1 and R2. This we call the "consistency property."

As we indicated in §2, it suffices to handle each of the schemata in turn, assuming equations for use with R2 which give the values of the given functions. The argument for consistency which we sketched in §3 for Schema (VI) applies as well to the other schemata. For Schema (IV) there is indeed a choice in the order in which the values of the several χ's are introduced, but it is without effect on the final result.

This very easy consistency proof was gained by restricting the replacement rule so that replacement is only performable on the right member of an equation, a part $f(\mathbf{x}_1, \cdots, \mathbf{x}_n)$ where f is a function symbol and $\mathbf{x}_1, \cdots, \mathbf{x}_n$ are numerals being replaced by a numeral $\mathbf{x}$. This eliminates the possibility of deriving an equation of the form $g(\boldsymbol{y}_1, \cdots, \boldsymbol{y}_m) = \boldsymbol{y}$, where g is a fixed function symbol, $\boldsymbol{y}_1, \cdots, \boldsymbol{y}_m$ are fixed numerals, and $\boldsymbol{y}$ is any numeral, along essentially different paths within the system, and therewith the possibility that such an equation should be derivable for different $\boldsymbol{y}$'s.

In some previous versions of the theories of general and partial recursive functions, the replacement rule was not thus restricted. The consistency proof which we gave in the version with the unrestricted replacement rule was based on the notion of verifiability of an equation(15). This notion makes presupposition of the values of the functions, and for the theory of partial recursive functions also of the determinateness whether or not the values are defined. In the latter case, it is not finitary. To give a constructive consistency proof for the theory of partial recursive functions with the stronger replacement rule seems to require the type of argument used in the Church-Rosser consistency proof for λ-conversion(16), and in the Ackermann-von Neumann consistency proof for a certain part of number theory in terms of the Hilbert ϵ-symbol(17).

It is easily shown, by using the method of proof of Theorem IV to obtain the same normal form with the stronger replacement rule, that every function partial recursive under the stronger replacement rule is such under the weaker.

Thus we find the curious fact that the main difficulty in showing the equivalence of the two notions of recursiveness comes in showing that the stronger rule suffices to define as many functions as the weaker. This is because the consistency of a stronger formalism is involved. The consistency of that formalism is of interest on its own account, but is extraneous for the theory of recursive definition, including the applications corresponding to those of Church in terms of the λ-notation which presuppose the complicated Church-Rosser consistency proof. All that is required for the theory of recursive definition is some consistent formalism sufficient for the derivation of the equations giving the values of the functions.

To this discussion we may add several supplementary remarks. We might in practice have a system E of equations and a method for deriving from E by R1 and the strong replacement rule, for all and only the n-tuples of a certain set, an equation of the form $\mathrm{f}(\mathbf{x}_1, \cdots, \mathbf{x}_n) = \mathbf{x}$ with a determinate $\mathbf{x}$, but lack the knowledge that unlimited use of the two rules could not lead to other such equations. In this situation, a function is defined intuitively for the n-tuples of the set, and undefined off the set. If we can characterize metamathematically our method of applying the two rules, we shall obtain a limited formalism known to be consistent, and the method used in establishing Theorem IV can then be applied to obtain equations defining the function recursively with the weak replacement rule.

For some types of equations which define a function recursively with the strong replacement rule (consistency being known), a more direct method may be available for obtaining a system defining the function recursively with the weak replacement rule. For example, consider (in informal language) the equation $\phi(\psi(x)) = \chi(x)$. To use this in deriving equations giving values of ϕ, we need to introduce values of ψ by replacement on the left. After

(15) Kleene [2, p. 731] and [4, §2, the bracketed portion of the fifth paragraph].

(16) Church and Rosser [1].

(17) Hilbert and Bernays [1, §2, part 4, pp. 93–130, and Supplement II, pp. 396–400].

241

expressing the equation in the form $\phi(y) = (\mu w[\psi((w)_1) = y \,\&\, \chi((w)_1) = (w)_2])_2$, and separating the latter into a series of equations without the μ-symbol by the method which the theory of the schemata affords, replacement will be required only on the right. This device is applicable to any equation of which the left member has the form $\mathrm{f}(\mathrm{g}_1(\mathrm{x}_1, \cdots, \mathrm{x}_n), \cdots, \mathrm{g}_m(\mathrm{x}_1, \cdots, \mathrm{x}_n))$.

The precise form of the restriction which is used to weaken the replacement rule is somewhat arbitrary, so long as it accomplishes its purpose of channelling the deductions of equations giving the values of the functions. The restriction as it was stated in the early Gödel version is now simplified, since we need to consider only equations having the forms appearing in the six schemata. Gödel provided for equations the left members of which could have the form $\mathrm{f}(\mathrm{g}_1(\mathrm{x}_1, \cdots, \mathrm{x}_n), \cdots, \mathrm{g}_m(\mathrm{x}_1, \cdots, \mathrm{x}_n))$ where f is the principal function symbol and $\mathrm{g}_1, \cdots, \mathrm{g}_m$ are given function symbols, and therefore allowed replacement on the left in the case of the g's.

9. **Predicates expressible in both one-quantifier forms.** By Theorem IV, for any general recursive predicate $P(x_1, \cdots, x_n)$,

(26) $$P(x_1, \cdots, x_n) \equiv (Ey)[T_n(e, x_1, \cdots, x_n, y) \,\&\, U(y) = 0],$$

(27) $$P(x_1, \cdots, x_n) \equiv (y)[T_n(e, x_1, \cdots, x_n, y) \rightarrow U(y) = 0],$$

where e is any Gödel number of the representing function of P.

Conversely, suppose that for a predicate P both $P(x_1, \cdots, x_n) \equiv (Ey)R(x_1, \cdots, x_n, y)$ and $P(x_1, \cdots, x_n) \equiv (y)S(x_1, \cdots, x_n, y)$ where R and S are general recursive. From the second of these equivalences, under classical interpretations, $\overline{P}(x_1, \cdots, x_n) \equiv (Ey)\overline{S}(x_1, \cdots, x_n, y)$. By the classical law of the excluded middle, $(Ey)[R(x_1, \cdots, x_n, y) \vee \overline{S}(x_1, \cdots, x_n, y)]$. Therefore

(28) $$P(x_1, \cdots, x_n) \equiv R(x_1, \cdots, x_n, \mu y[R(x_1, \cdots, x_n, y) \vee \overline{S}(x_1, \cdots, x_n, y)]),$$

where the second member is general recursive by Theorem III.

THEOREM V. *Every general recursive predicate $P(x_1, \cdots, x_n)$ is expressible in both of the forms $(Ey)R(x_1, \cdots, x_n, y)$ and $(y)R(x_1, \cdots, x_n, y)$ where the R for each is primitive recursive. Under classical interpretations, conversely, every predicate expressible in both of these forms where the R for each is general recursive is general recursive.*

Now consider any predicate expressible in one of the forms of Theorem II after the first. According as the innermost quantifier in this form is existential or universal, we can apply (26) or (27), and then absorb the extra quantifier by (15) or (16), respectively, to obtain the original form but with a primitive recursive R. For example,

$$(x)(Ey)R(a, x, y) \equiv (x)(Ey_1)(Ey_2)[T_3(e, a, x, y_1, y_2) \,\&\, U(y_2) = 0]$$
$$\equiv (x)(Ey)[T_3(e, a, x, (y)_1, (y)_2) \,\&\, U((y)_2) = 0].$$

COROLLARY. *The class of predicates expressible in a given one of the forms of Theorem* II *after the first (for a given n variables) is the same whether a primitive recursive or a general recursive R be allowed.*

This generalizes the observation of Rosser that a class enumerable by a general recursive function is also enumerable by a primitive recursive function([18]).

The formulas for the one-quantifier cases are

$$(29)\qquad (Ey)R(x_1, \cdots, x_n, y) \equiv (Ey)[T_{n+1}(e, x_1, \cdots, x_n, (y)_1, (y)_2) \,\&\, U((y)_2) = 0],$$

$$(30)\qquad (y)R(x_1, \cdots, x_n, y) \equiv (y)[T_{n+1}(e, x_1, \cdots, x_n, (y)_1, (y)_2) \rightarrow U((y)_2) = 0],$$

where e is any Gödel number of the representing function of R. These afford a new proof of the enumeration theorem of §4, with new enumerating predicates, and thence a new proof of Theorem II.

10. **Partial recursive predicates.** Let $P(x_1, \cdots, x_n)$ be a predicate which may not be defined for all n-tuples of natural numbers as arguments. By a *completion* of P we understand a predicate Q such that, if $P(x_1, \cdots, x_n)$ is defined, then $Q(x_1, \cdots, x_n)$ is defined and has the same value, and if $P(x_1, \cdots, x_n)$ is undefined, then $Q(x_1, \cdots, x_n)$ is defined. In particular, the completion $P^+(x_1, \cdots, x_n)$ which is false when $P(x_1, \cdots, x_n)$ is undefined, and the completion $P^-(x_1, \cdots, x_n)$ which is true when $P(x_1, \cdots, x_n)$ is undefined, we call the *positive completion* and *negative completion* of $P(x_1, \cdots, x_n)$, respectively. (In P and P^+, the "positive parts" coincide; in P and P^-, the "negative parts" coincide.)

If $P(x_1, \cdots, x_n)$ is a partial recursive predicate, then by Theorem IV,

$$(31)\qquad P^+(x_1, \cdots, x_n) \equiv (Ey)[T_n(e, x_1, \cdots, x_n, y) \,\&\, U(y) = 0],$$

$$(32)\qquad P^-(x_1, \cdots, x_n) \equiv (y)[T_n(e, x_1, \cdots, x_n, y) \rightarrow U(y) = 0],$$

where e is any Gödel number of the representing function of P.

Conversely, if $R(x_1, \cdots, x_n, y)$ is any general recursive predicate, then by Theorem III,

$$(33)\qquad (Ey)R(x_1, \cdots, x_n, y) \equiv \mu y R(x_1, \cdots, x_n, y) =^+ \mu y R(x_1, \cdots, x_n, y),$$

$$(34)\qquad (y)R(x_1, \cdots, x_n, y) \equiv \mu y \overline{R}(x_1, \cdots, x_n, y) \neq^- \mu y \overline{R}(x_1, \cdots, x_n, y).$$

THEOREM VI. *The positive completion $P^+(x_1, \cdots, x_n)$ of a partial recursive predicate $P(x_1, \cdots, x_n)$ is expressible in the form $(Ey)R(x_1, \cdots, x_n, y)$ where R is primitive recursive; and conversely, any predicate expressible in the form $(Ey)R(x_1, \cdots, x_n, y)$ where R is general recursive is the positive completion $P^+(x_1, \cdots, x_n)$ of a partial recursive predicate $P(x_1, \cdots, x_n)$.*

([18]) Rosser [1, Lemma I, Corollary I, p. 88].

i 232

Dually, for negative completions $P^{-}(x_1, \cdots, x_n)$ *and the predicate form* $(y)R(x_1, \cdots, x_n, y)$.

It follows that, for the predicate forms of Theorem II which have an existential quantifier (universal quantifier) innermost, we may, without altering the class of predicates expressible in that form, take R to be the positive completion (negative completion) of a partial recursive predicate.

Let us abbreviate $U(\mu y T_n(z, x_1, \cdots, x_n, y))$ as $\Phi_n(z, x_1, \cdots, x_n)$(19). Then Φ_n is a fixed partial recursive function of $n+1$ variables, from which any partial recursive function ϕ of n variables can be obtained thus (rewriting (25)),

$$(35) \qquad \phi(x_1, \cdots, x_n) \simeq \Phi_n(e, x_1, \cdots, x_n)$$

where e is any Gödel number of ϕ. Since for a constant z, $\Phi_n(z, x_1, \cdots, x_n)$ is always a partial recursive function of the remaining n variables, $\Phi_n(z, x_1, \cdots, x_n)$ therefore gives for $z=0, 1, 2, \cdots$ an enumeration (with repetitions) of the partial recursive functions of n variables. It follows that $\Phi_n(z, x_1, \cdots, x_n)=0$ is a partial recursive predicate of $n+1$ variables which enumerates (with repetitions) the partial recursive predicates of n variables.

This, seen in the light of Theorem VI, has as consequence the enumeration theorem of §2 (with other enumerating predicates), and thence by Cantor's diagonal method Theorem II.

Elsewhere, the enumeration theorem for partial recursive functions gave by Cantor's diagonal method what may be called the fundamental theorem for proofs of recursive definability(20).

This fundamental theorem, and the existence of partial recursive functions and predicates, no completions of which are general recursive(21), are what occasioned the introduction of the notion of a partial recursive function.

III. Incompleteness theorems in the foundations of number theory

11. **Introductory remarks.** We entertain various propositions about natural numbers. These propositions have meaning, independently of or prior to the consideration of formal postulates and rules of proof. We pose the problem of systematizing our knowledge about these propositions into a theory of some kind. For certain definitions of our objectives in constructing the theory, and certain classes of propositions, we shall be able to reach definite answers concerning the possibility of constructing the theory.

The naïve informal approach which we are adopting may be contrasted

(19) Using the notation of Kleene [4, bottom p. 152], but with the changes in the formulation of Theorem IV.

(20) Kleene [4, the last result in §2].

(21) Kleene [4, Footnote 3].

with that form of the postulational approach which consists in first listing formal postulates, which are then said to define the content of the theory based on them. In the case of number theory, the formal approach cannot render entirely dispensable an intuitive understanding of propositions of the kind which we commonly interpret the theory to be about. For the explicit statement of the postulates and characterization of the manner in which they are to determine the theory belong to a metatheory on another level of discourse; and the ultimate metatheory must be an intuitive mathematics unregulated by explicit postulates, and having the essential character of number theory.

Of course the informality of our investigation does not preclude the enumeration, from another level, of postulates which would suffice to describe it. Indeed, such regulation may perhaps be considered necessary from an intuitive standpoint for that part of it which belongs to the context of classical mathematics.

The propositions about natural numbers which we shall consider will contain parameters. We shall thus have infinitely many propositions of a given form, according to the natural numbers taken as values by the parameters. In other words, we have predicates, for which these parameters are the independent variables. Generally, in a theory, a number of predicates are dealt with simultaneously; but for our investigations it will suffice to consider a theory with respect to some one predicate without reference to other predicates which might be present. Usually, we shall write a one-variable predicate $P(a)$, though the discussion applies equally well to a predicate $P(a_1, \cdots, a_n)$ of n variables.

12. **Algorithmic theories.** As one choice of the objective, we can ask that the theory should give us an effective means for deciding, for any given one of the propositions which are taken as values of the predicate, whether that proposition is true or false. Examples of predicates for which a theoretical conquest of this kind has been obtained are: *a is divisible by b* (that is, in symbols, $(Ex)[a=bx]$), *$ax+by=c$ is solvable for x and y* (that is, $(Ex)(Ey)[ax+by=c]$). We shall call this kind of theory for a predicate a *complete algorithmic theory* for the predicate.

Let us examine the notion of this kind of theory more closely. In setting up a complete algorithmic theory, what we do is to describe a procedure, performable for each set of values of the independent variables, which procedure necessarily terminates and in such manner that from the outcome we can read a definite answer, "Yes" or "No," to the question, "Is the predicate value true?"

We can express this by saying that we set up a second predicate: *the procedure terminates in such a way as to give the affirmative answer.* The second predicate has the same independent variables as the first, is equivalent to the first, and the determinability of the truth or falsity of its values is guaranteed.

This last property of the second predicate we designate as the property of being *effectively decidable.*

Of course the original predicate becomes effectively decidable, in a derivative sense, as soon as we have its equivalence to the second; extensionally, the two are the same. But while our terminology is ordinarily extensional, at this point the essential matter can be emphasized by using the intensional language. The reader may if he wishes write in more explicit statements referring to the (generally) differing objects or processes with which the two predicates are concerned.

Now, the recognition that we are dealing with a well defined process which for each set of values of the independent variables surely terminates so as to afford a definite answer, "Yes" or "No," to a certain question about the manner of termination, in other words, the recognition of effective decidability in a predicate, is a subjective affair. Likewise, the recognition of what may be called *effective calculability* in a function. We may assume, to begin with, an intuitive ability to recognize various individual instances of these notions. In particular, we do recognize the general recursive functions as being effectively calculable, and hence recognize the general recursive predicates as being effectively decidable.

Conversely, as a heuristic principle, such functions (predicates) as have been recognized as being effectively calculable (effectively decidable), and for which the question has been investigated, have turned out always to be general recursive, or, in the intensional language, equivalent to general recursive functions (general recursive predicates). This heuristic fact, as well as certain reflections on the nature of symbolic algorithmic processes, led Church to state the following thesis[(22)]. The same thesis is implicit in Turing's description of computing machines[(23)].

THESIS I. *Every effectively calculable function (effectively decidable predicate) is general recursive.*

Since a precise mathematical definition of the term effectively calculable (effectively decidable) has been wanting, we can take this thesis, together with the principle already accepted to which it is converse, as a definition of it for the purpose of developing a mathematical theory about the term. To the extent that we have already an intuitive notion of effective calculability (effective decidability), the thesis has the character of an hypothesis—a point emphasized by Post and by Church[(24)]. If we consider the thesis and its converse as definition, then the hypothesis is an hypothesis about the application of the mathematical theory developed from the definition. For the acceptance of the hypothesis, there are, as we have suggested, quite compelling grounds.

---

(22) Church [1].

(23) Turing [1].

(24) Post [1, p. 105], and Church [2].

i 291

A full account of these is outside the scope of the present paper(25). We are here concerned rather to present the consequences.

In the intensional language, to give a complete algorithmic theory for a predicate $P(a)$ now means to find an equivalent effectively decidable predicate $Q(a)$. It would suffice that $Q(a)$ be given as a general recursive predicate; and by Thesis I, if $Q(a)$ is not so given, then at least there is a general recursive predicate $R(a)$ equivalent to $Q(a)$ and hence to $P(a)$. Thus to give a complete algorithmic theory for $P(a)$ means to find an equivalent general recursive predicate $R(a)$, or more briefly, to express $P(a)$ in the form $R(a)$ where R is general recursive. This predicate form is the one listed first in Theorem II; and Theorem II gives to each of the other forms a predicate not expressible in that form. Thus, while under our interpretations there is a complete algorithmic theory for each predicate of the form $R(a)$ where R is general recursive, to each of the other forms there is a predicate for which no such theory is possible. We state this in the following theorem, using the particular examples for the one-quantifier forms which were exhibited in the proof of Theorem II.

THEOREM VII. *There exists no complete algorithmic theory for either of the predicates* $(Ex)T_1(a, a, x)$ *and* $(x)\overline{T}_1(a, a, x)$.

Of course, once the definition of effective decidability is granted, which affords an enumeration of the effectively decidable predicates, Cantor's methods immediately give other predicates. This theorem, as additional content, shows the elementary forms which suffice to express such predicates.

Abstracting from the particular examples used here, the theorem is Church's theorem on the existence of an unsolvable problem of elementary number theory, and the corresponding theorem of Turing in terms of his machine concept(26). The unsolvability is in the sense that the construction called for by the problem formulation, which amounts to that of a recursive R with a certain property, is impossible. The theorem itself constitutes solution in a negative sense.

13. **Formal deductive theories.** A second possibility for giving theoretic cohesion to the totality of true propositions taken as values of a predicate $P(a)$ is that offered by the postulational or deductive method. We should like all and only those of the predicate values which are true to be deducible from given axioms by given rules of inference. To make the axioms and principles of inference quite explicit, according to modern standards of rigor, we shall suppose them constituted into a formal system (symbolic logic), in which the propositions taken as values of the predicate are expressible. Those and only those of the formulas expressing the true instances of the predicate

(25) For a resume, see Kleene [4, Footnote 2], where further references are given.

(26) Turing [1, §8].

should be provable. We call this kind of theory for a predicate $P(a)$ a *complete formal deductive theory* for the predicate.

This type of theory should of course not be confused with incompletely formalized axiomatic theories, such as the theory of natural numbers itself as based on Peano's axioms.

It is convenient in discussing a formal system to name collectively as the "postulates" the rules describing the formal axioms and the rules of inference.

Let us now examine more closely the concept of provability in a stated formal system. If the formalization does accomplish its purpose of making matters explicit, we should be able effectively to recognize each step of a formal proof as an application of a postulate of the system. Furthermore, if the system is to constitute a theory for the predicate $P(a)$, we should be able effectively to recognize, to each natural number a, a certain formula of the system which is taken as expressing the proposition $P(a)$. Together, these conditions imply that we should be able, given any sequence of formulas which might be submitted as a proof of $P(a)$ for a given a, to check it, thus determining effectively whether it is actually such or not.

Let us introduce a designation for the metamathematical predicate with which we deal in making this check, for a given formal system and predicate $P(a)$.

$\mathfrak{R}(a, \mathrm{X})$: X *is a proof in the formal system of the formula expressing the proposition* $P(a)$.

Then the concept of provability in the system of the formula expressing $P(a)$, or briefly, the provability of $P(a)$, is expressible as $(E\mathrm{X})\mathfrak{R}(a, \mathrm{X})$.

As we have just argued, the predicate $\mathfrak{R}(a, \mathrm{X})$ should be an effectively decidable metamathematical predicate. Here the formal objects over which X ranges, if the notation of the system is explicit, should be given in some manner which affords an effective enumeration of them. Using the indices in this enumeration, or generally any effective Gödel numbering of the formal objects, the metamathematical predicate $\mathfrak{R}(a, \mathrm{X})$ will be carried into a number-theoretic predicate $R(a, x)$, taken as false for any x not correlated to a formal object, which should then also be effectively decidable. By Thesis I, the effective decidability of the latter implies its general recursiveness. We are thus led to state a second thesis.

THESIS II. *For any given formal system and given predicate* $P(a)$, *the predicate that* $P(a)$ *is provable is expressible in the form* $(Ex)R(a, x)$ *where* R *is general recursive.*

This thesis corresponds to the standpoint that the role of a formal deductive system for a predicate $P(a)$ is that of making explicit the notion of what constitutes a proof of $P(a)$ for a given a. If a proposed "formal system" for $P(a)$ does not do this, we should say that it is not a formal system in the strict sense, or at least not one for $P(a)$. Taken this way, the thesis has a definitional character.

Presupposing, on the other hand, a prior conception of what constitutes a formal system for a given predicate in the strict sense, the thesis has the character of an hypothesis, to which we are led both heuristically and from Thesis I by general considerations.

Conversely, if a predicate of the form $(Ex)R(a, x)$ where R is general recursive is given, it is easily seen that we can always set up a formal system of the usual sort, with an explicit criterion of proof, in which all true instances of this predicate and only those are provable.

Using the thesis, and this converse, we can now say that to give a complete formal deductive theory for a predicate $P(a)$ means to find an equivalent predicate of the form $(Ex)R(a, x)$ where R is general recursive, or more briefly, to express the predicate in this form. By Theorem II, there are predicates of the other one-quantifier form, and of the forms with more quantifiers, not expressible in this form. Hence while there are complete formal deductive theories to each predicate of either of the forms $R(a)$ and $(Ex)R(a, x)$ where R is general recursive, to each of the other forms there is a predicate for which no such theory is possible. Specifically, using the one-quantifier example given in the proof of Theorem II:

THEOREM VIII. *There is no complete formal deductive theory for the predicate* $(x)\overline{T}_1(a, a, x)$.

This is the famous theorem of Gödel on formally undecidable propositions, in a generalized form. A proposition is formally undecidable in a given formal system if neither the formula expressing the proposition nor the formula expressing its negation is provable in the system. Gödel gave such a proposition for a certain formal system (by a method evidently applying to similar systems), subject to the assumptions of the consistency and ω-consistency of the system. Later Rosser gave another proposition, for which the latter assumption is dispensed with([27]).

In the present form of the theorem, we have a preassigned predicate $(x)\overline{T}_1(a, a, x)$ and a method which, to any formal system whatsoever for this predicate, gives a number f for which the following is the situation.

Suppose that the system meets the condition that the formula expressing the proposition $(x)\overline{T}_1(f, f, x)$ is provable only if that proposition is true. Then the proposition is true but the formula expressing it unprovable. This statement of results uses the interpretation of the formula, but if the system has certain ordinary deductive properties for the universal quantifier and recursive predicates, our condition on the system is guaranteed by the metamathematical one of consistency.

If the system contains also a formula expressing the negation of

([27]) Rosser [1].

$(x)\overline{T}_1(f, f, x)$, and if the system meets the further condition that this formula is provable only if true, then this formula cannot be provable, and we have a formally undecidable proposition. The further condition, if the system has ordinary deductive properties, is guaranteed by the metamathematical one of ω-consistency.

Moreover, we can incorporate Rosser's elimination of the hypothesis of ω-consistency into the present treatment. To do so, we replace the predicate $(Ex)R(a, x)$ for the application of Theorem II by $(Ex)[R(a, x) \& (y)[y<x \rightarrow \overline{S}(a, y)]]$ where $(Ex)S(a, y)$ is the predicate expressing the provability of the negation of $(x)\overline{T}_1(a, a, x)$. This changes the f for the system.

Thus we come out with the usual metamathematical results for a given formal system.

For the case that a formal system is sought which should not only prove the true instances of $P(a)$ but also refute the false ones, if the classical law of the excluded middle is applied to the propositions $P(a)$, then the Gödel theorem (Theorem VIII) comes under the Church theorem (Theorem VII). For had we completeness with respect both to $P(a)$ and to $\overline{P}(a)$, we could obtain a general recursive $R(a)$ equivalent to the given predicate by the method used in proving the second part of Theorem V. Informally, this amounts merely to the remark that we should have the algorithm for $P(a)$ which consists in searching through some list of the provable formulas until we encounter either the formula expressing $P(a)$ or the formula expressing $\overline{P}(a)$.

The connection between Gödel's theorem and the paradoxes has been much noted. The author gave a proof of Gödel's theorem along much the present lines but as a refinement of the Richard paradox rather than of the Epimenides(28). That gave the undecidable propositions as values of a predicate of the more complicated form $(x)(Ey)R(a, x, y)$ where R is general recursive. The Epimenides paradox now appears as the more basic. Currently, Curry has noted the same phenomenon in connection with the Kleene-Rosser inconsistency theorem(29).

14. **Discussion, incomplete theories.** In the present form of Gödel's theorem, several aspects are brought into the foreground which perhaps were not as clearly apparent in the original version.

Not merely, to any given formal system of the type considered, can a proposition be formulated with respect to which that system is incomplete, but all these propositions can be taken as values of a preassignable elementary predicate, with respect to which predicate therefore no system can be complete. This depends on the thesis giving a preassignable form to the concept of provability in a formal system.

---

(28) Kleene [2, XIII].
(29) Kleene and Rosser [1], Curry [2].

For the interpretation of the propositions we have required, as minimum, only the notions of effectively calculable predicates and of the quantifiers used constructively. It seems that lesser presuppositions, if one is to allow any mathematical infinite, are hardly conceivable.

Beyond that the system should fulfil the structural characteristic expressed in Thesis II, and should yield results correct under this modicum of interpretation, we have need of no reference whatsoever to its detailed constitution.

In particular, the nature of the intuitive evidence for the deductive processes which are formalized in the system plays no role.

Let us imagine an omniscient number theorist, whom we should expect, through his ability to see infinitely many facts at once, to be able to frame much stronger systems than any we could devise. Any correct system which he could reveal to us, telling us how it works without telling us why, would be equally subject to the Gödel incompleteness.

It is impossible to confine the intuitive mathematics of elementary propositions about integers to the extent that all the true theorems will follow from explicitly stated axioms by explicitly stated rules of inference, simply because the complexity of the predicates soon exceeds the limited form representing the concept of provability in a stated formal system.

We selected as the objective in constructing a formal deductive system that what constitutes proof should be made explicit in the sense that a proposed proof could be effectively checked, and either declared formally correct or declared formally incorrect.

Let us for the moment entertain a weaker conception of a formal system, under which, if we should happen to discover a correct proof of a proposition or be presented with one, then we could check it and recognize its formal correctness, but if we should have before us an alleged proof which is not correct, then we might not be able definitely to locate the formal fallacy. In other words, under this conception a system possesses a process for checking, which terminates in the affirmative case, but need not in the negative. Then the concept of provability would have the form $(Ex)P^+(a, x)$ where P^+ is the positive completion of a partial recursive predicate $P(a, x)$. By Theorem VI, $P^+(a, x)$ is expressible in the form $(Ey)R(a, x, y)$ where R is general recursive. Then the provability concept has the form $(Ex)(Ey)R(a, x, y)$, or by contraction of quantifiers $(Ex)R(a, (x)_1, (x)_2)$. This is of the form $(Ex)R(a, x)$ where R is general recursive. Thus the concept of provability has the usual form, and Gödel's theorem applies as before. If we take a new concept of proof based on $R(a, x)$, that is, if we redesignate the steps in the checking process as the formal proof steps, the concept of proof assumes the usual form.

We gave no attention, when we formulated the objectives both of an algorithmic and of a formal deductive theory, to the nature of the evidence for the correctness of the theory, or to various other practical considerations, simply because the crude structural objectives suffice to entail the corresponding incompleteness theorems. In this connection, it may be of some interest to give the corresponding definitions, although these may not take into account all the desiderata, for the case of incomplete theories of the two sorts. We shall state these for predicates of n variables $a_1, \cdots, a_n$, as we could also have done for the case of the complete theories.

To give an *algorithmic theory* (not necessarily complete) for a predicate $P(a_1, \cdots, a_n)$ is to give a general recursive function $\pi(a_1, \cdots, a_n)$, taking only 0, 1, and 2 as values, such that

$$\text{(36)} \qquad \begin{cases} \pi(a_1, \cdots, a_n) = 0 \rightarrow P(a_1, \cdots, a_n) \\ \pi(a_1, \cdots, a_n) = 1 \rightarrow \overline{P}(a_1, \cdots, a_n). \end{cases}$$

The algorithm always terminates, but if $\pi(a_1, \cdots, a_n)$ has the value 2 we can draw no conclusion about $P(a_1, \cdots, a_n)$.

To give a *formal deductive theory* (not necessarily complete) for a predicate $P(a_1, \cdots, a_n)$ is to give a general recursive predicate $R(a_1, \cdots, a_n, x)$ such that

$$\text{(37)} \qquad (Ex)R(a_1, \cdots, a_n, x) \rightarrow P(a_1, \cdots, a_n).$$

In words, to give a formal deductive theory for a predicate $P(a_1, \cdots, a_n)$ is to find a sufficient condition for it of the form $(Ex)R(a_1, \cdots, a_n, x)$ where R is general recursive. Here, according to circumstances, the sufficiency may be established from a wider context, or it may be a matter of postulation (hypothesis), or of conviction (belief).

From the present standpoint, the setting up of this sufficient condition is the essential accomplishment in the establishment of a so-called metatheory (in the constructive sense) for the body of propositions taken as the values of a predicate. We note that this may be accomplished without necessarily going through the process of setting up a formal object language, from which R is obtainable by subsequent arithmetization, although as remarked above, we can always set up the object language, if we have the R by some other means.

In the view of the present writer, the interesting variations of formal technique recently considered by Curry have the above as their common feature with formalization of the more usual sort(30). This is stated in our terminology, Curry's use of the terms "meta" and "recursive" being different. He gives examples of "formal systems," in connection with which he introduces some predicates by what he calls "recursive definitions," but what we should prefer to call "inductive definitions." This important type of definition, under suitable precise delimitation so that the individual clauses are constructive, can be shown to lead always to predicates expressible in the form $(Ex)R(a_1, \cdots, a_n, x)$ where R is recursive in our sense. Indeed, this fact

(30) Curry [1].

can be recognized by substantially the method indicated above for the case of the inductive definition establishing the notion of provability for a formal system of the usual sort.

Conversely, given any predicate expressible in the form $(Ex)R(a_1, \cdots, a_n, x)$ where R is recursive, we can set up an inductive definition for it.

15. **Ordinal logics.** In ordinal logics, studied by Turing(31), the requirement of effectiveness for the steps of deduction is relaxed to allow dependence on a number (or λ-formula) which represents an ordinal in the Church-Kleene theory of constructive ordinals(32). A presumptive proof in an ordinal logic cannot in general be checked objectively, since the proof character depends on the number which occupies the role of a Church-Kleene representative of an ordinal actually being such, for which there is no effective criterion. Nevertheless it was hoped that ordinal logics could be used to give complete orderings (with repetitions) of the true propositions of certain forms into transfinite series, by means of the ordinals represented in the proofs, in such a way that the proving of a proposition in the ordinal logic (and therewith the determination of a position for it in the series) would somehow make it easier to recognize the truth of the proposition.

Turing obtained a number of interesting results, largely outside the scope of this article, but among them the following. There are ordinal logics which are complete for the theory of a predicate of the form $(x)(Ey)R(a, x, y)$ where R is general recursive; however, for the example of such a logic which is given, its use would afford no theoretic gain, since the recognition that the number which plays the role of ordinal representative in a proof of the logic is actually such comes to the same as the direct recognition of the truth of the proposition proved.

Now let us approach the topic by inquiring whether, and if so where, the property of being provable in a given ordinal logic is located in the scale of predicate forms of Theorem II. First, it turns out that the property of a number a of being the representative of an ordinal is expressible in the form $(x)(Ey)R(a, x, y)$ where R is recursive(33). Now we may use the definition of ordinal logic in terms of λ-conversion, or we may take the notion in general terms as described above, and state the thesis that for a given predicate $P(a)$ and given ordinal logic the provability of $P(a)$ is expressible in the form $(E\alpha)(Ex)R(a, \alpha, x)$ where α ranges over the ordinal representatives and R is general recursive. In either case, it then follows that the provability of $P(a)$ is expressible in the form $(Ex)(y)(Ez)R(a, x, y, z)$ where R is general recursive. Conversely, to any predicate of the latter form, we can find an ordinal logic

(31) Turing [2]. Turing gave a somewhat restricted definition of "ordinal logic" in terms of the theory of λ-conversion for predicates expressible in the form $(x)(Ey)R(a, x, y)$ where R is recursive.

(32) Church and Kleene [1], Church [2], Kleene [4].

(33) Kleene [5].

in the more general sense such that provability in the logic expresses the predicate. Hence there is a complete ordinal logic to each predicate of each of the forms

$$R(a) \qquad \begin{matrix} (Ex)R(a, x) & (x)(Ey)R(a, x, y) & (Ex)(y)(Ez)R(a, x, y, z) \\ (x)R(a, x) & (Ex)(y)R(a, x, y) & \end{matrix}$$

where R is general recursive, but by Theorem II, classically there are predicates of the form $(x)(Ey)(z)R(a, x, y, z)$ and of each of the forms with more quantifiers, or classically and intuitionistically of the form $\overline{(Ex)}(y)(Ez)R(a, x, y, z)$ and of the negation of each of the forms with more quantifiers, for which no complete ordinal logic is possible. Specifically:

THEOREM IX. *There is no complete ordinal logic for the predicate* $\overline{(Ex)}(y)(Ez)T_3(a, a, x, y, z)$.

Ordinal logics form a class of examples of the systems of propositions which have recently come under discussion, in which more or less is retained of the ordering of propositions in deductive reasoning, but with an extension into the transfinite, or a sacrifice of constructiveness in individual steps. These may be called "non-constructive logics," in contrast to the formal deductive systems in the sense of §§13–14 which are "constructive logics." In general, the usefulness of a non-constructive logic may be considered to depend on the degree to which the statement of the non-constructive proof criterion is removed from the direct statement of the propositions.

Theorem IX is a "Gödel theorem" for the ordinal logics. The ordinal logics were at least conceived with somewhat of a constructive bias. Rosser has shown how Gödel theorems arise on going very far in the direction of non-constructiveness([34]), and Tarski has stated the Gödel argument for systems of sentences in general([35]). Incidental of Rosser's results for finite numbers of applications of the Hilbert "rule of infinite induction," also called "Carnap's rule," can easily be inferred from Theorem II, through the obvious correspondence of an application of this rule to a universal quantifier in the proof concept. However, the proof concepts for non-constructive logics soon outrun the scale of predicate forms of Theorem II. This appears to be the case even for the extension to protosyntactical definability given by Quine([36]). If one is going very far in the direction of non-constructiveness, and is not interested in considerations of the sort emphasized in §§12–14, there is no advantage in starting from the theory of recursive functions. But the more general results do not detract from the special significance which attaches to the Gödel theorems associated with provability criteria of the forms $R(a)$ and $(Ex)R(a, x)$

([34]) Rosser [2].
([35]) Tarski [2].
([36]) Quine [1].

where R is general recursive, that is, Church's theorem and Gödel's theorem, for which forms only it is true that a given proof is a finite object.

16. **Constructive existence proofs.** A proof of an existential proposition $(Ey)A(y)$ is acceptable to an intuitionist, only if in the course of the proof there is given a y such that $A(y)$ holds, or at least a method by which such a y could be constructed. Consider the case that $A(y)$ depends on other variables. Say that there is one of these, x, and rewrite the proposition as $(x)(Ey)A(x, y)$. The proposition asserts the existence of a y to each of the infinitely many values of x. In this case, the only way in which the constructivist demand could in general be met would be by giving the y as an effectively calculable function of x, that is, by giving the function. According to Thesis I, this function would have to be general recursive. Hence we propose the following thesis (and likewise for n variables $x_1, \cdots, x_n$):

THESIS III. *A proposition of the form $(x)(Ey)A(x, y)$ containing no free variables is provable constructively, only if there is a general recursive function $\phi(x)$ such that $(x)A(x, \phi(x))$.*

When such a ϕ exists, we shall say that $(x)(Ey)A(x, y)$ is *recursively fulfillable*(37).

This thesis expresses what seems to be demanded from the standpoint of the intuitionists. Whether such explicit rules of proof as they have stated do conform to the thesis is a further question which will be considered elsewhere(38). However, in its aspect as restriction on all intuitionistic existence proofs, the possibilities for which, as we know by Theorem VIII, transcend the limitations of any preassignable formal system, the thesis is more general than a metamathematical result concerning a given system.

We now examine the notion of recursive fulfillability as it applies to the values of a given predicate of the form $(x)(Ey)(z)R(a, x, y, z)$ where R is general recursive. Select any fixed value of a. Given a recursive ϕ which fulfils the corresponding proposition, by Theorem IV there is a number e such that $(x)(Ey)T_1(e, x, y)$ and $(x)(y)[T_1(e, x, y)\rightarrow(z)R(a, x, U(y), z)]$. Conversely, if such an e exists, the proposition is fulfilled by the general recursive function $U(\mu y T_1(e, x, y))$. Thus

$$(Ee)\{(x)(Ey)T_1(e, x, y) \;\&\; (x)(y)[T_1(e, x, y) \rightarrow (z)R(a, x, U(y), z)]\}$$

is a necessary and sufficient condition for recursive fulfillability. When the quantifiers are suitably brought to the front and contracted, this assumes the form $(Ex)(y)(Ez)R(a, x, y, z)$ with another general recursive R depending on the original R.

By Theorem II, classically, there is a predicate of the original form

(37) A further analysis of the implications of constructive provability is given in Kleene [6].
(38) Nelson [1].

$(x)(Ey)(z)R(a,x,y,z)$ which is not expressible in this form $(Ex)(y)(Ez)R(a,x,y,z)$, in which the condition of its recursive fulfillability is expressible.

Using the example of such a predicate given in the proof of Theorem II, we have then

$$\{(x)(Ey)(z)\overline{T}_3(a,\ a,\ x,\ y,\ z) \text{ rec. fulf.}\} \equiv (Ex)(y)(Ez)R(a,\ x,\ y,\ z) \tag{38}$$

for a certain general recursive R. Substituting the number f of (14) for a in (14) and (38),

$$(Ex)(y)(Ez)R(f,\ x,\ y,\ z) \equiv (Ex)(y)(Ez)T_3(f,\ f,\ x,\ y,\ z), \tag{39}$$

$$\{(x)(Ey)(z)\overline{T}_3(f,\ f,\ x,\ y,\ z) \text{ rec. fulf.}\} \equiv (Ex)(y)(Ez)R(f,\ x,\ y,\ z). \tag{40}$$

By the definition of recursive fulfillability,

$$\{(x)(Ey)(z)\overline{T}_3(f,\ f,\ x,\ y,\ z) \text{ rec. fulf.}\} \rightarrow (x)(Ey)(z)\overline{T}_3(f,\ f,\ x,\ y,\ z). \tag{41}$$

Suppose that $(x)(Ey)(z)\overline{T}_3(f,\ f,\ x,\ y,\ z)$ were recursively fulfillable. We could then conclude by (40) and (39), $(Ex)(y)(Ez)T_3(f,f,x,y,z)$, and by (41), $(x)(Ey)(z)\overline{T}_3(f,f,x,y,z)$. These results are incompatible. Therefore by reductio ad absurdum, $(x)(Ey)(z)\overline{T}_3(f,\ f,\ x,\ y,\ z)$ is not recursively fulfillable, and hence by Thesis III not constructively provable.

Now by (40) and (39), we have $\overline{(Ex)}(y)(Ez)T_3(f,\ f,\ x,\ y,\ z)$; and thence classically we can proceed to $(x)(Ey)(z)\overline{T}_3(f,\ f,\ x,\ y,\ z)$.

THEOREM X. *For a certain number f, the proposition* $(x)(Ey)(z)\overline{T}_3(f,f,x,y,z)$ *is true classically, but not constructively provable.*

Notice that we have here a fixed unprovable proposition for all constructive methods of reasoning, whereas in the preceding incompleteness theorems. we had only an infinite class of propositions, some of which must be unprovable in a given theory.

Intuitionistic number theory has been presented as a subsystem of the classical, so that the intuitionistic results hold classically, though many classical results are not asserted intuitionistically. The possibility now appears of extending intuitionistic number theory by incorporating Thesis III in the form

$$(x)(Ey)A(x,\ y) \rightarrow \{\text{for some general recursive } \phi,\ (x)A(x,\ \phi(x))\},$$

so that the two number theories should diverge, with the proposition of Theorem X true classically, and its negation true intuitionistically(39).

For the classical proof, an application of

$$\overline{(x)}A(x) \rightarrow (Ex)\overline{A}(x)$$

suffices as the sole non-intuitionistic step; therewith that law of logic would

(39) This is perhaps hinted in Church [1, first half of p. 363].

be refuted intuitionistically, for a certain A. Hitherto the intuitionistic refutations of laws of the classical predicate calculus have depended on the interpretation of the quantifiers in intuitionistic set theory(40).

The result of Theorem X, with another proposition as example, can be reached as follows. Consider the proposition,

$$(x)(Ey)\{[(Ez)T_1(x, x, z) \& y = 0] \vee [(z)\overline{T}_1(x, x, z) \& y = 1]\}.$$

This holds classically, by application of the law of the excluded middle in the form

$$(x)\{(Ez)A(x, z) \vee (z)\overline{A}(x, z)\},$$

or the form

$$(x)(A(x) \vee \overline{A}(x)),$$

from which the other follows by substituting $(Ez)A(x, z)$ for $A(x)$. But it is not recursively fulfillable, since it can be fulfilled only by the representing function of the predicate $(Ez)T_1(x, x, z)$, which, as we saw in the proof of Theorem II, is non-recursive.

17. **Non-elementary predicates.** The elementary predicates are enumerable. By Cantor's methods, there are therefore non-elementary number-theoretic predicates. However let us ask what form of definition would suffice to give such a predicate. Under classical interpretations, the enumeration of predicate forms given in Theorem II for n variables suffices for the expression of every elementary predicate of n variables. By defining relations of the form shown in the next theorem, we can introduce a predicate $M(a, k)$ so that it depends for different values of k on different numbers of alternating quantifiers. On the basis of Theorem II, it is possible to do this in such a way that the predicate will be expressible in none of the forms of Theorem II.

THEOREM XI. *Classically, there is a non-elementary predicate $M(a, k)$ definable by relations of the form*

$$\begin{cases} M(a, 0) \equiv R(a) \\ M(a, 2k+1) \equiv (Ex)M(\phi(a, x), 2k) \\ M(a, 2k+2) \equiv (x)M(\phi(a, x), 2k+1) \end{cases}$$

where R and ϕ are primitive recursive.

We are dealing here with essentially the same fact which Hilbert-Bernays discover by setting up a truth definition for their formal system (Z)(41).

The system (Z) has as primitive terms only $'$, $+$, $\cdot$, $=$ and the logical operations. The predicates expressible in these terms are elementary. Con-

(40) Heyting [1, p. 65].

(41) Hilbert and Bernays [1, pp. 328–340].

versely, using Theorem IV and Gödel's reduction of primitive recursive functions to these terms(42), every elementary predicate is expressible in (*Z*).

The Hilbert-Bernays result is an application to (*Z*) of Tarski's theorem on the truth concept(43), with the determination of a particular form of relations which give the truth definition for (*Z*). If (*Z*) is consistent, a formal proof that the relations do define a predicate is beyond the resources of (*Z*).

Bibliography

Alonzo Church

1. *An unsolvable problem of elementary number theory*, Amer. J. Math. vol. 58 (1936) pp. 345–363.[a]
2. *The constructive second number class*, Bull. Amer. Math. Soc. vol. 44 (1938) pp. 224–232.

Alonzo Church and S. C. Kleene

1. *Formal definitions in the theory of ordinal numbers*, Fund. Math. vol. 28 (1936) pp. 11–21.

Alonzo Church and Barkley Rosser

1. *Some properties of conversion*, Trans. Amer. Math. Soc. vol. 39 (1936) pp. 472–482.

H. B. Curry

1. *Some aspects of the problem of mathematical rigor*, Bull. Amer. Math. Soc. vol. 47 (1941) pp. 221–241.
2. *The inconsistency of certain formal logics*, J. Symbolic Logic vol. 7 (1942) pp. 115–117.

Kurt Gödel

1. *Über formal unentscheidbare Sätze der Principia Mathematica und verwandter Systeme* I, Monatshefte für Mathematik und Physik vol. 38 (1931) pp. 173–198.[a]
2. *On undecidable propositions of formal mathematical systems*, notes of lectures at the Institute for Advanced Study, 1934.[a]

David Hilbert and Paul Bernays

1. *Grundlagen der Mathematik*, vol. 2, Berlin, Springer, 1939.

Arend Heyting

1. *Die formalen Regeln der intuitionistischen Mathematik*, Preuss. Akad. Wiss. Sitzungsber, Phys.-math. Kl. 1930, pp. 57–71, 158–169.

S. C. Kleene

1. *A theory of positive integers in formal logic*, Amer. J. Math. vol. 57 (1935) pp. 153–173, 219–244.
2. *General recursive functions of natural numbers*, Math. Ann. vol. 112 (1936) pp. 727–742.[a]
3. *A note on recursive functions*, Bull. Amer. Math. Soc. vol. 42 (1936) pp. 544–546.
4. *On notation for ordinal numbers*, J. Symbolic Logic vol. 3 (1938) pp. 150–155.
5. *On the forms of the predicates in the theory of constructive ordinals*, to appear in Amer. J. Math. (Bull. Amer. Math. Soc. abstract 48-5-215).
6. *On the interpretation of intuitionistic number theory*, Bull. Amer. Math. Soc. abstract 48-1-85.

S. C. Kleene and Barkley Rosser

1. *The inconsistency of certain formal logics*, Ann. of Math. (2) vol. 36 (1935) pp. 630–636.

David Nelson

1. *Recursive functions and intuitionistic number theory*, under preparation.

E. L. Post

1. *Finite combinatory processes—formulation* I, J. Symbolic Logic vol. 1 (1936) pp. 103–105.[a]

---

(42) Gödel [1, Theorem VII]. See Kleene [3 (erratum: p. 544, line 11, "of" should be at the end of the line)].

(43) Tarski [1].

W. V. Quine
1. *Mathematical logic*, New York, Norton, 1940.

Barkley Rosser
1. *Extensions of some theorems of Gödel and Church*, J. Symbolic Logic vol. 1 (1936) pp. 87–91.[a]
2. *Gödel theorems for non-constructive logics*, ibid. vol. 2 (1937) pp. 129–137.

Alfred Tarski
1. *Der Wahrheitsbegriff in den formalisierten Sprachen*, Studia Philosophica vol. 1 (1936) pp. 261–405. (Original in Polish, 1933.)
2. *On undecidable statements in enlarged systems of logic and the concept of truth*, J. Symbolic Logic vol. 4 (1939) pp. 105–112.

A. M. Turing
1. *On computable numbers, with an application to the Entscheidungsproblem*, Proc. London Math. Soc. (2) vol. 42 (1937) pp. 230–265.[a]
2. *Systems of logic based on ordinals*, ibid. vol. 45 (1939) pp. 161–228.[a]

Addendum correcting an oversight noticed by J. C. E. Dekker: Footnote 21 cites only a function which is partial but not potentially recursive. For such a predicate, cf. Example 6, p. 332 of Kleene's Introduction to Metamathematics, New York and Toronto (Van Nostrand), Amsterdam (North Holland) and Groningen (Noordhoff), 1952.

FINITE COMBINATORY PROCESSES. FORMULATION I.

This paper gives an analysis of the computing process substantially identical to that given by Turing (this anthology, p. 116-154). Although this work is independent of Turing's, it is not independent of Church's, referring as it does to Church's paper, this anthology, pp. 89-107.

Note that what Turing refers to as internal configuration of a machine occurs in Post's treatment as instructions to be carried out by a human computer.

Emil L. Post

FINITE COMBINATORY PROCESSES. FORMULATION I.

The present formulation should prove significant in the development of symbolic logic along the lines of Gödel's theorem on the incompleteness of symbolic logics[1] and Church's results concerning absolutely unsolvable problems.[2]

We have in mind a *general problem* consisting of a class of *specific problems*. A solution of the general problem will then be one which furnishes an answer to each specific problem.

In the following formulation of such a solution two concepts are involved: that of a *symbol space* in which the work leading from problem to answer is to be carried out,[3] and a fixed unalterable *set of directions* which will both direct operations in the symbol space and determine the order in which those directions are to be applied.

In the present formulation the symbol space is to consist of a two way infinite sequence of spaces or boxes, i.e., ordinally similar to the series of integers · · · , −3, −2, −1, 0, 1, 2, 3, · · · . The problem solver or worker is to move and work in this symbol space, being capable of being in, and operating in but one box at a time. And apart from the presence of the worker, a box is to admit of but two possible conditions, i.e., being empty or unmarked, and having a single mark in it, say a vertical stroke.

One box is to be singled out and called the starting point. We now further assume that a specific problem is to be given in symbolic form by a finite number of boxes being marked with a stroke. Likewise the answer is to be given in symbolic form by such a configuration of marked boxes. To be specific, the answer is to be the configuration of marked boxes left at the conclusion of the solving process.

The worker is assumed to be capable of performing the following primitive acts:[4]

(a) *Marking the box he is in (assumed empty)*,

(b) *Erasing the mark in the box he is in (assumed marked)*,

(c) *Moving to the box on his right*,

(d) *Moving to the box on his left*,

(e) *Determining whether the box he is in, is or is not marked.*

The set of directions which, be it noted, is the same for all specific problems and thus corresponds to the general problem, is to be of the following form. It is to be headed:

Start at the starting point and follow direction 1.

Received October 7, 1936. The reader should compare an article by A. M. Turing, *On computable numbers*,[a] shortly forthcoming in the **Proceedings of the London Mathematical Society.** The present article, however, although bearing a later date, was written entirely independently of Turing's. *Editor.*

[1] Kurt Gödel, *Über formal unentscheidbare Sätze der Principia Mathematica und verwandter Systeme I*, **Monatshefte für Mathematik und Physik,** vol. 38 (1931), pp. 173–198.[a]

[2] Alonzo Church, *An unsolvable problem of elementary number theory*, **American Journal of Mathematics,** vol. 58 (1936), pp. 345–363.[a]

[3] Symbol space, and time.

[4] As well as otherwise following the directions described below.

Reprinted from THE JOURNAL OF SYMBOLIC LOGIC, vol. 1 (1936) pp. 103-105.

It is then to consist of a finite number of directions to be numbered 1, 2, 3, · · · n. The ith direction is then to have one of the following forms:

(A) *Perform operation* $O_i[O_i=$ (a), (b), (c), *or* (d)] *and then follow direction* j_i,

(B) *Perform operation* (e) *and according as the answer is yes or no correspondingly follow direction* j_i' *or* j_i'',

(C) *Stop.*

Clearly but one direction need be of type C. Note also that the state of the symbol space directly affects the process only through directions of type B.

A set of directions will be said to be *applicable* to a given general problem if in its application to each specific problem it never orders operation (a) when the box the worker is in is marked, or (b) when it is unmarked.[5] A set of directions applicable to a general problem sets up a deterministic process when applied to each specific problem. This process will terminate when and only when it comes to the direction of type (C). The set of directions will then be said to set up a *finite 1-process* in connection with the general problem if it is applicable to the problem and *if the process it determines terminates for each specific problem.* A finite 1-process associated with a general problem will be said to be a *1-solution* of the problem if the answer it thus yields for each specific problem is always correct.

We do not concern ourselves here with how the configuration of marked boxes corresponding to a specific problem, and that corresponding to its answer, symbolize the meaningful problem and answer. In fact the above assumes the specific problem to be given in symbolized form by an outside agency and, presumably, the symbolic answer likewise to be received. A more self-contained development ensues as follows. The general problem clearly consists of at most an enumerable infinity of specific problems. We need not consider the finite case. Imagine then a one-to-one correspondence set up between the class of positive integers and the class of specific problems. We can, rather arbitrarily, represent the positive integer n by marking the first n boxes to the right of the starting point. The general problem will then be said to be *1-given* if a finite 1-process is set up which, when applied to the class of positive integers as thus symbolized, yields in one-to-one fashion the class of specific problems constituting the general problem. It is convenient further to assume that when the general problem is thus 1-given each specific process at its termination leaves the worker at the starting point. If then a general problem is 1-given and 1-solved, with some obvious changes we can combine the two sets of directions to yield a finite 1-process which gives the answer to each specific problem when the latter is merely given by its number in symbolic form.

With some modification the above formulation is also applicable to symbolic logics. We do not now have a class of specific problems but a single initial finite marking of the symbol space to symbolize the primitive formal assertions of the logic. On the other hand, there will now be no direction of type (C). Consequently, assuming applicability, a deterministic process will be set up which is *unending.* We further assume that in the course of this process certain recognizable symbol groups, i.e., finite sequences of marked and unmarked boxes, will appear which are not further altered in the course of the process. These will be the derived assertions of the logic. Of course the set of directions corresponds to the deductive processes of the logic. The logic may then be said to be *1-generated.*

An alternative procedure, less in keeping, however, with the spirit of symbolic

---

[5] While our formulation of the set of directions could easily have been so framed that applicability would immediately be assured it seems undesirable to do so for a variety of reasons.

logic, would be to set up a finite 1-process which would yield the *n*th theorem or formal assertion of the logic given *n*, again symbolized as above.

Our initial concept of a given specific problem involves a difficulty which should be mentioned. To wit, if an outside agency gives the initial finite marking of the symbol space there is no way for us to determine, for example, which is the first and which the last marked box. This difficulty is completely avoided when the general problem is 1-given. It has also been successfully avoided whenever a finite 1-process has been set up. In practice the meaningful specific problems would be so symbolized that the bounds of such a symbolization would be recognizable by characteristic groups of marked and unmarked boxes.

The root of our difficulty however, probably lies in our assumption of an infinite symbol space. In the present formulation the boxes are, conceptually at least, physical entities, e.g., contiguous squares. Our outside agency could no more give us an infinite number of these boxes than he could mark an infinity of them assumed given. If then he presents us with the specific problem in a finite strip of such a symbol space the difficulty vanishes. Of course this would require an extension of the primitive operations to allow for the necessary extension of the given finite symbol space as the process proceeds. A final version of a formulation of the present type would therefore also set up directions for generating the symbol space.[6]

The writer expects the present formulation to turn out to be logically equivalent to recursiveness in the sense of the Gödel-Church development.[7] Its purpose, however, is not only to present a system of a certain logical potency but also, in its restricted field, of psychological fidelity. In the latter sense wider and wider formulations are contemplated. On the other hand, our aim will be to show that all such are logically reducible to formulation 1. We offer this conclusion at the present moment as a *working hypothesis*. And to our mind such is Church's identification of effective calculability with recursiveness.[8] Out of this hypothesis, and because of its apparent contradiction to all mathematical development starting with Cantor's proof of the non-enumerability of the points of a line, independently flows a Gödel-Church development. The success of the above program would, for us, change this hypothesis not so much to a definition or to an axiom but to a *natural law*. Only so, it seems to the writer, can Gödel's theorem concerning the incompleteness of symbolic logics of a certain general type and Church's results on the recursive unsolvability of certain problems be transformed into conclusions concerning all symbolic logics and all methods of solvability.

COLLEGE OF THE CITY OF NEW YORK

[6] The development of formulation 1 tends in its initial stages to be rather tricky. As this is not in keeping with the spirit of such a formulation the definitive form of this formulation may relinquish some of its present simplicity to achieve greater flexibility. Having more than one way of marking a box is one possibility. The desired naturalness of development may perhaps better be achieved by allowing a finite number, perhaps two, of physical objects to serve as pointers, which the worker can identify and move from box to box.

[7] The comparison can perhaps most easily be made by defining a 1-function and proving the definition equivalent to that of recursive function. (See Church, loc. cit., p. 350[i]) A 1-function $f(n)$ in the field of positive integers would be one for which a finite 1-process can be set up which for each positive integer n as problem would yield $f(n)$ as answer, n and $f(n)$ symbolized as above.

[8] Cf. Church, loc. cit., pp. 346, 356–358[ii] Actually the work already done by Church and others carries this identification considerably beyond the working hypothesis stage. But to mask this identification under a definition hides the fact that a fundamental discovery in the limitations of the mathematicizing power of Homo Sapiens has been made and blinds us to the need of its continual verification.

[i] 94 [ii] 90, 100–102

RECURSIVE UNSOLVABILITY OF A PROBLEM OF THUE

This paper contains the first unsolvability proof for a problem from classical mathematics - in this case the word problem for semigroups. The algebraically minded reader will readily note that what Post calls a Thue system determines a homomorphism on the free semigroup with the appropriate generators. Post thus constructs a particular semigroup for which it is recursively unsolvable to determine whether a given pair of elements of the free semigroup are mapped onto the same element by the homomorphism.

An independent proof of this result was given by A. A. Markov (C. R. (Doklady) Acad. Sci. U. S. S. R. (n. s.) 55(1947), pp. 583-586.)

The appendix to this paper has already been mentioned in the editorial remarks on p. 115 preceding Turing's paper, "On Computable Numbers, with an Application to the Entscheidungsproblem".

Emil Post

RECURSIVE UNSOLVABILITY OF A PROBLEM OF THUE

Alonzo Church suggested to the writer that a certain problem of Thue [6][1] might be proved unsolvable by the methods of [5]. We proceed to prove the problem recursively unsolvable, that is, unsolvable in the sense of Church [1], but by a method meeting the special needs of the problem.

Thue's (general) problem is the following. Given a finite set of symbols a_1, $a_2, \cdots, a_\mu$, we consider arbitrary *strings* (Zeichenreihen) on those symbols, that is, rows of symbols each of which is in the given set. Null strings are included. We further have given a finite set of pairs of corresponding strings on the a_i's, (A_1, B_1), (A_2, B_2), $\cdots$, (A_n, B_n). A string R is said to be a *substring* of a string S if S can be written in the form URV, that is, S consists of the letters, in order of occurrence, of some string U, followed by the letters of R, followed by the letters of some string V. Strings P and Q are then said to be *similar* if Q can be obtained from P by replacing a substring A_i or B_i of P by its correspondent B_i, A_i. Clearly, if P and Q are similar, Q and P are similar. Finally, P and Q are said to be *equivalent* if there is a finite set $R_1, R_2, \cdots, R_r$ of strings on $a_1, \cdots, a_\mu$ such that in the sequence of strings $P, R_1, R_2, \cdots, R_r, Q$ each string except the last is similar to the following string. It is readily seen that this relation between strings on $a_1, \cdots, a_\mu$, is indeed an equivalence relation. Thue's problem is then the problem of determining for arbitrarily given strings A, B on $a_1, \cdots, a_\mu$ whether, or no, A and B are equivalent.

This problem, at least for the writer, is more readily placed if it is restated in terms of a special form of the canonical systems of [3]. In that notation, strings C and D are similar if D can be obtained from C by applying to C one of the following operations:

$$PA_iQ \text{ produces } PB_iQ,\ PB_iQ \text{ produces } PA_iQ,\ i = 1, 2, \cdots, n. \tag{1}$$

In these operations the operational variables P, Q represent arbitrary strings. Strings A and B will then be equivalent if B can be obtained from A by starting with A, and applying in turn a finite sequence of operations (1). That is, A and B are equivalent if B is an assertion in the "canonical system"[2] with primitive assertion A and operations (1). Thue's general problem thus becomes the decision problem for the class of all canonical systems of this "Thue type."

This general problem could easily be proved recursively unsolvable if, instead of the pair of operations for each i of (1), we merely had the first operation of each pair.[3] In fact, by direct methods such as those of [3], we easily reduce the decision problem of an arbitrary "normal system" [3] to the decision problem of such a system of "semi-Thue type," the known recursive unsolvability of the

Reprinted from THE JOURNAL OF SYMBOLIC LOGIC, vol. 12 (1947) pp. 1-11.

[1] Numbers in brackets refer to the bibliography at the end of the paper.

[2] Null assertions, however, now being allowed.

[3] That is, using the language of propositions instead of operations, if we merely had an implication where (1) has an equivalence.

decision problem for the class of all normal systems then, no doubt, leading to the recursive unsolvability of the decision problem for the class of all semi-Thue systems. The crux of our method for handling the Thue systems themselves is to find such a reduction of a known unsolvable problem to a system of semi-Thue type that when, for each i, the second of the two operations in (1) is added to the semi-Thue system, no new assertions are thereby added to the system. The known unsolvable problem is thus reduced to the resulting Thue system, as desired. Such a reduction turns out to be possible for a certain unsolvable problem arising in the theory of Turing machines.

We shall adopt the following formulation of a Turing machine [7].[4] A two-way infinite linear tape is provided, ruled off into squares. Time is a one-way infinite sequence of discrete moments. A square will either be blank, or have at most one symbol printed upon it. At any moment the machine "scans" one of the squares. At such a moment the machine is capable of performing one of the following atomic acts: moving one square to the left, moving one square to the right, printing on the scanned square one of a given finite number of symbols $S_1, \cdots, S_m$, or a blank. Following Turing, we take "printing" here to mean "overprinting," that is, the letter or blank printed replaces any letter that may have been on the scanned square. Printing a blank is then equivalent to erasing, when the scanned square is not blank. The machine, furthermore, is capable of assuming but a finite number of internal states, internal configurations or m-configurations with Turing, $q_1, q_2, \cdots, q_R$. At any moment, the letter or blank on the scanned square together with the internal configuration of the machine determines the atomic act to be performed by the machine and the new internal configuration of the machine, or else, the machine then stops. At the initial moment a finite, possibly null, number of squares have S's printed on them, the machine scans a particular square and has a particular internal configuration.

Symbolically, the machine may be given as follows. Let S_o be used to represent a blank square. For the start of the action of the machine we need only consider the smallest unbroken piece of the tape containing the initially marked squares and the scanned square, replace these squares by their markings, or by S_o if blank, and insert the symbol q_{i_1} of the initial internal configuration prior to the S of the scanned square to yield the representation

$$S_{j_1}S_{j_2}\cdots S_{j_{k-1}}q_{i_1}S_{j_k}\cdots S_{j_K}. \tag{2}$$

A finite number of quadruplets of symbols of the three forms,

$$q_iS_jLq_l, \qquad q_iS_jRq_l, \qquad q_iS_jS_kq_l, \tag{3}$$

will then determine the behavior of the machine. Here, q_i and S_j represent the internal configuration of the machine, and the symbol or blank on the scanned

[4] Apart from the Turing convention, discussed in the appendix, this differs from Turing's formulation of an automatic machine in the nature of the tape, and in Turing's use, in his standard form [7, p. 240]$^{i}_{,}$ of the composite operation "print and move" where we just have "move." A number of comparisons with [2] will occur to a reader of that note.

^{i}126

square, at any moment; L, R, or S_k the correspondingly determined atomic act of motion left, right, or printing of S_k ; q_l the determined new internal configuration of the machine. It is fundamental that the pairs q_i , S_j of the several quadruplets are distinct, for they are to determine uniquely the consequent behavior of the machine. The machine will then continue acting deterministically from the initial moment on unless, and until, a q_iS_j is reached for which there is no quadruplet (3), in which case it will stop.

We can now readily set up a semi-Thue system whose assertions will represent the successive states of the tape, and the relation of the machine thereto, as (2) represented these at the initial moment. However, for simplicity, the portion of the tape represented, while including the marked squares and the scanned square, need not now be the smallest such portion.[5] Because of the particular needs of the semi-Thue form, we introduce a new symbol h. Each assertion of the semi-Thue system will then be in the form hPh with P free from h. If A represents the string (2) of S's and one q, the initial assertion of the semi-Thue system will be hAh. For each quadruplet (3) corresponding to moving one square to the left, we introduce the operations

$$PS_nq_iS_jQ \text{ produces } Pq_lS_nS_jQ,\ n = 0, 1, \cdots, m, \tag{4}$$

$$Phq_iS_jQ \quad \text{produces } Phq_lS_oS_jQ. \tag{5}$$

Note that (4) takes care of all cases where the scanned square is not the leftmost square of the part of the tape represented at the given moment, (5) where the scanned square is that leftmost square. Due to the form hPh of all assertions of the system, when (5) is applicable, the h of the premise thereof must be the leftmost of these two h's, so that P will be identified with the null string. The S_o of the conclusion then takes care of the necessary extension of the portion of the tape represented when the motion is one square to the left of that portion. Likewise, for each quadruplet (3) corresponding to motion of one square to the right we introduce

$$Pq_iS_jS_nQ \text{ produces } PS_jq_lS_nQ,\ n = 0, 1, \cdots, m, \tag{6}$$

$$Pq_iS_jhQ \quad \text{produces } PS_jq_lS_ohQ; \tag{7}$$

while for each quadruplet (3) corresponding to the printing of S_k over the scanned square we have

$$Pq_iS_jQ \text{ produces } Pq_lS_kQ. \tag{8}$$

Clearly both premise and conclusion of each operation thus introduced is of the form PBQ with fixed B, so that we do thus have a semi-Thue system. An obvious induction yields the form hPh with P free from h for each assertion. Likewise, each assertion has one and only one q therein. Finally, it is readily

[5] It could be made the smallest such portion by using more operations. There would then be a 1–1 correspondence between the intrinsic states of tape versus machine and the representations thereof.

verified from the deterministic character of the Turing machine, and from the forms of the above operations, that at most one of these operations is applicable to any string having no more than one occurrence of a q therein, and then in only one way.

The unsolvable problem that is to yield the unsolvability of the problem of Thue would seem to be furnished by the following result of Turing's [7, p. 248]:[ii] "There can be no machine $\mathfrak{E}$ which, when supplied with the S.D of an arbitrary machine $\mathfrak{M}$, will determine whether $\mathfrak{M}$ ever prints a given symbol (0 say)." There are, however, difficulties in using this result as given due to peculiarities of Turing's development. (The matter is discussed in the appendix.) We therefore proceed independently of Turing as follows.

We start with the known recursive unsolvability of the decision problem for the class of normal systems on two letters a, b.[6] It suffices here to think of this problem as consisting of a class of questions, each question Q being symbolized by a string on a given finite set of letters. By methods such as those used by Turing in setting up his universal computing machine [7], we then set up the quadruplets (3) of a fixed Turing machine with certain letters S_1, S_2, $\cdots$, S_m, and internal configurations q_1, q_2, $\cdots$, q_R, and give an effective method for translating each question Q into a Q' of form (2) to serve as the initial state of tape versus machine, the construction being such that the following is true. The answer to question Q is yes, or no, according as the constructed machine, when applied to Q', does, or does not, in the course of its operation print a certain fixed letter S_p. This letter S_p is not present in the Q' of any Q. Since such methods are fully exploited by Turing in [7], we do not give the details of this construction.[7]

Now, given Q, form the semi-Thue system T' with initial assertion $hQ'h$, and operations (4)–(8) corresponding to this Turing machine. Then, the answer to Q is yes, or no, according as some assertion of T' involves the letter S_p, or no assertion involves that letter. We now modify T' as follows. Delete all operations in T' such that the S_j of the premise, the symbol on the scanned square of the Turing machine, is S_p. Since, when S_p is first printed, it can appear so only as the S_k of (8), and thus would be the S_j of a next operation, the deductive processes of the semi-Thue system will now stop the first time S_p appears in an assertion. We now add operations which, in deterministic fashion, will erase all of this assertion except for the two h's and the q, while changing this q. For this purpose, we introduce two new "internal configurations"

[6] See [5, footnote 2]. The specific form of this problem, however, need not be known by the reader for an understanding of the present argument.

[7] This work was carried through before the definitive study of Turing's paper [7], referred to in the appendix, was made. As a result, some differences of method appear. A minor difference is that where Turing uses the method of "marking" a sequence of symbols [7, p. 235] to distinguish it, we introduce the *effect* of movable physical markers; two, indeed, suffice. A major difference is that instead of the m-configuration functions of Turing's skeleton tables [7, p. 236], we introduce a symbolism and technique based on the concept of a subset of directions of a given set of directions. Both differences were suggested by [2]. They may, perhaps, better be exploited in a more general setting.[i]

[i] For page references in this footnote, the reader should subtract 114 from the page given in order to obtain the corresponding page in this anthology.

[ii] 134

q_{R+1} and q_{R+2}. We further alter the operations of T' by changing the q_l of each operation (8) for which S_k is S_p to q_{R+1}, and add the following operations:

$$PS_n q_{R+1} Q \textit{ produces } Pq_{R+1}Q,\ n = 0, 1, \cdots, m;\ n \neq p. \quad (9)$$

$$Phq_{R+1}Q \quad \textit{produces } Phq_{R+2}Q. \quad (10)$$

$$Pq_{R+2}S_nQ \textit{ produces } Pq_{R+2}Q,\ n = 0, 1, \cdots, m. \quad (11)$$

Note that as a result of the previous changes, when S_p first appears in an assertion, the q therein is q_{R+1}. Operations (9) then serve to erase the S's of that assertion to the left of q_{R+1}, (10) then changes q_{R+1} to q_{R+2}, (11) erases the S's to the right of q_{R+2}. Finally, therefore, the assertion becomes $hq_{R+2}h$, to which no further operation is applicable. Call the resulting semi-Thue system T''. Clearly, for T'' it is also true that at most one of its operations is applicable to any string having no more than one occurrence of a q therein, and then in only one way. It follows that the answer to Q is yes, or no, according as $hq_{R+2}h$ is, or is not, an assertion in T'', the operations of T'' operating one by one in deterministic fashion, and, in the former case, terminating in $hq_{R+2}h$.

The proof of the reducibility of our initial unsolvable problem to the problem of Thue essentially becomes the proof of the following two lemmas.[8] By the *inverse* of an operation of the form *PAQ produces PBQ* we shall mean the operation *PBQ produces PAQ*. Let T''' be the semi-Thue system with primitive assertion $hq_{R+2}h$ and operations the inverses of those of T''. We then have:

LEMMA I. The primitive assertion $hq_{R+2}h$ of T''' is an assertion of T'' when, and only when, the primitive assertion $hQ'h$ of T'' is an assertion of T'''.

Proof. D is a result of applying "*PAQ produces PBQ*" to C when, and only when, C is a result of applying the inverse operation "*PBQ produces PAQ*" to D. For both statements are equivalent to the existence of strings P and Q such that $PAQ = C$, $PBQ = D$. If, then, operations $O_1, O_2, \cdots, O_n$ of T'' lead from its primitive assertion $hQ'h$ through assertions $C_1, C_2, \cdots, C_{n-1}$ to the assertion $hq_{R+2}h$, the inverses of these operations, all in T''', will in reverse order lead from $hq_{R+2}h$, the primitive assertion of T''', through $C_{n-1}, \cdots, C_2, C_1$ to $hQ'h$; and conversely.

As a result of Lemma I, the answer to question Q is yes, or no, according as $hQ'h$ is, or is not, an assertion of T'''. Note that while the initial assertion of T'' depended on Q, T''' is the same for all Q's. Now let T be the Thue system obtained from the semi-Thue system T''' by adding to the latter the inverse of each of its operations. We then have:

LEMMA II. The class of assertions of T is identical with the class of assertions of T'''.

Proof. Each assertion of T''' is, of course, an assertion of T. For the converse, let operations $O_1, O_2, \cdots, O_n$ of T lead from its primitive assertion $hq_{R+2}h$, through assertions $C_1, C_2, \cdots, C_{n-1}$, to an assertion C of T. If n is zero, C is $hq_{R+2}h$, and hence an assertion of T'''. Otherwise, note that the operations of T, being those of T''' and their inverses, are the combined operations of T'''

[8] These lemmas can be made more general.

and of T''. Now we saw that no operation of T'' is applicable to $hq_{R+2}h$, the deductive processes of T'' terminating in $hq_{R+2}h$ if leading thereto. Hence, operation O_1 must be in T'''. Assume O_{m+1} to be the first O not in T''', and hence in T''. Since O_m is in T''', its inverse is in T''. As O_m operates on $C_{m-1}(hq_{R+2}h$, if m is one) to yield C_m, the inverse of O_m is applicable to C_m yielding C_{m-1}. That is, both the inverse of O_m, and O_{m+1}, are operations in T'' applicable to C_m, and yielding C_{m-1} and C_{m+1} (C, if $m + 1 = n$) respectively. Now, since the premise and conclusion of each operation in T, and the primitive assertion of T, have exactly one occurrence of a q therein, the same is true of every assertion of T. But we saw that at most one operation of T'' is applicable to a string with a single occurrence of a q therein, and then in only one way. It follows that O_{m+1} is in fact that inverse of O_m, and hence C_{m+1} is C_{m-1} all over again. We may therefore delete operations O_m and O_{m+1} from the given sequence of operations, and still have a sequence of operations leading from $hq_{R+2}h$ to C. By repeating this process, we finally obtain such a sequence of O's with each O in T'''. The arbitrary assertion C of T is therefore also an assertion of T'''.[9]

Hence, $hQ'h$ is an assertion of T''' when and only when it is an assertion of T. Finally, then, the answer to Q is yes, or no, according as $hQ'h$ is, or is not, an assertion of the Thue system T. In terms of the language of Thue, we have then a fixed set of pairs of strings $(A_1, B_1), \cdots, (A_n, B_n)$ leading to a definition of equivalence of strings such that the answer to Q is yes, or no, according as $hQ'h$ is, or is not, equivalent to the fixed string $hq_{R+2}h$.[10] Certainly, then, a solution of the problem of Thue in its full generality would thus lead to a solution of the "decision problem for the class of normal systems on a, b." By the use of Gödel representations, the recursive unsolvability of the latter problem then easily leads to the recursive unsolvability of the problem of Thue.

A few concluding remarks may be in order. The methods of [5], and of the present paper, do have something in common, a something we may call *the method of the irrelevant modification*. Once an unsolvable problem has been obtained by a *reductio ad absurdum* argument based on the definition of solvability, the usual method of proving a new problem unsolvable is to reduce a known unsolvable problem to this given problem. In the method of the irrelevant modification, the known unsolvable problem is reduced to a problem which on modification becomes the given problem, while that modification does not affect the answers to the individual questions. In [5] the modification is a simplification, the existence of a solution of a certain string equation subject to

[9] Briefly, then, the effect of operations of T'' on deductive processes of T is to unravel work done by operations of T'''. Note that while the deductive processes of T'' give rise to a single sequence of assertions starting with $hQ'h$ and terminating in $hq_{R+2}h$, if leading thereto, the deductive processes of T''' give rise to a tree of assertions, elements not necessarily distinct, stemming from $hq_{R+2}h$, and containing the above sequence in reverse when that sequence terminates in $hq_{R+2}h$.

[10] By the method of the next to the last paragraph of [5], this definition of equivalence could be transformed into a definition of equivalence for strings on the two letters a, b. We have not paused to prove the recursive unsolvability of the resulting special case of the problem of Thue.

certain length conditions being equivalent to the mere existence of a solution of that string equation. In the present paper the modification is a complication, the answer to $hQ'h$ being or not being an assertion of T''' being unaffected by adding to the operations of T''' their inverses.

The writer has often felt that the multiplicity of equivalent formulations of recursiveness has been a deterrent to the general promulgation of this discipline. Yet, the writer's normal systems naturally lead to the unsolvable problem of [5], while the deterministic character of the Turing machine is basic to the above unsolvability proof. From this point of view, the several formulations of recursiveness are so many different instruments for tackling new unsolvability proofs.

Though we have not paused to verify this formally, it seems rather obvious that when the problem of [5], and the problem of Thue, are translated via positive integers as suggested in [4], they become decision problems of recursively enumerable sets of positive integers of the same degree of unsolvability as the complete set K, at worst, with respect to many-one reducibility [1]. This indicates how far practice lags behind theory in this field.

Appendix. The following critique of Turing's "computability" paper [7] concerns only pp. 230–248[ii] thereof. We have checked the work through the construction of the "universal computing machine" in detail;[11] but the proofs of the two theorems in the section following are there given in outline only, and we have not supplied the formal details. *We have therefore also left in intuitive form the proofs of the statements on recursiveness, and alternative procedures, we make below.*

Turing's definition of an arbitrary machine is not completely given in his paper, and, at a number of points, has to be inferred from his development. In the first instance his machine is a "computing machine" for obtaining the successive digits of a real number in dyadic notation, and, in that case, starts operating on a blank tape. Where explicitly stated, however, the machine may

[11] One major correction is needed. To the instructions for $\mathfrak{con}_1(\mathfrak{C}, \alpha)$ p. 244, add the line: None $PD, R, P\alpha, R, R, R$ $\mathfrak{C}$. This is needed to introduce the representation D of the blank scanned square when, as at the beginning of the action of the machine, or due to motion right beyond the rightmost previous point, the complete configuration ends with a q, and thus make the $\mathfrak{kmp}$ of p. 244 correct. We may also note the following minor slips and misprints in pp. 230–248. Page 236, to the instructions for $\mathfrak{f}(\mathfrak{C}, \mathfrak{B}, \alpha)$ add the line: None L $\mathfrak{f}(\mathfrak{C}, \mathfrak{B}, \alpha)$; p. 240 and p. 241, the S.D should begin, but not end, with a semicolon; p. 242, omit the first D in (C_2); p. 243, last paragraph, add ":" to the first list of symbols; pp. 244–246, replace $\mathfrak{g}$ by $\mathfrak{q}$; p. 245, in the instruction for $\mathfrak{mk}$, $\mathfrak{mk}$ should be $\mathfrak{mk}_1$; p. 215, in the second instruction for $\mathfrak{sim}_2$, replace the first R by L; p. 245, in the first instruction for $\mathfrak{sh}_2$, replace $\mathfrak{sh}_2$ by $\mathfrak{sh}_3$. A reader of the paper will be helped by keeping in mind that the "examples" of pages 236–239 are really parts of the table for the universal computing machine, and accomplish what they are said to accomplish not for all possible printings on the tape, but for certain ones that include printings arising from the action of the universal computing machine. In particular, the tape has ə printed on its first two squares, the occurrence of two consecutive blank squares insures all squares to the right thereof being blank, and, usually, symbols referred to are on "F-squares," and obey the convention of p. 235.[i]

[i] For page references in this footnote, the reader should subtract 114 from the page given in order to obtain the corresponding page in this anthology.

[ii] 116–134

start operating on a tape previously marked. From Turing's frequent references to the beginning of the tape, and the way his universal computing machine treats motion left, we gather that, unlike our tape, this tape is a one-way infinite affair going right from an initial square.

Primarily as a matter of practice, Turing makes his machines satisfy the following convention. Starting with the first square, alternate squares are called *F*-squares, the rest, *E*-squares. In its action the machine then never directs motion left when it is scanning the initial square, never orders the erasure, or change, of a symbol on an *F*-square, never orders the printing of a symbol on a blank *F*-square if the previous *F*-square is blank and, in the case of a computing machine, never orders the printing of 0 or 1 on an *E*-square. This convention is very useful in practice. However the actual performance, described below, of the universal computing machine, coupled with Turing's proof of the second of the two theorems referred to above, strongly suggests that Turing makes this convention part of the definition of an arbitrary machine. We shall distinguish between a Turing machine and a Turing convention-machine.

By a uniform method of representation, Turing represents the set of instructions, corresponding to our quadruplets,[12] which determine the behavior of a machine by a single string on seven letters called the standard description (S.D) of the machine. With the letters replaced by numerals, the S.D of a machine is considered the arabic representation of a positive integer called the description number (D.N) of the machine. If our critique is correct, a machine is said to be circle-free if it is a Turing computing convention-machine which prints an infinite number of 0's and 1's.[13] And the two theorems of Turing's in question are really the following. There is no Turing convention-machine which, when supplied with an arbitrary positive integer *n*, will determine whether *n* is the D.N of a Turing computing convention-machine that is circle-free. There is no Turing convention-machine which, when supplied with an arbitrary positive integer *n*, will determine whether *n* is the D.N of a Turing computing convention-machine that ever prints a given symbol (0 say).[14]

In view of [8], these "no machine" results are no doubt equivalent to the re-

---

[12] Our quadruplets are quintuplets in the Turing development. That is, where our standard instruction orders either a printing (overprinting) or motion, left or right, Turing's standard instruction always orders a printing and a motion, right, left, or none. Turing's method has certain technical advantages, but complicates theory by introducing an irrelevant "printing" of a symbol each time that symbol is merely passed over.

[13] "Genuinely prints," that is, a genuine printing being a printing in an empty square. See the previous footnote.

[14] Turing in each case refers to the S.D of a machine being supplied. But the proof of the first theorem, and the second theorem depends on the first, shows that it is really a positive integer *n* that is supplied. Turing's proof of the second theorem is unusual in that while it uses the unsolvability result of the first theorem, it does not "reduce" [4] the problem of the first theorem to that of the second. In fact, the first problem is almost surely of "higher degree of unsolvability" [4] than the second, in which case it could not be "reduced" to the second. Despite appearances, that second unsolvability proof, like the first, is a *reductio ad absurdum* proof based on the definition of unsolvability, at the conclusion of which, the first result is used.

cursive unsolvability of the corresponding problems.[15] But both of these problems are infected by the spurious Turing convention. Actually, the set of n's which are D.N's of Turing computing machines as such is recursive, and hence the condition that n be a D.N offers no difficulty. But, while the set of n's which are not D.N's of convention-machines is recursively enumerable, the complement of that set, that is, the set of n's which are D.N's of convention-machines, is not recursively enumerable. As a result, in both of the above problems, neither the set of n's for which the question posed has the answer yes, nor the set for which the answer is no, is recursively enumerable.

This would remain true for the first problem even apart from the convention condition. But the second would then become that simplest type of unsolvable problem, the decision problem of a non-recursive recursively enumerable set of positive integers [4]. For the set of n's that are D.N's of unrestricted Turing computing machines printing 0, say, is recursively enumerable, though its complement is not. The Turing convention therefore prevents the early appearance of this simplest type of unsolvable problem.

It likewise prevents the use of Turing's second theorem in the above unsolvability proof of the problem of Thue. For in attempting to reduce the problem of Turing's second theorem to the problem of Thue, when an n leads to a Thue question for which the answer is yes, we would still have to determine whether n is the D.N of a Turing convention-machine before the answer to the question posed by n can be given, and that determination cannot be made recursively for arbitrary n. If, however, we could replace the Turing convention by a convention that is recursive, the application to the problem of Thue could be made. An analysis of what Turing's universal computing machine accomplishes when applied to an arbitrary machine reveals that this can be done.

The universal computing machine was designed so that when applied to the S.D of an arbitrary computing machine it would yield the same sequence of 0's and 1's as the computing machine as well as, and through the intervention of, the successive "complete configurations"—representations of the successive states of tape versus machine—yielded by the computing machine. This it does for a Turing convention-machine.[16] For an arbitrary machine, we have to interpret a direction of motion left at a time when the initial square of the tape is scanned as meaning no motion.[17] The universal computing machine will then yield again the correct complete configurations generated by the given machine. But *the space sequence of* 0's *and* 1's *printed by the universal computing machine will now be identical with the time sequence of those printings of* 0's *and* 1's *by the given machine that are made in empty squares.* If, now, instead of Turing's

[15] Our experience with proving that "normal unsolvability" in a sense implicit in [3] is equivalent to unsolvability in the sense of Church [1], at least when the set of questions is recursive, suggests that a fair amount of additional labor would here be involved. That is probably our chief reason for making our proof of the recursive unsolvability of the problem of Thue independent of Turing's development.

[16] Granted the corrections given in footnote 11.

[17] This modification of the concept of motion left is assumed throughout the rest of the discussion, with the exception of the last paragraph.

convention we introduce the convention that the instructions defining the machine never order the printing of a 0 or 1 except when the scanned square is empty, or 0, 1 respectively, and never order the erasure of a 0 or 1, Turing's arguments again can be carried through. And this "(0, 1) convention," being recursive, allows the application to the problem of Thue to be made.[18] Note that if a machine is in fact a Turing convention-machine, we could strike out any direction thereof which contradicts the (0, 1) convention without altering the behavior of the machine, and thus obtain a (0, 1) convention-machine. But a (0, 1) convention-machine need not satisfy the Turing convention. However, by replacing each internal-configuration q_i of a machine by a pair q_i, q_i' to correspond to the scanned square being an F- or an E-square respectively, and modifying printing on an F-square to include testing the preceding F-square for being blank, we can obtain a "(q, q') convention" which is again recursive, and usable both for Turing's arguments and the problem of Thue, and has the property of, in a sense, being equivalent to the Turing convention. That is, every (q, q') convention-machine is a Turing convention-machine, while the directions of every Turing convention-machine can be recursively modified to yield a (q, q') convention-machine whose operation yields the same time sequence and spatial arrangement of printings and erasures as does the given machine, except for reprintings of the same symbol in a given square.

These changes in the Turing convention, while preserving the general outline of Turing's development and at the same admitting of the application to the problem of Thue, would at least require a complete redoing of the formal work of the proof of the second Turing theorem. On the other hand, very little added formal work would be required if the following changes are made in the Turing argument itself, though there would still remain the need of extending the equivalence proof of [8] to the concept of unsolvability. By using the above result on the performance of the universal computing machine when applied to the S.D of an arbitrary machine, we see that Turing's proof of his first theorem, whatever the formal counterpart thereof is, yields the following theorem. There is no Turing convention-machine which, when supplied with an arbitrary positive integer n, will determine whether n is the D.N of an arbitrary Turing machine that prints 0's and 1's in empty squares infinitely often. Now given an arbitrary positive integer n, if that n is the D.N of a Turing machine $\mathfrak{M}$, apply the universal computing machine to the S.D of $\mathfrak{M}$ to obtain a machine $\mathfrak{M}^*$. Since $\mathfrak{M}^*$ satisfies the Turing convention, whatever Turing's formal proof of his second theorem is, it will be usable intact in the present proof, and, via the new form of his first theorem, will yield the following usable result. There is no machine which, when supplied with an arbitrary positive integer n, will

[18] So far as recursiveness is concerned, the distinction between the Turing convention and the (0, 1) convention is that the former concerns the history of the machine in action, the latter only the instructions defining the machine. Likewise, despite appearances, the later (q, q') convention.

determine whether n is the D.N of an arbitrary Turing machine that ever prints a given symbol (0 say).[19]

These alternative procedures assume that Turing's universal computing machine is retained. However, in view of the above discussion, it seems to the writer that Turing's preoccupation with computable numbers has marred his entire development of the Turing machine. We therefore suggest a redevelopment of the Turing machine based on the formulation given in the body of the present paper. This could easily include computable numbers by defining a computable sequence of 0's and 1's as the *time sequence* of printings of 0's and 1's by an arbitrary Turing machine, provided there are an infinite number of such printings. By adding to Turing's complete configuration a representation of the act last performed, a few changes in Turing's method would yield a universal computing machine which would transform such a time sequence into a space sequence. Turing's convention would be followed as a matter of useful practice in setting up this, and other, particular machines. But it would not infect the theory of arbitrary Turing machines.

References

[1] Alonzo Church, *An unsolvable problem of elementary number theory*, ***American journal of mathematics***, vol. 58 (1936), pp. 345–363.[a]

[2] Emil L. Post, *Finite combinatory processes—formulation 1*, this Journal, vol. 1 (1936), pp. 103–105.[a]

[3] Emil L. Post, *Formal reductions of the general combinatorial decision problem*, ***American journal of mathematics***, vol. 65 (1943), pp. 197–215.

[4] Emil L. Post, *Recursively enumerable sets of positive integers and their decision problems*, ***Bulletin of the American Mathematical Society***, vol. 50 (1944), pp. 284–316.[a]

[5] Emil L. Post, *A variant of a recursively unsolvable problem*, ibid., vol. 52 (1946), pp. 264–268.

[6] Axel Thue, *Probleme über Veränderungen von Zeichenreihen nach gegebenen Regeln*, ***Skrifter utgit av Videnskapsselskapet i Kristiania***, I. Matematisk-naturvidenskabelig klasse 1914, no. 10 (1914), 34 pp.

[7] A. M. Turing, *On computable numbers, with an application to the Entscheidungsproblem*, ***Proceedings of the London Mathematical Society***, ser. 2 vol. 42 (1937), pp. 230–265.[a]

[8] A. M. Turing, *Computability and λ-definability*, this Journal, vol. 2 (1937), pp. 153–163.

[19] It is here assumed that the suggested extension of [8] includes a proof to the effect that the existence of an arbitrary Turing machine for solving a given problem is equivalent to the existence of a Turing convention-machine for solving that problem.

RECURSIVELY ENUMERABLE SETS OF POSITIVE INTEGERS AND THEIR DECISION PROBLEMS

This paper initiated the classification theory of recursively enumerable sets. Because of its clear informal style, the early part of this paper can be recommended as an introduction to the theory of recursive functions and to its relation to Gödel's incompleteness theorem. The later part contains ingenious constructions and raises the question which came to be known as Post's problem: Can there be two recursively enumerable but non-recursive sets such that the first is recursive relative to the second (i.e. using the second as an "oracle") but not vice versa? This question was answered much later in the affirmative by Richard Friedberg (Proceedings of the National Academy of Sciences (U. S. A.), vol. 43(1957), pp. 236-238) and independently by A. A. Mucnik (Doklady Akademii Nauk S. S. S. R., n.s. vol. 108(1956), pp. 194-197).

Emil Post

RECURSIVELY ENUMERABLE SETS OF POSITIVE INTEGERS AND THEIR DECISION PROBLEMS

Introduction. Recent developments of symbolic logic have considerable importance for mathematics both with respect to its philosophy and practice. That mathematicians generally are oblivious to the importance of this work of Gödel, Church, Turing, Kleene, Rosser and others as it affects the subject of their own interest is in part due to the forbidding, diverse and alien formalisms in which this work is embodied. Yet, without such formalism, this pioneering work would lose most of its cogency. But apart from the question of importance, these formalisms bring to mathematics a new and precise mathematical concept, that of the general recursive function of Herbrand-Gödel-Kleene, or its proved equivalents in the developments of Church and Turing.[1] It is the purpose of this lecture to demonstrate by example that this concept admits of development into a mathematical theory much as the group concept has been developed into a theory of groups. Moreover, that stripped of its formalism, such a theory admits of an intuitive development which can be followed, if not indeed pursued, by a mathematician, layman though he be in this formal field. It is this intuitive development of a very limited portion of a sub-theory of the hoped for general theory that we present in this lecture. We must emphasize that, with a few exceptions explicitly so noted, we have obtained formal proofs of all the consequently mathematical theorems here developed informally. Yet the real mathematics involved must lie in the informal development. For in every instance the informal "proof" was first obtained; and once gotten, transforming it into the formal proof turned out to be a routine chore.[2]

We shall not here reproduce the formal definition of *recursive function of positive integers*. A simple example of such a function is an

[1] For "general recursive function" see [9] ([8] a prerequisite), [12] and [11]; for Church's "λ-defineability," [1] and [6]; for Turing's "computability," [24] and the writer's related [18]. To this may be added the writer's method of "canonical systems and normal sets" [19]. See pp. 39–42 and bibliography of [6] for a survey of the literature and further references. Numbers in brackets refer to the bibliography at the end of the paper.

[2] Our present formal proofs, while complete, will require drastic systematization and condensation prior to publication.

arbitrary polynomial $P(x_1, x_2, \cdots, x_n)$, with say non-negative integral coefficients, and not identically zero. If the x's are assigned arbitrary positive integral values expressed, for example, in the arabic notation, the algorithms for addition and multiplication in that notation enable us to calculate the corresponding positive integral value of the polynomial. That is, $P(x_1, x_2, \cdots, x_n)$ is an *effectively calculable function of positive integers*. The importance of the technical concept recursive function derives from the overwhelming evidence that it is coextensive with the intuitive concept effectively calculable function.[3]

A set of positive integers is said to be *recursively enumerable* if there is a recursive function $f(x)$ of one positive integral variable whose values, for positive integral values of x, constitute the given set. The sequence $f(1), f(2), f(3), \cdots$ is then said to be a *recursive enumeration* of the set. The corresponding intuitive concept is that of an *effectively enumerable* set of positive integers. To prepare us in part for our intuitive approach, consider the following three examples of recursively enumerable sets of positive integers.

(a): $1^2, 2^2, 3^2, \cdots.$

(b): $1, 2, 2^{1+2}, 2^{1+2+2^{1+2}}, \cdots.$

(c):
$$\begin{array}{l} 1^2, 2^2, 3^2, \cdots \\ 1^3, 2^3, 3^3, \cdots \\ 1^4, 2^4, 3^4, \cdots \\ \cdot \quad \cdot \quad \cdot \quad \cdots \\ \cdot \quad \cdot \quad \cdot \quad \cdots \\ \cdot \quad \cdot \quad \cdot \quad \cdots . \end{array}$$

In the first example, the set is given by a recursive enumeration thereof via the recursive function x^2. In the second example, the set is generated in a linear sequence, each new element being effectively obtained from the elements previously generated, in this case by raising 2 to the power the sum of the preceding elements. The set is effectively enumerable, since the nth element of the sequence can be found, given n, by regenerating the sequence through its first n elements. In the third example, we rather imagine the positive integers $1, 2, 3, \cdots$ generated in their natural order, and, as each positive integer n is generated, a corresponding process set up which generates $n^2, n^3, n^4, \cdots$, all these to be in the set. Actually, the standard method for proving that an enumerable set of enumerable sets is enumerable yields an effective enumeration of the set.

---

[3] See Kleene [**13**, footnote 2]. In the present paper, "recursive function" means "general recursive function."

Several more examples would have to be given to convey the writer's concept of a *generated set*, in the present instance of positive integers. Suffice it to say that each element of the set is at some time written down, and earmarked as belonging to the set, as a result of predetermined effective processes. It is understood that once an element is placed in the set, it stays there. The writer elsewhere has referred to a generalization which may be restated *every generated set of positive integers is recursively enumerable.*[4] For comparison purposes this may be resolved into the two statements: every generated set is effectively enumerable, every effectively enumerable set of positive integers is recursively enumerable. The first of these statements is applicable to generated sets of arbitrary symbolic expressions; their converses are immediately seen to be true. We shall find the above concept and generalization very useful in our intuitive development. But while we shall frequently say, explicitly or implicitly, "set so and so of positive integers is a generated, and hence recursively enumerable set," as far as the present enterprise is concerned that is merely to mean "the set has intuitively been shown to be a generated set; it can indeed be proved to be recursively enumerable." Likewise for other identifications of informal concepts with corresponding mathematically defined formal concepts.

At a few points in our informal development we have to lean upon the formal development. The latter is actually yet another formalism, due to the writer [**19**] but proved completely equivalent to that of general recursive function. It will suffice to give the equivalent of "recursively enumerable set of positive integers" in this development.

A positive integer n is represented in the most primitive fashion by a succession $11 \cdots 1$ of n strokes. For working purposes, we introduce the letter b, and consider "strings" of 1's and b's such as $11b1bb1$. An operation on such strings such as "*b1bP produces P1bb1*" we term a normal operation. This particular normal operation is applicable only to strings starting with $b1b$, and the derived string is then obtained from the given string by first removing the initial $b1b$, and then tacking on $1bb1$ at the end. Thus $b1bb$ becomes $b1bb1$. "*gP produces Pg′*" is the form of an arbitrary normal operation. A system in normal form, or normal system, is given by an initial string A of 1's and b's, and a finite set of normal operations "g_iP *produces* Pg_i'," $i=1, 2, \cdots, \mu$. The derived strings of the system are A and all strings obtainable from A be repeated applications of the μ normal

[4] See [**19**, p. 201 and footnote 18]. In this connection note Kleene's use of the word "Thesis" in [**14**, p. 60].ⁱ We still feel that, ultimately, "Law" will best describe the situation [**18**].

ⁱ 274

operations. Each normal system uniquely defines a set, possibly null, of positive integers, namely the integers represented by those derived strings which are strings of 1's only. It can then be proved that every recursively enumerable set of positive integers is the set of positive integers defined by some normal system, and conversely.[5] *We here, as below, arbitrarily extend the concept recursively enumerable set to include the null set.*

By the *basis B* of a normal system, and of the recursively enumerable set of positive integers it defines, we mean the string of letters and symbols here represented by

$$A;\ g_1P \textit{ produces } Pg_1',\ \cdots,\ g_\mu P \textit{ produces } Pg_\mu'.$$

When meaningfully interpreted, B determines the normal system, and recursively enumerable set of positive integers, in question. Each basis is but a finite sequence of the symbols 1, b, P, the comma, semicolon and the letters of the word "*produces*." The set of bases is therefore enumerably infinite, and can indeed be effectively generated in a sequence of distinct elements

$$O: \qquad B_1,\ B_2,\ B_3,\ \cdots.$$

Since each B_i defines a unique recursively enumerable set of positive integers and each such set is defined by at least one B_i, O is also an ordering of all recursively enumerable sets of positive integers, though each set will indeed recur an infinite number of times in O. We may then say, in classical terms, that whereas there are $2^{\aleph_0}$ arbitrary sets of positive integers, there are but $\aleph_0$ recursively enumerable sets.

By the *decision problem* of a given set of positive integers we mean the problem of effectively determining for an arbitrarily given positive integer whether it is, or is not, in the set. While, in a certain sense, the theory of recursively enumerable sets of positive integers is potentially as wide as the theory of general recursive functions, the decision problems for such sets constitute a very special class of decision problems. Nevertheless they are important, as is shown by the following special and general examples.

One of the problems posed by Hilbert in his Paris address of 1900 [**10**, problem 10] is the problem of determining for an arbitrary diophantine equation with rational integral coefficients whether it has, or has not, a solution in rational integers. If the variables in a

[5] We have thus restricted the normal operations and normal systems of [**19**] because of the following result. If in the initial string and in the normal operations of a normal system with primitive letters 1, a_1', $\cdots$, $a_{\mu'}'$, each a_i', $i=1, \cdots, \mu'$, is replaced by $b1 \cdots 1b$ with i 1's, a normal system with primitive letters 1, b results, defining the same set of strings on 1 only as the original normal system.

diophantine equation be chosen from a given enumerably infinite set of variables, it is clear that the set of diophantine equations is enumerably infinite. Indeed they can be effectively put into one-one correspondence with the set of positive integers. Since for any one diophantine equation, and assignment of rational integral values to its variables, it can be effectively determined whether or no the equation is satisfied by those values, the set of diophantine equations having rational integral solutions can be generated. The corresponding integers under the above one-one correspondence can then also be generated, and, indeed, constitute a recursively enumerable set of positive integers.[6] And under that correspondence, Hilbert's problem is transformed into the decision problem of that recursively enumerable set.

The assertions of an arbitrary symbolic logic[7] constitute a generated set A of what may be called symbol-complexes or formulas. We assume that A is a subset of an infinite generated set E of symbol-complexes, which in one case may be the set of meaningful enunciations of the logic, in another the set of all symbol-complexes of a given mode of symbolization. The decision problem of the logic, more precisely its deducibility problem [**3**], is then the problem of determining of an arbitrary member of E whether it is, or is not, in A. Granting that every generated set is effectively enumerable, the members of E can be effectively set in one-one correspondence with the set of positive integers. The positive integers corresponding to the members of A then constitute a generated, and hence, under our generalization, a recursively enumerable set of positive integers. And under that correspondence the decision problem of the symbolic logic is transformed into the decision problem of this recursively enumerable set of positive integers.

Closely related to the technical concept recursively enumerable set of positive integers is that of a *recursive set* of positive integers. This is a set for which there is a recursive function $f(x)$ such that $f(x)$ is say 2 when x is a positive integer in the set, 1 when x is a positive integer not in the set. We may also make this the definition of the decision problem of the set being *recursively solvable*. For 2 and 1 may be regarded as the two possible truth-values, true, false, of the proposition "positive integer x is in the set," and the definition of recursive set is equivalent to this truth-value being recursively calculable for all positive integers x. If then recursive function is coextensive with

[6] In view of [**17**] we inadvertantly carried through our formal verification with "rational integral solution" replaced by "positive integral solution."

[7] See Church [**5**, p. 225] for our omitting the qualifying "finitary."

effective calculability, recursive solvability is coextensive with solvability in the intuitive sense. In particular, the decision problem of a recursively enumerable set would be solvable or unsolvable according as the set is, or is not, recursive. More generally than in our two illustrations, through the more precise mechanism of Gödel representations [8], a wide variety of decision and other problems are transformed into problems about positive integers; and whether those problems are, or are not, solvable in the intuitive sense would be equivalent to their being, or not being, recursively solvable in the precise technical sense.

Gödel's classic theorem on the incompleteness and extendibility of symbolic logics [8] in all but wording led him to the recursive unsolvability of a generalization of the above problem of Hilbert [8, 9, 22]. Church explicitly formulated the concept of recursive unsolvability, and arrived at the unsolvability of a number of problems; certainly he proved them recursively unsolvable [1–4]. The above problem of Hilbert begs for an unsolvability proof (see [17]). Like the classic unsolvability proofs, these proofs are of unsolvability by means of given instruments. What is new is that in the present case these instruments, in effect, seem to be the only instruments at man's disposal.

Related to the question of solvability or unsolvability of problems is that of the reducibility or non-reducibility of one problem to another. Thus, if problem P_1 has been reduced to problem P_2, a solution of P_2 immediately yields a solution of P_1, while if P_1 is proved to be unsolvable, P_2 must also be unsolvable. For unsolvable problems the concept of reducibility leads to the concept of *degree of unsolvability*, two unsolvable problems being of the same degree of unsolvability if each is reducible to the other, one of lower degree of unsolvability than another if it is reducible to the other, but that other is not reducible to it, of incomparable degrees of unsolvability if neither is reducible to the other. A primary problem in the theory of recursively enumerable sets is the problem of determining the degrees of unsolvability of the unsolvable decision problems thereof. We shall early see that for such problems there is certainly a highest degree of unsolvability. Our whole development largely centers on the single question of whether there is, among these problems, a lower degree of unsolvability than that, or whether they are all of the same degree of unsolvability. Now in his paper on *ordinal logics* [26, section 4], Turing presents as a side issue a formulation which can immediately be restated as the general formulation of the "recursive reducibility" of one problem to another, and proves a result which immediately generalizes to the result that for any "recursively given" unsolvable problem there is another of higher degree of unsolvability.[8] While his theorem does not help us in our search for that lower degree of unsolvability, his formulation makes our problem precise. It remains a problem at the end of this paper. But on the way we do obtain a number of special results, and towards the end obtain some idea of the difficulties of the general problem.

1. **Recursive versus recursively enumerable sets.** The relationship between these two concepts is revealed by the following

THEOREM. *A set of positive integers is recursive when and only when both it and its complement with respect to the set of all positive integers are recursively enumerable.*[9]

For simplicity, we assume both the set S and its complement $\bar{S}$ to be infinite. If, then, S is recursive, there is an effective method for telling of any positive integer n whether it is, or is not, in S. Generate the positive integers 1, 2, 3, $\cdots$ in their natural order, and, as a positive integer is generated, test its being or not being in S. Each time a positive integer is thus found to be in S, write it down as belonging to S. Thus, an effective process is set up for effectively enumerating the elements of S. Hence, S is recursively enumerable. Likewise $\bar{S}$ can be shown to be recursively enumerable.

Conversely, let both S and $\bar{S}$ be recursively enumerable, and let $n_1, n_2, n_3, \cdots$ be a recursive enumeration of S; $m_1, m_2, m_3, \cdots$, of $\bar{S}$. Given a positive integer n, generate in order n_1, m_1, n_2, m_2, n_3, m_3, and so on, comparing each with n. Since n must be either in S or in $\bar{S}$, in a finite number of steps we shall thus come across an n_i or m_j identical with n, and accordingly discover n to be in S, or $\bar{S}$. An effective method is thus set up for determining of any positive integer n whether it is, or is not, in S. Hence, S is recursive.

COROLLARY. *The decision problem of a recursively enumerable set is recursively solvable when and only when its complement is recursively enumerable.*

For then and only then is the recursively enumerable set recursive.

It is readily proved that the logical sum and logical product of two

[8] Both our generalization of his formulation and of his theorem have been carried through, rather hastily, by the formalism of [**19**], without, as yet, an actual equivalence proof. It may be that Tarski's Theorem 9.1 [**23**] can be transformed into a like absolute theorem.

[9] The only portion of this theorem we can find in the literature is Rosser's Corollary II [**20**, p. 88].

i 232

recursively enumerable sets are recursively enumerable, the complement of a recursive set, and the logical sum, and hence logical product, of two recursive sets are recursive.

Clearly, *any finite set of positive integers is recursive.* For if $n_1, n_2, \cdots, n_\nu$ are the integers in question, we can test n being, or not being, in the set by directly comparing it with $n_1, n_2, \cdots, n_\nu$.[10] Likewise for a set whose complement is finite. For arbitrary infinite sets we have the following result of Kleene [12]. *An infinite set of positive integers is recursive when and only when it admits of a recursive enumeration without repetitions in order of magnitude.* Indeed, if $n_1, n_2, n_3, \cdots$ is a recursive enumeration of S without repetitions in order of magnitude, all n_i's beyond the nth must exceed n. Hence we can test n being, or not being, in S by generating the first n members of the given recursive enumeration of S, and seeing whether n is, or is not, one of them. Conversely, if infinite S is recursive, the recursive enumeration thereof we set up in the proof of our first theorem is of the elements of S without repetition, and in order of magnitude.

A direct consequence of the first half of the last result is the following

THEOREM. *Every infinite recursively enumerable set contains an infinite recursive set.*

For, if $n_1, n_2, n_3, \cdots$ is a recursive enumeration of an infinite set S, for each n_i there must be, in this sequence, a later $n_j > n_i$. Hence, generate the elements $n_1, n_2, n_3, \cdots$ in order, and let $m_1 = n_1$, $m_2 = n_{i_2}$, the first n_i greater than n_1, $m_3 = n_{i_3}$, the first n_i beyond n_{i_2} greater than n_{i_2}, and so on. The sequence $m_1, m_2, m_3, \cdots$ is then a recursive enumeration of a subset of S without repetitions in order of magnitude. That subset is therefore infinite, and recursive.

Basic to the entire theory is the following result we must credit to Church, Rosser, Kleene, jointly [1, 20, 12].

THEOREM. *There exists a recursively enumerable set of positive integers which is not recursive.*[11]

By our first theorem this is equivalent to the existence of a recursively enumerable set of positive integers whose complement is

[10] The mere existence of a general recursive function defining the finite set is in question. Whether, given some definition of the set, we can actually discover what the members thereof are, is a question for a theory of proof rather than for the present theory of finite processes. For sets of finite sets the situation is otherwise, as seen in §11.

[11] In each of our existence theorems we show how to set up the basis of the set in question—at least, the corresponding formal proof does exactly that.

not recursively enumerable. Generate in order the distinct bases $B_1, B_2, B_3, \cdots$ of all recursively enumerable sets of positive integers as mentioned in the introduction, and keep track of these bases as the first, second, third, and so on, in this enumeration O. As the nth basis B_n is generated, with $n=1, 2, 3, \cdots$, set going the processes whereby the corresponding recursively enumerable set is generated, and whenever n is thus generated by B_n, place n in a set U. Being a generated set of positive integers, U is recursively enumerable. A positive integer n, then, is, or is not, in U according as it is, or is not, in the nth recursively enumerable set in O considered as an ordering of all recursively enumerable sets. Hence, n is, or is not, in $\overline{U}$, the complement of U, according as it is not, or is, in the nth set in O. We thus see that $\overline{U}$ differs from each recursively enumerable set in the presence or absence of at least one positive integer. Hence $\overline{U}$ is not recursively enumerable.

COROLLARY. *There exists a recursively enumerable set of positive integers whose decision problem is recursively unsolvable.*

Taken singly, finite sets, or sets whose complements are finite, are rather trivial examples of recursive sets. On the other hand, if we define two sets of positive integers to be *abstractly* the same if one can be transformed into the other by a recursive one-one transformation of the set of all positive integers into itself, then all infinite recursive sets with infinite complements are abstractly the same. Our theory being essentially an abstract theory of recursively enumerable sets, our interest therefore centers in recursively enumerable sets that are not recursive. Such sets, as well as their complements, are always infinite. We do not further pursue the question of two sets being abstractly the same, for that is but a special case of each set being one-one reducible to the other (§4).

2. A form of Gödel's theorem. Given any basis B, and positive integer n, the couple (B, n) may be used to represent the proposition, true or false, "n is in the set generated by B." By interlacing the process for generating the distinct bases in the sequence $B_1, B_2, B_3, \cdots$ and the process for generating the positive integers in the sequence $1, 2, 3, \cdots$ by the addition of 1's, we can effectively generate the distinct couples (B, n) in the single infinite sequence

$$O': \quad (B_1, 1), (B_2, 1), (B_1, 2), (B_3, 1), (B_2, 2), (B_1, 3), \cdots .$$

On the one hand, the set of all couples (B, n) is thus a generated set of expressions which we shall call E. On the other hand, O' leads to an effective 1-1 correspondence between the members of E and the set of positive integers, (B, n) corresponding to m if (B, n) is the mth member of O'. We may call m the Gödel representation[12] of (B, n). Given a generated subset of E, the Gödel representations of its members will constitute a generated set of positive integers, and conversely. Thus, in the former case we can generate the members of the subset of E, and, as a couple (B, n) is generated, find its Gödel representation m by regenerating O'. The set of these m's is thus a generated set. Likewise for the converse. If, therefore, we formally define a subset of E to be recursively enumerable if the set of Gödel representations of its members is recursively enumerable,[13] we can conclude that every generated subset of E is recursively enumerable, and, of course, conversely. Similarly for a like formal definition of a recursive subset of E.

While E is just the set of couples (B, n), it may be interpreted as the set of enunciations "n is in the set generated by B." The subset T of E consisting of those couples (B, n) for which n *is* in the set generated by B may then be interpreted as the set of true propositions in E, while $\overline{T}$, the complement of T with respect to E, consists of the false propositions in E.

Actually, T itself can be generated as follows. Generate B_1, B_2, B_3, $\cdots$ in order. As a B is generated, set up the process for generating the set of positive integers determined by B, and, whenever a positive integer n is thus generated, write down the couple (B, n). Each (B, n) for which n is in the set generated by B will thus be written down, and conversely. This generated set of (B, n)'s is then T. We therefore conclude that *T is recursively enumerable.*

Now let F be any recursively enumerable subset of $\overline{T}$. If (B, n) is in F, it is in $\overline{T}$, and hence n is certainly not in the set generated by B. Now generate the members of F, and if (B, n) is thus generated, find the nth member B_n of $O: B_1, B_2, B_3, \cdots$, and if B_n is B, place n in a set of positive integers S_0. Since S_0 is thus a generated set of positive integers, it is recursively enumerable. It will therefore be determined by some basis B. Let this basis be in the νth in O, that is, let the basis be B_ν, and form the couple (B_ν, ν). Now by construction, S_0 consists of those members of F of the form (B_n, n). Suppose that (B_ν, ν) is in F. Then, on the one hand, proposition (B_ν, ν) being false,

[12] Rather is the Gödel representation in [8] not just an effectively corresponding positive integer, but one which, when expressed according to a specific algorithm, is "formally similar," in the sense of Ducasse [7, p. 51], to the symbolic expression represented.

[13] In our own development [19], "recursively enumerable subset of E" is defined directly as a normal subset of E, or rather of the set of symbolic representations of the members of E.

ν would not be in the set generated by B_ν, that is (1): ν would not be in S_0. But (B_ν, ν) being of the form (B_n, n), (2): ν would be in S_0. Our assumption thus leading to a contradiction, it follows that (B_ν, ν) *is not in* F. But ν can only be in S_0 by (B_ν, ν) being in F. Hence, ν *is not in* S_0. Finally, (B_ν, ν) as proposition says that ν is in S_0. The proposition (B_ν, ν) is therefore false, that is (B_ν, ν) *is in* $\overline{T}$.

For any recursively enumerable subset F of $\overline{T}$ there is then always this couple (B_ν, ν) in $\overline{T}$, but not in F. On the one hand, then, $\overline{T}$ can never be F. Hence, $\overline{T}$ *is not recursively enumerable.* By the definitions of this section, and the first theorem of the last, it follows that T, *while recursively enumerable, is not recursive.* By the decision problem of T we mean the problem of determining for an arbitrarily given member of E whether it is, or is not, in T. But that can be interpreted as the decision problem for the class of recursively enumerable sets of positive integers, that is, the problem of determining for any arbitrarily given recursively enumerable set, that is, arbitrarily given basis B of such a set, and arbitrary positive integer n whether n is, or is not, in the set generated by B. We may therefore say that the *decision problem for the class of all recursively enumerable sets of positive integers is recursively unsolvable,* and hence, in all probability, unsolvable in the intuitive sense.

On the other hand, since (B_ν, ν) of $\overline{T}$ is not in F, T and F together can never exhaust E. Now T, or any recursively enumerable subset T' of T, in conjunction with F may be called a recursively generated logic relative to the class of enunciations E. For the appearance of (B, n) in T' assures us of the truth of the proposition "n is in the set generated by B," while its presence in F would guarantee its falseness. We can then say *that no recursively generated logic relative to E is complete,* since F alone will lead to the (B_ν, ν) which is neither in T' nor in F. That is, (B_ν, ν) is undecidable in this logic. Moreover, if, with a given "basis" for F, the above argument is carried through formally,[14] the recursively enumerable S_0 obtained above will actually be given by a specific basis B which can be constructed by that formal argument. Having found this B, we can then re-generate $O: B_1, B_2, B_3, \cdots$, until B is reached, and thus determine the ν such that $B = B_\nu$. That is, given the basis of F, the (B_ν, ν) in $\overline{T}$ and not in F can actually be found. If then we add this (B_ν, ν) to F, a wider recursively enumerable subset F' of $\overline{T}$ results. We may then say that *every recursively generated logic relative to E can be extended.* Outwardly, these two results, when formally developed, seem to be

[14] Here, the basis of F may be taken to be the basis of the recursively enumerable set of Gödel representations of the members of F. But see the preceding footnote.

Gödel's theorem in miniature. But in view of the generality of the technical concept general recursive function, they implicitly, in all probability, justify the generalization that every symbolic logic is incomplete and extendible relative to the class of propositions constituting E.[15] The conclusion is unescapable that even for such a fixed, well defined body of mathematical propositions, *mathematical thinking is, and must remain, essentially creative.* To the writer's mind, this conclusion must inevitably result in at least a partial reversal of the entire axiomatic trend of the late nineteenth and early twentieth centuries, with a return to meaning and truth as being of the essence of mathematics.

3. **The complete set K; creative sets.** Return now to the effective 1-1 correspondence between the set E of distinct (B, n)'s and the set of positive integers obtained via the effective enumeration O' of E. Since T is a recursively enumerable subset of E, the positive integers corresponding to the elements of T constitute a recursively enumerable set of positive integers, K. We shall call K the *complete set.*[16] Since $\overline{T}$ is not recursively enumerable, $\overline{K}$, which consists of the positive integers corresponding to the elements of $\overline{T}$, is not recursively enumerable. Now let B be the basis of a recursively enumerable subset α of $\overline{K}$. The elements of E corresponding to the members of α constitute, then, a recursively enumerable subset F of $\overline{T}$. Find then the (B_ν, ν) of $\overline{T}$ not in F, and, via O', the positive integer n corresponding to (B_ν, ν). This n will then be an element of $\overline{K}$ not in α.

Actually, we have no general method of telling when a basis B defines a recursively enumerable subset of $\overline{K}$. Indeed, the above method will yield a unique positive integer n for any basis B of a recursively enumerable set α of positive integers. However, when α is a subset of $\overline{K}$, n will also be in $\overline{K}$, but not in α.

Furthermore, even the formal proof of this result merely gives an effective method for finding n, given B. But this method itself can be formalized, so that, as a result, n is given as a "recursive function of B." This can mean that a recursive function $f(m)$ can be set up such that $n = f(m)$ where $B = B_m$. We now isolate this property of K by setting up the

DEFINITION. *A creative set C is a recursively enumerable set of positive integers for which there exists a recursive function giving a unique*

[15] See Kleene's Theorem XIII in [12] for a mathematically stateable theorem approximating the generality of our informal generalization.

[16] "A complete set" might be better. Just how to abstract from K the property of completeness is not, at the moment, clear. By contrast, see "creative set" below.

positive integer n for each basis B of a recursively enumerable set of positive integers α such that whenever α is a subset of $\bar{C}$, n is also in $\bar{C}$, but not in α.

THEOREM. *There exists a creative set; to wit, the complete set K.*

Actually, the class of creative sets is infinite, and very rich indeed as shown by the following easily proved results.[17] If C is a creative set, and E a recursively enumerable set of positive integers, then if E contains $\bar{C}$, CE is creative, if $\bar{C}$ contains E, $C+E$ is creative. Results of §1 enable us actually to construct creative sets according to the first method by using E's which are the complements of recursive subsets of C. Results of the rest of this section lead to constructions using the second method.

It is convenient to talk as if the n in the definition of a creative set were determined by the α thereof instead of by the basis B of α. Clearly *every creative set C is a recursively enumerable set which is not recursive.* For were $\bar{C}$ recursively enumerable, there could be no n in $\bar{C}$ not in the recursively enumerable subset $\bar{C}$ of $\bar{C}$. *The decision problem of each creative set is therefore recursively unsolvable.* On the other hand, the *complement $\bar{C}$ of any creative set C contains an infinite recursively enumerable set.* Recall that every finite set is recursive, and hence recursively enumerable. With, then, α of the definition of creative set as the null set, find the $n=n_1$ of $\bar{C}$ "not in α." With α the unit set having n_1 as sole member, $n=n_2$ will be in $\bar{C}$, and distinct from n_1. With α consisting of n_1 and n_2, $n=n_3$ will be in $\bar{C}$, and distinct from n_1 and n_2, and so on. The set of positive integers $n_1, n_2, n_3, \cdots$ is then an infinite generated, and hence recursively enumerable, subset of $\bar{C}$.

Actually, with this subset of $\bar{C}$ as α, a new element n_ω of $\bar{C}$ is obtained, and so on into the constructive transfinite. But this process is essentially creative. For any mechanical process could only yield n's forming a generated, and hence recursively enumerable, subset α of $\bar{C}$, and hence could be transcended by finding that n of $\bar{C}$ not in α.

4. One-one reducibility, to K; many-one reducibility. Let S_1 and S_2 be any two sets of positive integers. One of the simplest ways in which the decision problem of S_1 would be reduced to the decision problem of S_2 would arise if we had an effective method which would determine for each positive integer n a positive integer m such that n is, or is not, in S_1 according as m is, or is not, in S_2. For if we could

[17] Of course, all sets abstractly the same as a given creative set, in the sense of §1, are creative. Likewise for our later simple and hyper-simple sets.

somehow determine whether m is, or is not, in S_2, we would determine n to be, or not be, in S_1 correspondingly. If "effective method" be replaced by "recursive method," we shall say, briefly, that S_1 is then *many-one reducible* to S_2. If, furthermore, different n's always lead to different m's, we shall say that S_1 is *one-one reducible* to S_2.[18] "Recursive method" here can mean that $m=f(n)$, where $f(n)$ is a recursive function.

THEOREM. *The decision problem of every recursively enumerable set of positive integers is one-one reducible to the decision problem of the complete set K.*

For let B' be a basis of any one recursively enumerable set S'. The effective one-one correspondence between all (B, n)'s and all positive integers yielded by the effective enumeration O' of E, the set of all (B, n)'s, then yields a unique positive integer m for each (B', n), B' fixed, and thus a unique m for each n, different n's yielding different m's. Now n is, or is not, in S' according as (B', n) is in T, or $\overline{T}$, and hence according as m is in K, or $\overline{K}$, whence our result.

Since K itself is recursively enumerable, we may say that for recursively enumerable sets of positive integers with recursively unsolvable decision problems there is a *highest degree of unsolvability relative to one-one reducibility*, namely, that of K. Actually, one-one reducibility is a special case of all the more general types of reducibility later introduced, and, though the proof of this is still in the informal stage, these latter are special cases of general recursive, that is, Turing reducibility. The same result then obtains relative to these special types of reducibility and, more significantly, for reducibility in the general sense.[19]

We have thus far explicitly obtained two recursively enumerable sets with recursively unsolvable decision problems, the U of our first section, and K. We may note that a certain necessary and sufficient condition for the many-one reducibility of K to a recursively enumerable set, the proof of which is still in the informal stage, has as an immediate consequence that K is many-one reducible to U. It would then follow that K and U are of the same degree of unsolvability relative to many-one reducibility.

---

[18] The resulting one-to-one correspondence is then between $S_1+\overline{S}_1$ and a subset, recursively enumerable indeed, of $S_2+\overline{S}_2$. Of course, both $S_1+\overline{S}_1$ and $S_2+\overline{S}_2$ constitute the set of all positive integers.

[19] It seems rather obvious that K and the problem of Church [**1**] are each at least many-one reducible to the other; likewise for the problem of [**1**] and of [**2**, **3**]. Had we verified this in detail, we would have called this highest degree of unsolvability of decisions problems of recursively enumerable sets the *Church degree of unsolvability*.

5. **Simple sets.** It is readily proved that the necessary and sufficient condition that every recursive set be one-one reducible to a given recursively enumerable set of positive integers S is that S is infinite, and $\bar{S}$ contains an infinite recursively enumerable set. We are thus led to ask if there exist sets satisfying the following

DEFINITION. *A simple set is a recursively enumerable set of positive integers whose complement, though infinite, contains no infinite recursively enumerable set.*

We now prove the

THEOREM. *There exists a simple set.*

Recall the set T of all couples (B, n) such that positive integer n is in the recursively enumerable set of positive integers determined by basis B. Since T is recursively enumerable, we can set up an effective enumeration

$$O'': \quad (B_{i_1}, n_1), (B_{i_2}, n_2), (B_{i_3}, n_3), \cdots$$

of its members. The subscript of each B is its subscript in the effective enumeration $O: B_1, B_2, B_3, \cdots$ of all distinct B's. Now the complement of a set containing no infinite recursively enumerable set is equivalent to the set itself having an element in common with each infinite recursively enumerable set. Generate then the distinct bases $B_1, B_2, B_3, \cdots$, and as a B_i is generated, regenerate the sequence O'' of (B, n)'s in T, and the first time, if ever, B is B_i, and *n is greater than $2i$*, place n in a set S. The resulting set S is then a generated, and hence recursively enumerable, set of positive integers. We proceed to prove it simple.

If S' is an infinite recursively enumerable set of positive integers, it will be determined by some basis B_i, and will have some element m greater than $2i$. Since (B_i, m), being then in T, will appear in O'', our construction will place m in S, if some earlier (B_i, n) of O'' has not already contributed an element of S' to S. That is, S has an element in common with each infinite recursively enumerable S'. As for $\bar{S}$ being infinite, note that each B_i contributes at most one element to S. The first n B's in O therefore contribute at most n elements to S. Each B_i with $i \geqq n+1$ can only contribute to S an element greater than $2n+2$. Of the first $2n+2$ positive integers, at most n are therefore in S, and hence at least $n+2$ are in the consequently infinite $\bar{S}$.[20]

---

[20] $n>i$ can replace $n>2i$ in the above construction, but the proof will then depend on there being an infinite number of bases defining the null set.

Having one simple set, the method of our succeeding §8 can be modified to yield a rich infinite class of simple sets. Clearly, *every simple set S is a recursively enumerable set that is not recursive.* For were S recursive, $\overline{S}$ would be an infinite recursively enumerable subset of $\overline{S}$. *The decision problem of each simple set is therefore recursively unsolvable.* We thus have obtained two infinite mutually exclusive classes of recursively enumerable sets with recursively unsolvable decision problems, the class of creative sets, and the class of simple sets. They are poles apart in that the complements of creative sets have a creative infinity of infinite recursively enumerable subsets, those of simple sets, not one.

In passing, we may note that every recursively enumerable set of positive integers S with recursively unsolvable decision problem leads to an incompleteness theorem for symbolic logics relative to the class of propositions $n \in S$, n an arbitrary positive integer. Creative sets S are then exactly those recursively enumerable sets of this type each of which admits a universal extendibility theorem as well, simple sets S those for which, given S, each logic can prove the falsity of but a finite number of the infinite set of false propositions $n \in S$.

It is readily seen that no creative set C can be one-one reducible to a simple set S. For under such a reduction, each infinite recursively enumerable subset of $\overline{C}$, proved above to exist, would be transformed into an infinite recursively enumerable subset of $\overline{S}$, contradicting the simplicity of S. Simple sets thus offer themselves as *candidates* for recursively enumerable sets with decision problems of lower degree of unsolvability than that of the complete set K. Even for many-one reducibility the situation is no longer immediately obvious; for an infinite recursively enumerable subset of $\overline{C}$ could thus be transformed into a finite subset of $\overline{S}$, the complement of simple S, without contradiction. However we can actually go much further than that.

6. Reducibility by truth-tables. If S_1 is many-one reducible to S_2, positive integer n being, or not being, in S_1 may be said to be determined by its correspondent m being, or not being, in S_2 in accordance with the truth-table

(S_2) m	n (S_1)
+	+
−	−

Here, the two signs $+$, $-$ under m represent the two possibilities m is in S_2, m is not in S_2, respectively. And by the sign under n in the same horizontal row as the corresponding sign under m the table in the same language tells whether n correspondingly is $(+)$, or is not $(-)$, in S_1. The table then says that when m is in S_2, n is in S_1, when m is not in S_2, n is not in S_1, as required by many-one reducibility. Now there are altogether four ways in which n being, or not being, in S_1 can be made to depend solely on m being, or not being, in S_2, the signs under n being $+$, $-$ as above; or $+$, $+$; $-$, $-$; $-$, $+$. If then we have an effective method which for each positive integer n will not only determine a unique corresponding positive integer m, but also one of these four "first order" truth-tables, and if in each case the table is such that for the correct statement of membership or non-membership of m in S_2, it gives the correct statement of membership or non-membership of n in S_1, then the decision problem of S_1 will thus be reduced to the decision problem of S_2. For here also, given n, if we could somehow determine whether m is, or is not, in S_2, we could thereby determine which row of the corresponding table correctly describes the membership or non-membership of m in S_2, and from that row correctly determine whether n is, or is not, in S_1.

More generally, let there be an effective method which for each positive integer n determines a finite sequence of positive integers $m_1, m_2, \cdots, m_\nu$, ν as well as the m's depending on n. Let that method correspondingly determine for each n a "νth order" truth-table of the form

$$\begin{array}{cccccc|cc}
(S_2) & m_1 & m_2 & \cdots & m_\nu & & n & (S_1) \\
\hline
 & + & + & \cdots & + & & - & \\
 & + & + & \cdots & - & & + & \\
 & \cdot & \cdot & \cdots & \cdot & & \cdot & \\
 & \cdot & \cdot & \cdots & \cdot & & \cdot & \\
 & \cdot & \cdot & \cdots & \cdot & & \cdot & \\
 & - & - & \cdots & - & & - & \\
\end{array}$$

Each horizontal row, to the left of the vertical bar, specifies one of the 2^ν possible ways in which the ν m_i's may, or may not, be in S_2, to the right of the bar correspondingly commits itself to one of the statements n is in S_1, n is not in S_1. If then for each n that row of the corresponding table which gives the correct statements for the m's being or not being in S_2 also gives the correct statement regarding the membership or non-membership of n in S_1, the decision problem of S_1 is again thereby reduced to the decision problem of S_2.

If such a situation obtains with "effective method" replaced by "recursive method," we shall say that *S_1 is reducible to S_2 by truth-tables*. "Recursive method" here can mean that a suitable Gödel representation of the couple consisting of the sequence $m_1, m_2, \cdots,$

m_ν and the truth-table of order ν is a recursive function of n. If the orders of the truth-tables arising in such a reduction are bounded, we shall say that S_1 *is reducible to* S_2 *by bounded truth-tables.* Since there are 2^{2^ν} distinct truth-tables of order ν, reducibility by bounded truth-tables is equivalent to reducibility by truth-tables in which but a finite number of distinct tables arise.

7. **Non-reducibility of creative sets to simple sets by bounded truth-tables.** Let us suppose that creative set C is reducible to simple set S by bounded truth-tables. Let $T_1, T_2, \cdots, T_\kappa$ be the finite set of distinct truth-tables entering into such a reduction. That reduction then effectively determines for each positive integer n a finite sequence of positive integers $m_1, m_2, \cdots, m_\nu$, and a unique T_i, $1 \leqq i \leqq \kappa$.

The gist of our reductio-ad-absurdam proof consists in showing that under the assumed reduction we can obtain for each natural number p a sequence of m's at least p of which are in S. We then immediately have our desired contradiction. For in each case $p \leqq \nu$. The finite set of ν's, the orders of the T_i's, being bounded, p cannot then be arbitrarily large as stated.

More precisely we prove by mathematical induction that under the assumed reduction the following would be true. *For each natural number p an effective process Π_p can be set up which will determine for each recursively enumerable subset α of $\overline{C}$ an element n of $\overline{C}$ not in α, and which for the corresponding $m_1, m_2, \cdots, m_\nu$ and T_i yielded by the assumed reduction will correctly designate p of these m's as belonging to S.* The mode of designation may be assumed to be by specifying the sequence of subscripts, $i_1, i_2, \cdots, i_p$, of the m's to be designated, with say $i_1 < i_2 < \cdots < i_p$. With the assumed reduction adjoined to this process, Π_p then determines for each α in question the quadruplet (n, M, T_i, I), M being the sequence of m's, I the sequence of subscripts of the p designated m's.

For $p=0$, Π_p is immediately given by the creative character of C. For that immediately gives us for each recursively enumerable subset α of $\overline{C}$ a definite element n of $\overline{C}$ not in α. The assumed reduction yields the corresponding M and T_i; and with no members of M *designated* as being in S, I is the null sequence.

Inductively, assume that we have the process Π_p for $p=k$. Let α be any given recursively enumerable subset of $\overline{C}$, and let $(n', M', T_{i'}, I')$ be the corresponding quadruplet yielded by Π_k. Now suppose n is a positive integer for which the assumed reduction yields the same table $T_{i'}$ as it did for n', and a sequence of m's, M, consequently of the same length as M', having the following property. For each undesignated element of M', the correspondingly placed element of M is identical with that of M'; for each element of M' designated as being in S, the corresponding element of M is also in S. Such an n must then be in $\overline{C}$ along with n'. For that row of $T_{i'}$ which correctly tells of the m's of M' whether they are, or are not, in S will also be the correct row for M. And since in the former case that row must say that n' is in $\overline{C}$, in the latter case it will say that n is in $\overline{C}$, and correctly so. We proceed to show how all such n's may be generated.

We first show how to generate all M's obtainable from M' by replacing the designated elements of M' by arbitrary elements of S. For any one such M, the replacing elements, being finite in number, will be among the first N elements, for some positive integer N, of a given recursive enumeration of S. Generate then the positive integers $1, 2, 3, \cdots$, and as a positive integer N is generated, generate the first N elements of the given recursive enumeration of S. For each N place in a set β the at most N^k sequences M that can be obtained from M' by replacing the designated elements of M' by elements chosen from the first N elements of S. The generated set of sequences β then consists of all M's obtainable from M' by replacing the designated elements of M' by arbitrary elements of S.

The n's we wish to generate are then those positive integers for which the assumed reduction yields the table $T_{i'}$ and a sequence of m's, M, such that M is a member of β. Generate then the elements of β. As an element M of β is generated, generate the positive integers $1, 2, 3, \cdots$, and as a positive integer n is generated, find the corresponding sequence of m's and table yielded by the reduction of C to S. If then that sequence of m's is M, and the table is $T_{i'}$, *add n to the given set α.* As seen above, each such n will be in $\overline{C}$. Hence the resulting generated, and hence recursively enumerable, set α' is a subset of $\overline{C}$ containing α. Our reason for thus adding the desired n's to α instead of just forming the class thereof is that the iterative process we are about to set up requires a cumulative effect.

As a result of our hypothesis and construction we thus have a derived process Π_k' which for every recursively enumerable subset α of $\overline{C}$ yields a definite recursively enumerable subset α' of $\overline{C}$ containing α. Starting with α, we may then iterate the process Π_k' to obtain the infinite sequence $A:\alpha_1, \alpha_2, \alpha_3, \cdots$, where $\alpha_1=\alpha$, $\alpha_{n+1}=(\alpha_n)'$. Each member of A is thus a recursively enumerable subset of $\overline{C}$, and contained in the next member of A. By applying the original process Π_k to the members of A we correspondingly obtain the infinite sequence $\Sigma:\sigma_1, \sigma_2, \sigma_3, \cdots$, where σ_j is the quadruplet $(n^{(j)}, M^{(j)}, T_i^{(j)}, I^{(j)})$ yielded by Π_k for α_j. We then observe the following. If for $j_1 \neq j_2$

the T's of σ_{j_1} and σ_{j_2} are the same, and the I's are the same, then the sequences obtained from the M's by deleting the designated m's cannot be identical. For if they also were identical, then, with say $j_1 < j_2$, $n^{(j_2)}$ would have been assigned to $\alpha^{(j_1+1)}$, whereas it actually is outside of $\alpha^{(j_2)}$ which contains $\alpha^{(j_1+1)}$. Hence, the infinite sequence Σ', obtained from Σ by deleting from each σ_j the integer $n^{(j)}$ and the designated m's of $M^{(j)}$, itself consists of distinct elements.

It follows that *there are an infinite number of distinct undesignated m's appearing in Σ*. Indeed, the distinct $T_i^{(j)}$'s of Σ are at most κ in number. With $T_i^{(j)}$ fixed, the order $\nu^{(j)}$ of $T_i^{(j)}$ is fixed; and since $1 \leqq i_1^{(j)} < i_2^{(j)} < \cdots < i_k^{(j)} \leqq \nu^{(j)}$, the number of distinct I's is finite. Finally, with T and I fixed, were the total number of distinct undesignated m's finite, the number of distinct ways in which those $\nu^{(j)} - k$ undesignated m's could assume values would be finite. Hence Σ' would be finite, not infinite.

Now were each of this infinite set of undesignated m's in $\overline{S}$, we could regenerate the elements of Σ, and as an element σ_j thereof is generated, place all of its undesignated m's in a set γ, and thus obtain an infinite generated, and hence recursively enumerable, subset of $\overline{S}$. As this contradicts the simplicity of S, it follows that *at least one undesignated m arising in Σ is in S*.

We can then find a unique such m, as well as a σ in which it occurs, as follows. With $N = 1, 2, 3, \cdots$, generate the first N elements of the given recursive enumeration of S, and the first N elements of Σ, and test the latter in order to see if any undesignated m is among those first N elements of S. If a particular undesignated m of Σ in S, proved above to exist, is the Lth member of S, and in the Kth member σ_K of Σ, then an affirmative answer to the above test will certainly be obtained for $N = \max(L, K)$. Find then the first N for which an affirmative answer is obtained, and let (m, M, T_i, I) be the first σ to yield the affirmative answer for this N, $m_{i'}$ the first undesignated m of M thus found to be in S. We can then add $m_{i'}$ to the designated m's of M, thus obtaining a quadruplet (n, M, T_i, I_1), where I_1 designates $(k+1)$ of the m's of M as being in S, and where n is certainly a member of $\overline{C}$ not in the originally given α. But the whole process leading up to (n, M, T_i, I_1) is determined by that α. It is therefore the desired process Π_p for $p = k+1$.

Under the assumed reduction of C to S, Π_p would therefore exist for every natural number p. With α say the null set, we would thus obtain for every natural number p a quadruplet (n_p, M_p, T_{i_p}, I_p) such that p of the members of the sequence M_p are in S. Yet the total length of M_p is the order of T_{i_p}, and hence bounded. Hence the

THEOREM. *No creative set is reducible to a simple set by bounded truth-tables.*

We recall that every recursively enumerable set of positive integers is one-one reducible to the creative set K, the complete set. Hence the

COROLLARY. *Every simple set is of lower degree of unsolvability than the complete set K relative to reducibility by bounded truth-tables.*

8. **Counter-example for unbounded truth-tables.** We recall that for the particular simple set S constructed in §5, of the first $2m+2$ positive integers at most m were in S, m being any positive integer. Hence, of the $m+1$ integers $m+2, m+3, \cdots, 2m+2$, at least one is in $\bar{S}$. By setting $m=2^n-1$, with $n=1, 2, 3, \cdots$, we can effectively generate the infinite sequence of mutually exclusive finite sequences

$$\sigma: \quad (3, 4), (5, 6, 7, 8), \cdots, (2^n+1, 2^n+2, \cdots, 2^{n+1}), \cdots$$

such that each sequence in σ has at least one member thereof in $\bar{S}$. An effective one-one correspondence between the positive integers 1, 2, 3, $\cdots$ and the elements of σ is then obtained by making the positive integer n correspond to the sequence $(2^n+1, 2^n+2, \cdots, 2^{n+1})$ constituting the nth element of σ.

Given a creative set C, regenerate the elements of S, placing each in a set S_1. Furthermore, regenerate the elements of C, and as an element n thereof is generated, place all of the positive integers in the nth sequence of σ in S_1. The resulting set S_1 is a generated, and hence recursively enumerable, set of positive integers. Since S_1 contains S, $\bar{S}$ contains $\bar{S}_1$. As S is simple, $\bar{S}$, and hence $\bar{S}_1$, does not have an infinite recursively enumerable subset. Moreover, $\bar{S}_1$ is also infinite. For $\bar{C}$ is infinite. And, for each element of $\bar{C}$, the corresponding sequence in σ has only those of its members that are already in S also in S_1, and hence at least one element in $\bar{S}_1$. Hence, S_1 is simple.

Likewise we see that a positive integer n is in C, or $\bar{C}$, according as all of the integers in the nth sequence of σ are in S_1, or at least one is in $\bar{S}_1$. If then we make correspond to each positive integer n the sequence of 2^n positive integers $(2^n+1, 2^n+2, \cdots, 2^{n+1})$, and the truth-table of order 2^n in which the sign under n is $+$ in that row in which the signs under the 2^n "m's" are all $+$, and in every other row the sign under n is $-$, we have a reduction of C to S_1 by truth-tables. Hence the

THEOREM. *For each creative set C a simple set S can be constructed such that C is reducible to S by unbounded truth-tables.*

COROLLARY. *A simple set S can be constructed which is of the same degree of unsolvability as the complete set K relative to reducibility by truth-tables unrestricted.*

Simple sets as such do not therefore give us the absolutely lower degree of unsolvability than that of K we are seeking.

9. **Hyper-simple sets.** The counter-example of the last section suggests that we seek a set satisfying the following

DEFINITION. *A hyper-simple set H is a recursively enumerable set of positive integers whose complement $\overline{H}$ is infinite, while there is no infinite recursively enumerable set of mutually exclusive finite sequences of positive integers such that each sequence has at least one member thereof in $\overline{H}$.*[21]

In this definition we may use the original Gödel method for representing a finite sequence of positive integers $m_1, m_2, \cdots, m_\nu$ by the single positive integer $2^{m_1} 3^{m_2} \cdots p^{m_\nu}$, where $2, 3, \cdots, p_\nu$ are the first ν primes in order of magnitude. A set of finite sequences of positive integers is then recursively enumerable if the set of Gödel representations of those sequences is recursively enumerable.

THEOREM. *A hyper-simple set exists.*

Our intuitive argument must again draw upon the formal development to the effect that each recursively enumerable set of finite sequences of positive integers will be determined by a "basis" B^*, and that all such bases can be generated in a single infinite sequence of distinct bases

$$O^*: \qquad B_1^*, \ B_2^*, \ B_3^*, \cdots .$$

As in §2, generate the elements of O^*, and as an element B^* is generated, set up the process for generating the set of sequences determined by B^*, and as a sequence s is thus generated, write down the couple (B^*, s). The resulting set of couples is then a generated set, and can indeed be effectively ordered in a sequence of distinct couples

$$O_1^*: \qquad (B_{i_1}^*, s_1), \ (B_{i_2}^*, s_2), \ (B_{i_3}^*, s_3), \cdots .$$

[21] Mutually exclusive sequences here mean no element of one sequence is an element of another. Curry suggests that "hyper-simple" is linguistically objectionable, and should be replaced by "super-simple." But we would not then know what to use in place of the letter H.

O_1^* then consists of all distinct couples (B^*, s) such that finite sequence s is a member of the recursively enumerable set of finite sequences of positive integers determined by basis B^*.

Now the condition that no infinite recursively enumerable set of mutually exclusive finite sequences of positive integers has the property that each sequence has at least one positive integer thereof in $\overline{H}$ is equivalent to each such set of sequences having at least one sequence all of whose members are in H. Our method of constructing the desired hyper-simple set H will then consist in placing in H for certain B^*'s in O^* all of the positive integers in a sequence in the set of sequences determined by B^*. For purposes of presentation we shall call each such basis B^* a *contributing basis*, while every B^* determining an infinite recursively enumerable set of mutually exclusive sequences will be called a *relevant basis*. Set H, if recursively enumerable, will then be hyper-simple if each relevant basis is a contributing basis, and $\overline{H}$ is infinite.

If B^* is a relevant basis, then among the infinite number of mutually exclusive sequences generated by B^* there must be a sequence each of whose elements exceeds an arbitrarily given positive integer N. For did every sequence generated by B^* have as element one of the integers $1, 2, \cdots, N$, for any $N+1$ of these sequences at least two would have one of these integers in common. We shall then generate H by regenerating sequence O_1^*, and, as an element $(B_{i_n}^*, s_n)$ thereof is generated, we shall place all the elements of s_n in H if $B_{i_n}^*$ has not thus been made a contributing basis earlier in the process, while the elements of s_n are all greater than a certain positive integer N_n, about to be determined; otherwise none. Inductively, assume N_m to have been determined for $1 \leqq m < n$, and thus the entire process up to the time $(B_{i_n}^*, s_n)$ was brought up for consideration. Let $B_{j_1}^*, B_{j_2}^*, \cdots, B_{j_\nu}^*$ be the bases that have thus far contributed to H, and in the order in which they became contributing bases. These bases are then distinct, and hence their subscripts, which give their position in the sequence O^* of all distinct bases, are distinct. Let $k_1, k_2, \cdots, k_\nu$ be the largest integer placed in H by the first contributing basis, by the first two, $\cdots$, by the first ν. The result being cumulative, $k_1 \leqq k_2 \leqq \cdots \leqq k_\nu$. The crux of our construction is to make N_n depend not on the history of all these ν contributions to H, but only on that part of that history up to and including the last contribution, if any, made by a B^* preceding $B_{i_n}^*$ in O^*. Specifically, if $B_{j_\mu}^*$ is the last of the above ν contributing bases preceding $B_{i_n}^*$ in O^*, that is, with $j_\mu < i_n$, N_n is to be one more than the largest integer present in H as a result of all the contributions made up to and including the contribution made by $B^*_{j_\mu}$. That is, $N_n = k_\mu + 1$. Actually, if none of the ν contributing bases precede $B^*_{i_n}$ in O^*, no condition is to be placed on s_n, and all of its elements are placed in H so long as $B^*_{i_n}$ is distinct from the ν contributing bases obtained thus far.

Furthermore, in our induction assume that we have been able to keep a record of the sequence $B^*_{j_1}, B^*_{j_2}, \cdots, B^*_{j_\nu}$, of $k_1, k_2, \cdots, k_\nu$, and also of $j_1, j_2, \cdots, j_\nu$ up to the time $(B^*_{i_n}, s_n)$ was about to be generated. We then generate $(B^*_{i_n}, s_n)$, and by regenerating O^* find the place of $B^*_{i_n}$ in O^* thus determining the subscript i_n. Our criterion for determining whether, or no, the elements of s_n are to be placed in H then becomes effective. In the latter case, the record is unchanged as we generate $(B^*_{i_{n+1}}, s_{n+1})$. In the former, $B^*_{i_n}$ is written into the record as $B^*_{j_{\nu+1}}$, i_n as $j_{\nu+1}$ while we can write in for $k_{\nu+1}$ the maximum of k_ν and the largest integer in s_n. The entire process is thus effective at each stage, and H is thus a generated, and hence recursively enumerable, set of positive integers. We proceed to prove it hyper-simple.

Let B^* be any relevant basis. Of the finite number of bases preceding B^* in O^*, but a finite number can be contributing bases. Let $B^*_{j_\mu}$ be the last of these contributing bases, if any, appearing in the sequence $B^*_{j_1}, B^*_{j_2}, B^*_{j_3}, \cdots$ of distinct contributing bases determined by the above generation of H. There will then be a sequence s generated by B^* each of whose elements is greater than $k_\mu + 1$. When then (B^*, s), a definite element of O^*_1, is generated in the course of generating H, B^* will contribute each element of s to H unless it became a contributing basis earlier in the process. Hence, every relevant basis is a contributing basis.

It also follows, or is easily seen directly, that the number of contributing bases is infinite. Consider then the infinite sequence of contributing bases $B^*_{j_1}, B^*_{j_2}, B^*_{j_3}, \cdots$, the corresponding infinite sequence of subscripts $j_1, j_2, j_3, \cdots$, and the associated infinite sequence $k_1, k_2, k_3, \cdots$. Since the contributing bases are distinct, so are their subscripts. Hence, for each j_m, among the infinite set of j's following j_m there is a unique least j, $j_{m'}$. Consider then the resulting infinite sequence $j_{\lambda_1}, j_{\lambda_2}, j_{\lambda_3}, \cdots$, where j_{λ_1} is the least j in the whole infinite sequence of j's, while $\lambda_2 = (\lambda_1)'$, $\lambda_3 = (\lambda_2)'$, $\cdots$. Now k_{λ_n} is the largest integer contributed to H through the contributing basis with subscript j_{λ_n}. Since j_{λ_n} is the smallest j following $j_{\lambda_{n-1}}$ it is less than all succeeding j's. Hence B^* with subscript j_{λ_n} precedes in O^* all bases following that B^* in the above infinite sequence of contributing bases. Hence, each element added to H by contributing bases thus following B^* with subscript j_{λ_n} must exceed $k_{\lambda_n}+1$. It follows, on the one hand, that for each positive integer n, $k_{\lambda_n}+1$ is in $\bar{H}$. On the other hand, $k_{\lambda_{n+1}}$ itself exceeds $k_{\lambda_n}+1$ so that $k_{\lambda_{n+1}}+1>k_{\lambda_n}+1$. These members of $\bar{H}$ therefore constitute an infinite subset of the consequently infinite $\bar{H}$. Hence, H is hyper-simple.

Clearly, *every hyper-simple set H is simple.* For an infinite recursively enumerable subset of $\bar{H}$, as set of unit sequences, would contradict H being hyper-simple. Our construction of §6, in view of §8, gives us, however, *a simple set which is not hyper-simple.* Hyper-simple sets thus constitute a third class of recursively enumerable sets with recursively unsolvable decision problems—a class which is a proper subclass of the class of simple sets.

10. **Non-reducibility of creative sets to hyper-simple sets by truth-tables unrestricted.** Let creative set C be reducible by truth-tables to a recursively enumerable set of positive integers H. The given reduction will again determine for each positive integer n a finite sequence of positive integers $m_1, m_2, \cdots, m_\nu$, and a truth-table T of order ν such that that row of the table which correctly tells of the m's whether they are, or are not, in H will correctly tell of n whether it is, or is not, in C. Of course ν and T as well as the m's depend on n, and the set of distinct T's now entering into our reduction may be infinite, and hence the set of distinct ν's unbounded.

Let $l_1, l_2, \cdots, l_\mu$ be any given finite sequence of distinct positive integers. A particular hypothesis on the l's being, or not being, in H may then be symbolized by a sequence of μ signs, each $+$ or $-$, such as $+-\cdots+$, such that the ith sign is $+$, or $-$, according as the hypothesis says that l_i is in H, or $\bar{H}$, respectively. We shall speak of such a sequence of signs as a *truth-assignment* for the l's, the ith sign in that sequence as the *sign of l_i* in that truth-assignment. Of the 2^μ possible truth-assignments for the l's, constituting a set V_1, one and only one correctly tells of each l_i whether it is, or is not, in H. Every set V of truth-assignments for the l's is then a subset of V_1, and will be called a *possible set* of truth-assignments if it includes this *correct* truth-assignment.

Let then V be any given possible set of truth-assignments for the l's. Let n be a positive integer with corresponding $m_1, m_2, \cdots, m_\nu$, T yielded by the given reduction of C to H such that *each m not an l is in H.* The correct row of table T must then have the following two properties. First, the sign under each m not an l must be $+$. Second, the signs under those m's which are l's must be the same as the signs of those integers in some one and the same truth-assignment for the l's in V, in fact, as in the correct truth assignment for the l's. Any row of T having these two properties, given the l's, m's and V, will be called a *relevant row* of T. Since for our n the correct row of T is thus a relevant row, it follows that n will surely be in $\overline{C}$ if for each relevant row of T the sign under n is $-$.

Generate then the positive integers 1, 2, 3, $\cdots$, and as a positive integer N is generated, generate the first N members of a given recursive enumeration of H, and for each n, with $1 \leqq n \leqq N$, find the corresponding $m_1, m_2, \cdots, m_\nu, T$ yielded by the given reduction of C to H. Of those m's, if any, which are not l's, see if each is one of those first N members of H. If they all are, see if for each relevant row of T the sign under n is $-$. If that also is true, place n in a set α_V. Since each such n must be in $\overline{C}$, as seen above, α_V is a subset of $\overline{C}$. And being a generated set, α_V is therefore a recursively enumerable subset of $\overline{C}$.

C being creative, we can therefore find a definite positive integer n' in $\overline{C}$ but not in α_V, and, by the given reduction, the corresponding $m_1', m_2', \cdots, m_{\nu'}', T'$. Let $p_1, p_2, \cdots, p_\lambda$ be those m''s, if any, which are not l's. Now suppose that each p is in H. Then for at least one relevant row of T' the sign under n' must be $+$. For otherwise, if p_i is say the k_ith element in the given recursive enumeration of H, n' would have been placed in α_V in the above generation thereof for $N = \max(k_1, k_2, \cdots, k_\lambda, n')$. Since n' is in $\overline{C}$, such a relevant row cannot be the correct row. But, with each p in H, the signs in that row under m's that are not l's are correctly $+$. Hence the sign under at least one m' that is an l must be incorrect. But, by our definition of a relevant row, the signs under all such m''s are the same as the signs of those integers in at least one truth-assignment in V. Such a truth-assignment in V cannot therefore be the correct truth-assignment for the l's, and hence may be deleted from V. Perform this deletion for all such truth-assignments in V, and for all such relevant rows of T', to obtain the set of truth-assignments V'. Under our hypothesis that each p is in H, V' will then be a proper subset of V, and yet a possible set of truth-assignments for the l's.

Actually, let V be any given set of truth-assignments for the l's, possible or not. Each step of the above construction can then still be carried out, though the constructed entities need not now have all the properties they otherwise possess.[22] In particular, the set of integers, possibly null, $p_1, p_2, \cdots, p_\lambda$ can be found, all different from any l. Likewise, whether the p's are, or are not, all in H, the subset V' of V can be found. What we can say is that if V is a possible set,

[22] Recall that in the definition of creative set, §3, each B determines an n, whether the α determined by B is, or is not, a subset of $\overline{C}$.

and if furthermore each p is in H, then V' is a proper subset of V, and itself is also a possible set of truth-assignments for the l's.

For the given sequence of l's, start then with $V=V_1$, the possible set of all 2^μ truth-assignments for the l's, obtain the corresponding p's, p_1', p_2', $\cdots$, $p_{\lambda'}'$ and corresponding V', $V_2=(V_1)'$. With $V=V_2$, likewise find p_1'', p_2'', $\cdots$, $p_{\lambda''}''$, and $V_3=(V_2)'$, and so on. Now each V_{i+1} is a subset of V_i, while V_1 is but a finite set of 2^μ members. Hence in at most 2^μ steps we shall come across a V_κ such that either $V_{\kappa+1}$ is identical with V_κ, or is null. But if all the p''s, p'''s, $\cdots$, $p^{(\kappa)}$'s were in H, V_1 being a possible set, V_2, $\cdots$, V_κ as well as $V_{\kappa+1}$ would all be possible sets, each a proper subset of the preceding. $V_{\kappa+1}$ could not then either be identical with V_κ, or null. It follows that at least one of the $p_i^{(j)}$'s with $1\leqq j\leqq\kappa$ is in $\overline{H}$. Each $p_i^{(j)}$ is an integer that is not one of the l's. If then we take this finite set of $p_i^{(j)}$'s and arrange them in a sequence of distinct elements in say order of magnitude, we obtain for our arbitrarily given sequence of distinct positive integers $l_1, l_2, \cdots, l_\mu$ a sequence of distinct positive integers $k_1, k_2, \cdots, k_\nu$ having no element in common with the former sequence, and having at least one element in $\overline{H}$.

Starting with the null sequence as the sequence of l's, we can thus find the sequence of k's, $(k_1', k_2', \cdots, k_{\nu'}')$ of distinct positive integers at least one of which is in $\overline{H}$. Inductively, let us have thus generated the sequences $(k_1', k_2', \cdots, k_{\nu'}')$, $\cdots$, $(k_1^{(\mu)}, k_2^{(\mu)}, \cdots, k_{\nu^{(\mu)}}^{(\mu)})$, mutually exclusive, of distinct positive integers, each having at least one element in $\overline{H}$. With the single sequence $k_1', \cdots, k_{\nu^{(\mu)}}^{(\mu)}$ as the sequence of l's, we can find the corresponding sequence of k's, $(k_1^{(\mu+1)}, k_2^{(\mu+1)}, \cdots, k_{\nu^{(\mu+1)}}^{(\mu+1)})$ of distinct positive integers with no element in common with any of the preceding sequences, and having at least one element in $\overline{H}$.

With creative C reducible to recursively enumerable H by truth-tables we can thus obtain an infinite generated, and hence recursively enumerable, set of mutually exclusive finite sequences of positive integers each having an element in $\overline{H}$. The set H is therefore not hyper-simple. Hence the

THEOREM. *No creative set is reducible to a hyper-simple set by truth-tables.*

COROLLARY. *Every hyper-simple set is of lower degree of unsolvability than the complete set K relative to reducibility by truth-tables.*

Despite this result, the brief discussion of Turing reducibility, still in the informal stage, entered into in the next section makes it dubious that hyper-simple sets as such will give us the desired absolutely lower degree of unsolvability than that of K. But, in the absence of a counter-example, they remain candidates for this position.

11. General (Turing) reducibility. The process envisaged in our concept of a generated set may be said to be *polygenic*. In a *monogenic* process act succeeds act in one time sequence. The intuitive picture is that of a machine grinding out act after act (Turing [**24**]) or a set of rules directing act after act (Post [**18**]). The actual formulations are in terms of "atomic acts," the first leading to a development proved by Turing [**25**] equivalent to those arising from general recursive function or λ-definability, and hence of the same degree of generality. In our intuitive discussion the acts may be "molecular."

An effective solution of the decision problem for a recursively enumerable set S_1 of positive integers may therefore be thought of as a machine, or set of rules, which, given any positive integer n, will set up a monogenic process terminating in the correct answer, "yes" or "no," to the question "is n in S_1." Now suppose instead, says Turing [**26**] in effect, this situation obtains with the following modification. That at certain times the otherwise machine determined process raises the question is a certain positive integer in a given recursively enumerable set S_2 of positive integers, and that the machine is so constructed that were the correct answer to this question supplied on every occasion that arises, the process would automatically continue to its eventual correct conclusion.[23] We could then say that the machine effectively reduces the decision problem of S_1 to the decision problem of S_2. Intuitively this should correspond to the most general concept of the reducibility of S_1 to S_2. For the very concept of the decision problem of S_2 merely involves the answering for an arbitrarily given single positive integer m of the question is m in S_2; and in finite time but a finite number of such questions can be asked. A corresponding formulation of "Turing reducibility" should then be the same degree of generality for effective reducibility as say general recursive function is for effective calculability.[24]

We may note that whereas in reducibility by truth-tables the posi-

---

[23] Turing picturesquely suggests access to an "oracle" which would supply the correct answer in each case. The "if" of mathematics is however more conducive to the development of a theory.

[24] A reading of McKinsey [**16**] suggested generalizing the reducibility of a recursively enumerable set S to a recursively enumerable set S' to the reducibility of S to a finite set of recursively enumerable sets $S_1, S_2, \cdots, S_n$. However, no absolute gain in generality is thus achieved, as a single recursively enumerable set S' can be constructed such that reducing S to $(S_1, S_2, \cdots, S_n)$ is equivalent to reducing S to S'. Points of interest, however, do arise.

tive integers m of which we ask the questions "is m in S_2" are effectively determined, for given n, by the reducibility process, in Turing reducibility, except for the first such m, the very identity of the m's for which this question is to be asked depends, in general, on the correct answers having been given to these questions for all preceding m's. The mode of this dependence is, however, effective, hence we still have effective reducibility in the intuitive sense.

Let now creative set C be Turing reducible to a recursively enumerable set S of positive integers. We shall talk as if our intuitive discussion has already been formalized. Generate the positive integers 1, 2, 3, $\cdots$, and as a positive integer N is generated, for each n with $1 \leqq n \leqq N$ proceed as follows. Set going the reducibility process of C to S for n. Each time a question of the form "is m in S" is met, see if m is among the first N integers in a given recursive enumeration of S. If it is, supply the answer "yes," thus enabling the reducibility process to continue. Finally, if under these circumstances the process terminates in a "no" for the initial question of n being in C, place n in a set α_0. This α_0 is then a recursively enumerable subset of $\overline{C}$ consisting of all members thereof for which the given Turing reduction of C to S leads only to questions of the form "is m in S" whose answer is "yes."

Find then n_0 of $\overline{C}$ not in α_0, and set the reducibility process going for n_0. Now if at any time a wrong answer is supplied to a question "is m in S," we can nevertheless expect our machine for reducing C to S either to effectively pick up the wrong answer and operate on it to give a next step in the process, or to cease operating. Generate then the positive integers 1, 2, 3, $\cdots$, and as a positive integer N is generated, generate the first N members of the given recursive enumeration of S, and make the reducibility process for n_0 *effective* though perhaps *incorrect* as follows. Each time a question of the form "is m in S" is reached, see if m is among the first N members of S. If it is, answer the question "yes," and correctly so; if not, answer the question "no," whether that answer be correct or no. If now this *pseudo-reduction* terminates in a "no," place the finite number of m's thus arising in a set β_{n_0}. Note that β_{n_0} consists of all such m's for all such pseudo-reductions for the given n_0. Being a generated set of positive integers, β_{n_0} *is recursively enumerable.*

Now let the correct, though possibly non-effective, reducibility process for n_0 involve the μ questions "is m_i in S," $i=1, 2, \cdots, \mu$. Let $m_{i_1}, m_{i_2}, \cdots, m_{i_\nu}$ be those of these m's actually in S, and let them be the n_1st, n_2nd, $\cdots$, n_νth members of the given recursive enumeration of S. If then $N \geqq M = \max(n_1, n_2, \cdots, n_\nu)$, or $M=1$ if

$\nu = 0$, the corresponding psuedo-reduction for n becomes the correct reduction. For, inductively, if that be so through the time a question "is m in S" is raised, m will be $m_1, m_2, \cdots$, or m_μ, hence will, or will not, be in S according as it is, or is not, one of the first N members of S. The answer is then correctly given by that pseudo-reduction, which therefore continues to be correct through the raising of the next question. Finally, since n_0 is in $\overline{C}$, the correct reduction, now the pseudo-reduction, must terminate with a "no."

It follows that all N's with $N > M$ merely repeat the contribution to β_{n_0} made by $N = M$, that is, of the integers $m_1, m_2, \cdots, m_\mu$. Since but a finite number of m's are contributed by N's with $N < M$, it follows that *β_{n_0} is a finite set.* Finally, were each of the integers $m_1, m_2, \cdots, m_\mu$ in S, n_0 would be in α_0. Hence, *at least one member of β_{n_0} is in $\overline{S}$.*

Formally, we would thus obtain a basis for a finite recursively enumerable set of positive integers at least one of whose members is in $\overline{S}$. Instead of recursively enumerable sets of finite sequences of positive integers, we would thus be led to consider recursively enumerable sets of bases for finite recursively enumerable sets of positive integers. Though, in the last analysis, each sequence in the former case must be generated atom by atom, there will come a time for each sequence when the process will say "this sequence is completed." In the latter case, in general, we cannot have an effective method which, for each basis, will give a point in the ensuing process at which it can say all members of the finite set in question have already been obtained, even though, with the process made monogenic, there always is such a stage in the process.

This suggests, then, that we strengthen the condition of hyper-simplicity still further by replacing "infinitive recursively enumerable set of mutually exclusive finite sequences of positive integers" in the definition of §9 by "infinite recursively enumerable set of bases defining mutually exclusive finite recursively enumerable sets of positive integers." Whether such a "hyper-hyper-simple" set exists, or whether, if it exists, it will lead to a stronger non-reducibility result than that of the last section we do not know.

On the other hand, an equivalent definition of hyper-simple set is obtained if, for example, we replace the quoted phrase by "recursively enumerable set of finite sequences of positive integers having for each positive integer n a member each of whose elements exceeds n." We now can say that with this as the definition of a hyper-simple set, the corresponding extension to a hyper-hyper-simple set cannot be made. For we prove the

THEOREM. *For any recursively enumerable set of positive integers S, with infinite $\bar{S}$, there exists a recursively enumerable set of bases defining finite recursively enumerable sets of positive integers, each set having at least one element in $\bar{S}$, and at least one set having each of its elements greater than an arbitrarily given positive integer n.*

Briefly, with n given, for each positive integer N, and each positive integer m, place all of the integers $n+1, n+2, \cdots, n+m$ in a set α_n if all, or all but the last, are among the first N members of a given recursive enumeration of S. It is readily seen that α_n is a generated, and hence recursively enumerable, set of positive integers. A corresponding basis $B^{(n)}$ can actually be found, and the set of $B^{(n)}$'s, $n=1, 2, 3, \cdots$, being a generated set, is therefore recursively enumerable. Moreover, if ν_n is the smallest integer in the infinite $\bar{S}$ greater than n, α_n will consist of exactly the integers $n+1, n+2, \cdots, \nu_n$, and hence will be finite, with indeed ν_n as the only element in $\bar{S}$, and with each element greater than n.

As a result we are left completely on the fence as to whether there exists a recursively enumerable set of positive integers of absolutely lower degree of unsolvability than the complete set K, or whether, indeed, all recursively enumerable sets of positive integers with recursively unsolvable decision problems are absolutely of the same degree of unsolvability. On the other hand, if this question can be answered, that answer would seem to be not far off, if not in time, then in the number of special results to be gotten on the way.[25]

Such then is the portion of "Recursive theory" we have thus far developed. In fixing our gaze in the one direction of answering the lower degree of unsolvability question, we have left unanswered many questions that stud even the short path we have traversed. Moreover, both our special, and the general Turing, definitions of reducibility are applicable to arbitrary decision problems whose questions in symbolic form are recursively enumerable, and indeed to problems with recursively enumerable set of questions whose answers belong to a recursively enumerable set. Thus, only partly leaving the field of decisions problems of recursively enumerable sets, work of Turing [26] suggests the question is the problem of determining of an arbitrary basis B whether it generates a finite, or infinite, set of positive

[25] This is a matter of practical concern as well as of theoretical interest. For according as the second or first of the above alternatives holds will the method of reducing new decision problems to problems previously proved unsolvable be, or not be, the general method for proving the unsolvability of decision problem either of recursively enumerable sets of positive integers or of problems equivalent thereto.

integers of absolutely higher degree of unsolvability than K. And if so, what is its relationship to that decision problem of absolutely higher degree of unsolvability than K yielded by Turing's theorem.

Actually, the theory of recursive reducibility can be but one chapter in the theory of recursive unsolvability, and that, but one volume of the theory and applications of general recursive functions. Indeed, if general recursive function is the formal equivalent of effective calculability, its formulation may play a rôle in the history of combinatory mathematics second only to that of the formulation of the concept of natural number.

Bibliography

1. Alonzo Church, *An unsolvable problem of elementary number theory*, Amer. J. Math. vol. 58 (1936) pp. 345–363.[a]

2. ———, *A note on the Entscheidungsproblem*, Journal of Symbolic Logic vol. 1 (1936) pp. 40–41.[a]

3. ———, *Correction to a note on the Entscheidungsproblem*, ibid. pp. 101–102.[a]

4. ———, *Combinatory logic as a semi-group*, Preliminary report, Bull. Amer. Math. Soc. abstract 43-5-267.

5. ———, *The constructive second number class*, ibid. vol. 44 (1938) pp. 224–232.

6. ———, *The calculi of lambda-conversion*, Annals of Mathematics Studies, no. 6, Princeton University Press, 1941.

7. C. J. Ducasse, *Symbols, signs, and signals*, Journal of Symbolic Logic vol. 4 (1939) pp. 41–52.

8. Kurt Gödel, *Über formal unentscheidbare Sätze der Principia Mathematica und verwandter Systeme* I, Monatshefte für Mathematik und Physik vol. 38 (1931) pp. 173–198.[a]

9. ———, *On undecidable propositions of formal mathematical systems*, mimeographed lecture notes, The Institute for Advanced Study, 1934.[a]

10. David Hilbert, *Mathematical problems*. Lecture delivered before the International Congress of Mathematicians at Paris in 1900. English translation by Mary Winston Newsom, Bull. Amer. Math. Soc. vol. 8 (1901–1902) pp. 437–479.

11. David Hilbert and Paul Bernays, *Grundlagen der Mathematik*, vol. 2, Julius Springer, Berlin, 1939.

12. S. C. Kleene, *General recursive functions of natural numbers*, Math. Ann. vol. 112 (1936) pp. 727–742.[a]

13. ———, *On notation for ordinal numbers*, Journal of Symbolic Logic vol. 3 (1938) pp. 150–155.

14. ———, *Recursive predicates and quantifiers*, Trans. Amer. Math. Soc. vol. 53 (1943) pp. 41–73.[a]

15. C. I. Lewis, *A survey of symbolic logic*, Berkley, 1918, chap. 6, §3.

16. J. C. C. McKinsey, *The decision problem for some classes of sentences without quantifiers*, Journal of Symbolic Logic vol. 8 (1943) pp. 61–76.

17. Rozsa Péter, *Az axiomatikus módszer korlátai* (The bounds of the axiomatic method), Review of, Journal of Symbolic Logic vol. 6 (1941) pp. 111.

18. Emil L. Post, *Finite combinatory processes—formulation* 1, ibid. vol. 1 (1936) pp. 103–105.[a]

19. ———, *Formal reductions of the general combinatorial decision problem*, Amer. J. Math. vol. 65 (1943) pp. 197–215.

20. J. B. Rosser, *Extensions of some theorems of Gödel and Church*, Journal of Symbolic Logic vol. 1 (1936) pp. 87–91.[a]

21. ———, *An informal exposition of proofs of Gödel's theorems and Church's theorem*, ibid. vol. 4 (1939) pp. 53–60.[a]

22. Th. Skolem, *Einfacher beweis der unmöglichkeit eines allgemeinen losungsverfahrens für arithmetische probleme*, Review of, Mathematical Reviews vol. 2 (1941) p. 210.

23. Alfred Tarski, *On undecidable statements in enlarged systems of logic and the concept of truth*, Journal of Symbolic Logic vol. 4 (1939) pp. 105–112.

24. A. M. Turing, *On computable numbers, with an application to the Entscheidungsproblem*, Proc. London Math. Soc. (2) vol. 42 (1937) pp. 230–265. [a]

25. ———, *Computability and λ-definability*, Journal of Symbolic Logic vol. 2 (1937) pp. 153–163.

26. ———, *Systems of logic based on ordinals*, Proc. London Math. Soc. (2) vol. 45 (1939) pp. 161–228. [a]

THE CITY COLLEGE,
NEW YORK CITY.

ABSOLUTELY UNSOLVABLE PROBLEMS AND RELATIVELY UNDECIDABLE PROPOSITIONS

ACCOUNT OF AN ANTICIPATION

This paper was submitted by Post to a mathematics periodical in 1941 and was rejected. Certainly the speculative (not to say metaphysical) "Appendix" is not what one ordinarily expects to find in a mathematics paper. So, it appears in print for the first time in this anthology.

The paper recounts Post's anticipation by a decade of the fundamental conclusions of Gödel, Church, and Turing with which the papers in this anthology are concerned. For Post, these conclusions were based on an assumption which is equivalent to Church's thesis. Hence, Post's further efforts were directed towards verification of the assumption, which to him was a matter of "psychological analysis of the mental processes involved in combinatory mathematics." He proposed to publish his work only after he had completed this analysis. In the meanwhile, the more complete development by Gödel appeared.

In the present period of increasing interest in logic, it may not be amiss to recall that during the twenties the field suffered from virtually total neglect in the United States. Post was faced not only with a lack of understanding of what he was trying to do but also with the difficulty of earning his living by teaching in the high schools.

Of the technical material in the present paper, only the material on reduction of systems in Canonical form C (nowadays called "canonical systems") to normal systems has appeared in print. The equivalence proofs given between Canonical forms A, B, and C are of historical interest, showing as they do the evolution of Post's ideas from his early work on the propositional calculus.

The reader should note that what Post calls "enunciations" would now be called "w. f. f. 's". Similarly, his "assertions" would be called "theorems", his "primitive assertions" would be called "axioms", and his "finiteness problem" would be called "decision problem".

The problem of tag, which for Post was such a frustrating initial step in his effort to solve the decision problem for Principia Mathematica, has recently been proved unsolvable by Marvin Minsky (Annals of Mathematics, vol. 74 (1961) pp. 437-455).

Emil Post

ABSOLUTELY UNSOLVABLE PROBLEMS AND RELATIVELY UNDECIDABLE PROPOSITIONS ACCOUNT OF AN ANTICIPATION[1]

There would be little point in publicizing the writer's anticipation of the existence of absolutely unsolvable problems in the sense of Church,[2] and, as a corollary thereof, of Gödel's

1. The phrase "absolutely unsolvable" is due to Church who thus described his problem in answer to a query of the writer as to whether the unsolvability of his elementary number theory problem was relative to a given logic (see footnote 2). By contrast, the undecidable propositions in this and related papers are undecidable only with respect to a given logic. A fundamental problem is the question of the existence of absolutely undecidable propositions, that is, propositions which in some a priori fashion can be said to have a determined truth-value, and yet cannot be proved or disproved by any valid logic. An attempt at formulating such a proposition appears in the appendix. The ideal candidate therefor would be a suitable arithmetic proposition in the sense of Gödel. For to the writer it is axiomatic that if the truth-value of $\phi(n)$ "is determined" for each natural number n, the truth-value of $(\exists n)\phi(n)$ and of $(n)\phi(n)$, n restricted to natural numbers, "is determined"- whether determinable by us, or not.

The writer cannot overemphasize the fundamental importance to mathematics of the existence of absolutely unsolvable combinatory problems. True, with a specific criterion of solvability under consideration, say recursiveness (see footnote 6), the unsolvability in question, as in the case of the famous problems of antiquity, becomes merely unsolvability by a given set of instruments. And, indeed, the corresponding proofs for combinatory problems are almost trivial in comparison with the classic unsolvability proofs. The fundamental new thing is that for the combinatory problems the given set of instruments is in effect the only humanly possible set.

2. Alonzo Church, "An unsolvable problem of elementary number theory." American Journal of Mathematics, vol. 58 (1936), pp. 345-363 [this anthology, pp. 89-107] .

theorem on the incompleteness of symbolic logics,[3] merely as a claim to unofficial priority. Indeed, the present development is but fragmentary by comparison. But with the Principia Mathematica of Whitehead and Russell as a common starting point, the roads followed towards our common conclusions are so different that much may be gained from a comparison of these parallel evolutions. Less important is the bridge the present paper would provide between the writer's past and it is hoped future work.

The point of departure of the present paper is the formulation presented in the third part of the writer's dissertation of 1921 (see §1). Within its scope the development of Part I is complete, and, indeed, whatever persuasiveness there is in the actual anticipation presented in Part II derives from it. The formative work which largely led to this anticipation falls within the writer's tenure as Proctor Fellow in Princeton for the academic year 1920-21, and the summers immediately preceding and following. Actual dates are, for §1 and §2, October 1920, §4 and §5 August and September 1921; Part II, September and October 1921, the last entry of the appendix, February 24, 1922.[4]

3. Kurt Gödel, "Uber formal unentscheidbare Sätze der Principia Mathematica und verwandter Systeme I." Monatshefte für Mathematik und Physik, vol 38 (1931), pp. 173-198 [this anthology, pp. 5-38].

4. This work was significantly carried forward during June-July 1924. The first half of this month of work was devoted to developing an "operational logic" for sequences (see the next to the last paragraph of §8 and footnote 8). This development was completely detailed as far as it went. The consideration of quantifiers led to the second half of the month's work where a hierarchy of "mandates" (directions), "super mandates" etc. were introduced. At the time, this hierarchy was thought to exhaust all possibilities, but, at least by 1929, was later seen to but include mandates of finite type, admitting of extension into the transfinite. The more detailed discussion of mandates of lowest type all but anticipated the writer's published note of 1936 (see footnote 6). We might add that since February 1938 we have given an occasional week to a continuation of this work, and largely

In keeping with the parallel evolution idea the present account tries to keep as close as it can to our original notes, even to the terminology employed. The footnotes then serve to give us a running critique. Because the work covered in Part I aimed at completeness as far as it went, we have there taken the liberty of correcting a number of errors, more or less serious, occurring in the originally unchecked notes, as well as adding more explanatory material than is there given. These departures are noted in the footnotes, and in every case attempt to go no further than a re-check on the notes at the time would have yielded. On the other hand, in the case of Part II we have deemed it essential not to alter or supplement the essential content of the notes, but merely to connect and occasionally smooth out their presentations. Finally, except for the introductory paragraphs, the appendix consists entirely of quotations from notes and concurrent diary,[5] corrected only for spelling, obvious slips in words, and flagrant crimes against grammar.

Perhaps the chief difference in method between the present development and its more complete successors is its preoccupation with the outward forms of symbolic expressions, and possible operations thereon, rather than with logical concepts as clothed in, or reflected by, correspondingly particularized symbolic expressions, and operations thereon. While this in part is perhaps responsible for the fragmentary nature of our development, it also allows greater freedom of method and technique. In particular, it reveals the "Gödel Representation" to be merely a case of resymbolization, and suggests that with the growth of mathematics such resymbolization will have to be effected again and again.

in the spirit of the appendix. Our goal, however, is now an analysis of proof, perhaps leading to an absolutely undecidable proposition, rather than an analysis of finite process.

5. This diary, under the title "Time Accounts", was begun in the spring of 1916 and continued without interruption to the spring of 1922.

Apart from the indirect role the present paper may play as a different, if imperfect, pair of lenses with which to view recent developments, it has a direct contribution to make to present day literature in adding still another precise formulation to the list of general recursiveness, λ-definability, computability.[6] Where in these formulations the informal basic idea is that of effective calculability, our own is that of a generated set. This derives from the idea of a symbolic logic rather than that of an algorithm, and may be described by saying that each member of the set is at some time generated[7] by the continued application of a given method, while that method will at no time yield an individual on the primitives of the set which is not in the set. Our emphasis is on generated sets of sequences on a fixed set of symbols $a_1, a_2, \ldots, a_\mu$.[8] The precise

6. For the first two see footnote 2, for the third see A. M. Turing, "On computable numbers, with an application to the Entscheidungsproblem." Proc. London Math. Soc. (2), vol. 42 (1937), pp. 230-265 [this anthology, pp. 116-154] later referred to as computable numbers. We might also add, E.L. Post, "Finite combinatory processes - formulation I." Journal of Symbolic Logic, vol. 1 (1936), pp. 103-105 [this anthology, pp. 289-291].

In this connection we must emphasize the distinction between a formulation which includes an equivalent for every possible "finite process", and a description which will cover every possible method for setting up finite processes. It is towards the latter goal that the appendix of the present paper strained, the first having been achieved in the generalization ending §7. While the Turing simplifications referred to in footnote 9 may make the detailed development envisioned in the appendix unnecessary for the analysis of process, though retaining an intrinsic interest as added description, it is doubtful if Turing considerations can replace such a development in the analysis of proof.

7. Produced, created- in practice, written down.

8. More exactly, "strings" of such symbols to use a term of C.I. Lewis (A Survey of Symbolic Logic. Berkeley, Cal. 1918; chapter VI, sec. III). On the other hand, in the 1924 "operational logic" for sequences, referred to in footnote 4, after many pros and cons, this concept was given up for essentially the Principia Mathematica concept of sequences of repeatable elements. Though this avoided

formulation is that of a normal system on a fixed set of symbols (see the beginning of §9), and our identification, not as definition, but as at least partially verified conclusion,[9] is of generated set of sequences on $a_1, a_2, \ldots, a_\mu$ with augmented normal system[10] on $a_1, a_2, \ldots, a_\mu$, i.e., the subset of sequences on $a_1, a_2, \ldots, a_\mu$ only, of a normal system on $a_1, a_2, \ldots, a_\mu$, and additional letters $a_1', a_2', \ldots, a_\mu'$. The reductions of §§4,5, coupled with certain private correspondence with Church, make it certain that generated set in this precise sense is equivalent to "recursively enumerable set", while a "recursive set" of sequences on $a_1, a_2, \ldots, a_\mu$ would be one for which both it and its negative, i.e., complement with respect to the set of all finite sequences on $a_1, a_2, \ldots, a_\mu$, are generated sets.[11] We thus reverse the usual order of these two concepts; and since our "generated set" is not burdened with the added ordering superimposed on the "recursively enumerable set", there may be certain advantages in making it the primary concept.

the difficulties of "identity" for sequences as strings, and allowed a considerable development prior to the introduction of quantifiers, the ultimate wisdom of this decision is questionable.

9. In this connection see the last paragraph of the writer's note referred to in footnote 6. However, should Turing's finite number of mental states hypothesis (computable numbers, p. 250) [this anthology, p. 136] bear up under adverse criticism, and an equally persuasive analysis be found for all humanly possible modes of symbolization, then the writer's position, while still tenable in an absolute sense, would become largely academic.

10. Though we have had no occasion to introduce this phrase in our account proper, it occurs frequently in our notes, especially in the work of 1924. The concept, of course, is present in the generalization ending §7. In describing an augmented normal system as a set of sequences, we continue the phraseology of §9 where a normal system is so considered. Actually, apart from the purposes of §9, the notes concept of normal system included also the methods of generating the sequences in question, and the same is true of augmented normal system. Thus, in 1924, the notes repeat that every set of finite sequences on $a_1, a_2, \ldots, a_\mu$ that can be generated, can be generated by an augmented normal system.

11. In accordance with footnote 96, the concept of recursive set of sequences

But perhaps the greatest service the present account could render would stem from its stressing of its final conclusion that mathematical thinking is, and must be, essentially creative.[12] It is to the writer's continuing amazement that ten years after Gödel's remarkable achievement current views on the nature of mathematics are thereby affected only to the point of seeing the need of many formal systems, instead of a universal one. Rather has it seemed to us to be inevitable that these developments will result in a reversal of the entire axiomatic trend of the late nineteenth and early twentieth centuries, with a return to meaning and truth. Postulational thinking will then remain as but one phase of mathematical thinking. While in the appendix we may be but following a will-o'-the-wisp, its very gropings may in some measure give an inkling of the creative unity that is Mathematics.

appeared in our work in the following equivalent form: a generated set of sequences for which there is a finite-normal-test (see §9). But it was not then positively stressed. By 1924 its importance was beginning to be recognized, the formulation given in the body of the introduction then appearing. But its use was postponed for later applications, with generated set remaining the basic concept.

12. Yet, as this account emphasizes, the creativeness of human mathematics has as counterpart inescapable limitations thereof— witness the absolutely unsolvable (combinatory) problems. Indeed, with the bubble of symbolic logic as universal logical machine finally burst, a new future dawns for it as the indisputable means for revealing and developing those limitations. For, in the spirit of the appendix, Symbolic Logic may be said to be Mathematics become self-conscious. Actually, the old dream of symbolic logic is finding partial realization in Tarski's recent positive work on decision problems .

PART I. FORMAL TRANSFORMATIONS

1. Canonical form A and its reduction to a canonical form B.

In a previous paper[13] the writer proposed the following formal postulational generalization of the $(\sim, \vee)$ system of Principia Mathematica.[14] For an arbitrary finite set of primitive functions of propositions

$$f_1(p_1, p_2, \ldots, p_{m_1}), \ \ldots, \ f_\mu(p_1, p_2, \ldots, p_{m_\mu}),$$

of an arbitrary finite number of arguments each, assume a set of postulates of the following form.

I. If $p_1, \ldots, p_{m_1}$ are elementary propositions, so is $f_1(p_1, \ldots, p_{m_1})$.

. . . .

If $p_1, \ldots, p_{m_\mu}$ are elementary propositions, so is $f_\mu(p_1, \ldots, p_{m_\mu})$.

II. The assertion of a function involving a variable p produces the assertion of any function found from the given one by substituting for p any other variable q, or $f_1(q_1, \ldots, q_{m_1}), \ldots,$ or $f_\mu(q_1, \ldots, q_{m_\mu})$.

III. "$\vdash g_{11}(P_1, P_2, \ldots, P_{k_1})$" . . . "$\vdash g_{\kappa 1}(P_1, P_2, \ldots, P_{k_\kappa})$"

. . . .

"$\vdash g_{1\kappa_1}(P_1, P_2, \ldots, P_{k_1})$" . . . "$\vdash g_{\kappa\kappa_\kappa}(P_1, P_2, \ldots, P_{k_\kappa})$"

produce . . . produce

"$\vdash g_1(P_1, P_2, \ldots, P_{k_1})$", . . . , "$\vdash g_\kappa(P_1, P_2, \ldots, P_{k_\kappa})$",

13. E.L. Post, "Introduction to a general theory of elementary propositions." American Journal of Mathematics, vol. 43 (1921) pp. 163-185. See §8 p. 176 thereof.

14. That is, its propositional calculus. See A.N. Whitehead and B. Russell, Principia Mathematica. 2nd ed. vol. I. Cambridge, England, 1925: part I, section A.

where the P's are any combination of f's including the special case of the unmodified variable, while the g's are particular combinations of this kind which need not have all the indicated arguments.

IV. $\vdash h_1(p_1, p_2, \ldots, p_{l_1})$,
$\vdash h_2(p_1, p_2, \ldots, p_{l_2})$,
. . . .

$\vdash h_\lambda(p_1, p_2, \ldots, p_{l_\lambda})$,

where the h's are particular combinations of the f's.[15] As in the description of the $(\sim, \vee)$ system given in that paper, we may observe that I may be said to determine the enunciations of the system under consideration[16]. On the other hand, the repeated application of operations II and III to the primitive assertions in IV, and all derived assertions that may thus result, yield a subset of the set of enunciations consisting of the assertions of the system[17].

15. By a combination of f's is meant any expression built up out of the primitive functions and variables by the operation of substitution — strictly so for the P's of III, with added abstraction of the final variables for the g's and h's. Where our notes always talk of "the capital P's", we have allowed ourselves occasionally the phrase "the operational variables". These occur only in the basis of a system, while "variables" refer to the p's, q's, etc. which are in the system itself. Note that our formulation is really that of a class of systems, each explicitly given basis yielding a corresponding system.

16. "Enunciation" is at least approximately equivalent to the more recent "well-formed formula".

17. We take this opportunity to point out an error in §10 of the above paper. Lemma 1 thereof requires the added condition that the expressions replacing the r's do not involve any letter upon which a substitution is made in the given deductive process. This necessitates several minor changes in the proof of the basic theorem that follows. Actually, a little further analysis than was effected at the time allows the lemma 2 of the paper to be strengthened to

> Lemma 2'. The most general process of obtaining an assertion from a given set of assertions in accordance with II and III can be reduced to first asserting a number of functions in accordance with II, and then applying only operations III.

In the case of the $(\sim, \vee)$ system the writer proved, in the paper referred to, that by introducing the truth-table concept a finite method was thereby afforded for determining of any enunciation in the system whether it was or was not an assertion of the system. We shall say that we thus solved the finiteness problem[18] for the $(\sim, \vee)$ system. While this solution was purely formal, nevertheless it was suggested by the intuitive interpretation of "$\sim$" and "$\vee$". For the above generalizations such interpretations are not at hand. Nevertheless, even before the publication of the above paper, the writer solved the finiteness problem for those of the above systems in which the primitive functions are all functions of one variable, the resulting relative simplicity of the systems allowing a direct analysis of the formal processes involved[19]. While considerable further labor produced but minor dents in the problem for the above

It then suffices to replace lemma 1 by the simpler

> Lemma 1'. If a given set of functions gives rise to some other function solely through the use of operations III, then the same deductive process will be valid if we have given the original functions with an arbitrary substitution on their letters as described in II, provided this substitution is also made throughout the process.

With these simpler lemmas the proof of the theorem is valid as it stands. In this connection see footnotes 21 and 24.

18. That is, the "deducibility problem" in the sense of Church. "Decision problem", if a translation of "Entscheidungsproblem", seems to have this as but one of two distinct meanings. (See Alonzo Church, "Correction to 'A note on the Entscheidungsproblem." Journal of Symbolic Logic, vol. 1 (1936) pp. 101-102) [115]. An alternative name for the "finiteness problem" in our notes is "the fundamental problem". This name occurs in an unpublished "Note on a fundamental problem in postulate theory. "of the writer bearing the date June 4, 1921. This note was then left at Princeton, presumably for publication, but was withdrawn the following fall as a result of the nullifying of its program by the anticipation recorded in Part II.

19. An abstract of this as yet unpublished paper appears under E.L. Post, "On a simple class of deductive systems." Bull. Amer. Math. Soc., vol. 27 (1921) pp. 396-7.

systems not so restricted, impetus was lent to the work by our formally reducing the subsystem of Principia Mathematica treated in *10 and *11 thereof[20] to a system of the above type. For thereby a solution of the finiteness problem for all of the above systems would immediately lead to a solution of that problem for this important subsystem of Principia Mathematica. Actually, it is necessary to replace the above formulation by an equivalent formulation, the reduction in question being to a system in this modified form. To avoid the lengthy interruption of the direct proof of the equivalence of these two formulations, but one half of this equivalence is established in the present section, the rest following from sections 4 and 5 with the help of the short additional section 6.

Before this modification can be introduced, we must re-examine a certain dubious feature of the above original formulation. Formally, I should have been recast in operational form à la II, III, and IV to constitute a method for generating all enunciations of the system. A symbol for enunciation corresponding to "⊢" for assertion could have been introduced. Of course, a vital difference between the two iterative processes thus set up would have been that whereas enunciations thus arise with, as it were, complexities monotonically increasing, not so, in general, for assertions. Now II is precisely enough stated to be independent of this assumed generation of the enunciations of the system. But III, as given, is so dependent. Now actually, the description of the P's occurring in III could have been omitted for any P occurring in a premise therein. For once that premise has arisen as an assertion, and is written in the assumed form, the P's thereof can only be variables or enunciations. But not so for a P occurring in the conclusion of a production, but in no premise thereof; and III as stated allows such productions. Furthermore, when such a production is used, it has the theoretical disadvantage that the premises do not determine the conclu-

20. That is, its restricted functional calculus.

sion, but a class of conclusions made precise only by the precise generation of enunciations. In our new formulation we therefore replace III by

III': III with the added restriction that each capital *P* of a conclusion is present in at least one premise of the corresponding production.[21]

Our principal aim however is to weaken operation II; for a formal system to which *10 - *11 Principia Mathematica is to be reduced must not allow, for example, the replacing of a variable x in an expression involving $f(x)$ by anything except a variable, and that distinct from the variable f. We therefore replace II by the very weak

II': II restricted to the replacing of a variable by any other variable, and that not present in the given assertion.

By iterating II' we can therefore first perform any 1-1 replacement of the variables in an assertion by variables not in the assertion, and hence, finally, any 1-1 replacement of variables by variables.

In saying that the formulations ***A***:I, II, III, IV, and ***B***:I, II', III', IV, are equivalent we mean that the solution of the finiteness problem for all systems coming under either formulation would

21. This change in III is not made in our notes; but, in point of fact, all of the productions occurring in our notes corresponding to the work of §2 on do satisfy the restriction present in III'. The notes attempt in part to remedy this defect in III by suggesting the convention that the general description given for the ***P***'s only apply to ***P***'s in the premises of a production, while a ***P*** in the conclusion thereof not present in any premise represent a variable only. Due to the presence of II, this convention does not alter the effectiveness of II and III combined. On the other hand, the notes explicitly use lemma 2' as given in footnote 17, but this strengthening of lemma 2 is not possible under the proposed convention. It may be that an implicit assumption of this convention was responsible for the use of lemmas 1 and 2 instead of 1' and 2' in the paper referred to. We finally note that with III' replacing III, the proofs of the emended and modified lemmas are greatly simplified.

lead to the solution of the finiteness problem for all systems coming under the other formulation. More specifically, a system S_1 will be said to be reduced to a system S_2 if a method is presented which would transform a solution of the finiteness problem for S_2 into one for S_1.[22] We shall say that a system falling under a formulation X is in canonical form X. A system S_1 will be said to be reduced to canonical form X if it is reduced to a system in canonical form X, and canonical form Y will be said to be reduced to canonical form X if a method is presented whereby each system in canonical form Y is reduced to canonical form X. Canonical forms X and Y are then equivalent if each is reducible to the other.

We proceed to prove that canonical form A is reducible to canonical form B.[23]

Let S_1 be any system in canonical form A. By a lemma stated in our earlier paper,[24] an arbitrary proof in S_1 can

22. Actually, no definition is given in the notes. But the reference therein to transforming one system into part of another, coupled with the actual developments given, suggest that the following more precise definition was tacitly assumed. A system S_1 is reduced to a system S_2 if a 1-1 correspondence is (effectively) set up between the enunciations of S_1, and certain of the enunciations of S_2, so that an enunciation of S_1 is asserted when and only when its correspondent in S_2 is asserted.

23. This, the easier part of the equivalence proof, was given incorrectly in the notes and probably for that very reason. The error was subsequently noted, but, while saying that it could easily be overcome, the further error noted in footnote 21 arose. Actually, the notes weaken operation II in two successive stages, the first allowing a variable to be replaced by any other variable. The bulk of the work concerns the reducibility of each of the resulting canonical forms to canonical form A. In the case of the intermediate form, the proof depends on an unpublished method of the writer (the L.C.M. process referred to early in §3), but the method appears in simplified form in the second reduction of §6. In the case of the present canonical form B (except for III'), the method depended on replacing the variables p_1, $p_2, p_3, \ldots$ by $a(p)$, $a(a(p))$, $a(a(a(p)))$, ..., $a(p)$ being a new primitive function, and thus was a precursor of the method fully presented in §4.

24. So say the notes. But while the paper states lemma 2, this restates lemma

always be replaced by one which first merely obtains a set of assertions by the repeated use of II starting with assertions in IV, and then merely repeats the use of III starting with the set thus found. Now the assertions of S_1 that can be thus found via II and IV only are all enunciations of the form $h_i(P_1, P_2, \ldots, P_{l_i})$ $i=1,2,\ldots,\lambda$, where the P's are arbitrary variables or enunciations in S_1. It will be convenient in this connection to include variables in the class of enunciations of a system. We therefore introduce a new primitive function $e(p)$, and set up a system in canonical form B such that $e(P)$ will be asserted in that system when and only when P is an enunciation in S_1. For that system, I is to be the I of S_1 with the additional postulate corresponding to the primitive function $e(p)$, II′ the one II′ of all systems in canonical form B. Our immediate purpose is then achieved by taking for III′ and IV the following.

III.′ "$\vdash e(P_1)$"

. . . .

"$\vdash e(P_{m_i})$" $i=1,2,\ldots,\mu$.

produce

"$\vdash e(f_i(P_1,\ldots,P_{m_i}))$".

IV. $\vdash e(p)$.

That the desired result is thus achieved may be seen from the fact that this IV and II′ yield all "$\vdash e(P)$"'s where P is a variable, that is, where P is any enunciation of rank zero in S_1;[25] and, assuming inductively that the same is true for all enunciations in S_1 of rank $\leqslant \rho$. III′ insures the result for all enunciations in S_1 of rank $\rho+1$. Hence every "$\vdash e(P)$" with P an enunciation in S_1 appears in our new system; and a similar induction shows

2′ (see footnotes 17 and 21). Whether this was an oversight on the part of the notes, or the result of further analysis, is not clear.

25. The rank of an arbitrary enunciation $f_i(P_1,\ldots,P_{m_i})$ is then inductively defined as one more than the maximum of the ranks of $P_1,\ldots,P_{m_i}$.

that no other "$\vdash e(P)$"'s thus appear than with P an enunciation in S_1.

We now add to the above III$'$, keeping I, II$'$ and IV unchanged. These additions are such that no conclusion of a production added to III$'$ is of the form $e(P)$. Hence no new assertions of the form "$\vdash e(P)$" result.

First add to III$'$ the productions

III: " $e(P_1)$"

. . . .

" $e(P_{l_i})$" $\quad i=1,2,\ldots,\lambda$.

produce

" $h_i(P_1,\ldots,P_{l_i})$".

We shall then have asserted in our system all assertions in S_1 obtained solely by the use of the II and IV of S_1. The effect of III of S_1 is then easily reproduced in our new system as follows. Let the i-th production in III of S_1 have $P_{j_{i,1}},\ldots,P_{j_{i,\nu_i}}$ for those of its operational variables that are present in the conclusion of that i-th production, but in no premise thereof. The full effect of III of S_1 will then be achieved if we add to III$'$,

III:$'$ "$\vdash g_{i1}(P_1,\ldots,P_{k_i})$"

. . . .

"$\vdash g_{i\kappa_i}(P_1,\ldots,P_{k_i})$"

"$\vdash e(P_{j_{i,1}})$" $\quad i=1,2,\ldots,\kappa$,

. . . .

"$\vdash e(P_{j_{i,\nu_i}})$"

produce

"$\vdash g_i(P_1,\ldots,P_{j_{i,1}},\ldots,P_{j_{i,\nu_i}},\ldots,P_{k_i})$",

the g's being those of the III of S_1. In fact, with "$\vdash e(P)$" read meaningfully, as is justified by our proved result concerning its occurrence in our new system, these additions to III′ become exactly the III of S_1.

Let then S_2 be the last found system, i.e., having for basis the I, II′, IV as first given for our new system, and the III′ consisting of the several parts thereof listed above. S_2 is then in canonical form *B*. Clearly the enunciations of S_1 are among the enunciations of S_2. By the restatement of the way in which an arbitrary assertion of S_1 can be found, it also follows from the above that every assertion in S_1 is also an assertion in S_2. In fact, it is immediately seen that apart from assertions of the form "$\vdash e(P)$" the assertions in S_2 are the assertions in S_1. The solution of the finiteness problem for S_2 thus immediately becomes a solution of the finiteness problem for S_1. That is, S_1 in canonical form *A* has been reduced to S_2 in canonical form *B*.

2. Reduction of *10-*11 Principia Mathematica to canonical form *B*.[26]

Leaving aside for the moment the question of real versus apparent variables,[27] we observe that *10 allows for three distinct classes of variables: propositional variables p, q, r, . . . , individual variables x, y, z, . . . , functional variables

26. Since, with the exception of the explanatory *11.07, the basis for this subsystem is given in *10, we shall henceforth refer to this subsystem as *10, it being understood that in its verbal title "one apparent variable" is replaced by "an arbitrary finite number of apparent variables". Actually, our version of this subsystem is narrower than that of Principia Mathematica, since we assume the variables x, y, z,... thereof to have a common range. This explains the concluding remark of the notes, "This completes the discussion (except for some questions on type)". At the start the notes state, "We shall now attempt to prove that the development of *10 is equivalent to the entire set *A* or *B* or *C*", but at this stage only the reducibility of *10 is considered. Concerning the other half of the equivalence see the second paragraph of footnote 79.

27. Free versus bound variables, in more recent terminology.

$f, g, h, \ldots$. On the other hand, a system in canonical form B makes no distinction among its variables, permitting, indeed, any 1-1 replacement of variables in an assertion. Actually this distinction may also be disregarded in *10. For a propositional variable p always appears in some context $\sim p$, $p \vee P$, $P \vee p$, a functional variable f in some context $f(x_{i_1}, x_{i_2}, \ldots, x_{i_n})$, an individual variable x in some context $f(\ldots, x, \ldots)$, so that the type of the variable is determined by the context, and needs no other distinguishing mark. For purposes of presentation we shall continue to use the symbol differentiations suggested by Principia Mathematica. But theoretically, $x \vee p(f)$ is a valid enunciation of *10, x being now a propositional variable, f an individual variable, p a functional variable. With this understanding, any 1-1 replacement of variables in an enunciation of *10 leaves it an enunciation of *10.

In order to make *10 of Principia Mathematica correspond as closely as possible to *9 thereof, we shall assume that in an enunciation of *10 all apparent variables will be distinct, and distinct from any real variables occurring therein.[28] No loss of generality is thus incurred, since *10 was built up on the assumption that the particular symbols used for apparent variables is irrelevant. On the other hand, as a result of this convention, the enunciations of *10 will consist of those of its unrestricted enunciations which could appear in *9 as definitions.[29]

28. This contrasts strongly with the treatment of apparent, i.e., bound variables in the critical treatments appearing in the literature, and is one of the features to be noted in the "parallel evolution" idea referred to in the introduction.

29. Despite the theoretical advantages of an ideal *9 development over that of *10, the writer was forced to decide in favor of the latter because he found the basis of *9 to be incapable of yielding the complete development desired. To be specific, the proofs to the effect that the primitive propositions of the ($\sim$, $\vee$) system are valid in *9 when p, q, and r are replaced by arbitrary enunciations of *9 do not universally go over when more than one of these enunciations involve an apparent variable, or when any of them involves more than one apparent variable. A lengthy communication to this effect was sent to one of the authors of Principia

With this understood, we may then completely determine the enunciations of *10 as follows.[30] It is convenient to allow an unmodified variable, thus identified as propositional, to be an enunciation of *10. The enunciations of rank 0 of *10 are all unmodified variables p, and all simple functional expressions[31] $f(x_{i_1}, x_{i_2}, \ldots, x_{i_n})$ where the individual variables x_{i_j} are arbitrarily distinct or repeated, but the functional variable f is distinct from all the x_{i_j}'s. The remaining enunciations of *10 are then inductively determined by the following three rules.

(a). If P is an enunciation of *10, $\sim P$ is an enunciation of *10.
(b). If P is an enunciation of *10 involving a real individual variable x, $(x)P$ is an enunciation of *10.
(c). If P and Q are enunciations of *10, then $P \vee Q$ is an enunciation of *10 provided the following conditions are satisfied. (α) Any variable occurring in both P and Q either appears as a propositional variable in both P and Q, or as an individual variable in both, or as a functional variable in both. (β) In the third case of the preceding condition the corresponding simple functional expressions involve the same total number of arguments. (γ) Each apparent variable of P, and of Q, is distinct from all of the individual variables of Q, and of P respectively.

Note that $(\exists x)f(x)$ need not be referred to explicitly, since it is definable in *10. As a consequence of these rules of formation, consistency of context is preserved for each separate enunciation. Hence the need of (α) in (c); also of (β), for while

Mathematica. However, the heavy handed modification of *9 proposed by the writer to remove the inadequacy of the original formulation was quite out of keeping with the finesse of Principia Mathematica. Whether the new *8 of the revised edition (appendix, vol. I) overcomes this difficulty, the writer cannot say. At any rate, that is not its expressed purpose.

30. While the formulation given in this paragraph does not explicitly appear in the notes, it is clearly tacit in all of the development that is there given.

31. This phrase does not occur in the notes.

we may have $f(x)$ and $f(x, y)$ occurring in different contexts, not so for one and the same enunciation. Condition (γ) of (c) embodies the above convention on apparent variables. It follows inductively that given any part of an enunciation P constituting a constituent of P of the form $(x)Q$ then x appears nowhere else in P than in that $(x)Q$.[32]

We have already defined the enunciations of rank 0 of *10. A unique rank can then be inductively assigned to each enunciation of *10 as follows. If P is of rank ϱ, $\sim P$ and $(x)P$ are of rank $\varrho+1$; if P and Q are of ranks ϱ_1 and ϱ_2, $P \vee Q$ is of rank $\varrho+1$, where $\varrho = \max(\varrho_1, \varrho_2)$.

The enunciations of *10, while employing the parenthesis notation of canonical form B, cannot themselves be considered enunciations of the system in canonical form B to which *10 is to be reduced. Thus, in the latter but a finite number of primitive functions can be allowed, while in *10 we must allow for an infinite number of functional variables. We therefore first give a method of translating the enunciations of *10 into certain enunciations of the system to which it is to be reduced.

Variables of *10 are to be their own correspondents in the new system. To obtain the correspondents of the simple functional expressions of *10 we introduce two primitive functions, $a(F, X)$, $b(X, Y)$. The correspondent of $f(x)$ will then be $a(f, x)$, of $f(x_{i_1}, x_{i_2}, \ldots, x_{i_n})$, $a(f, b(x_{i_1}, b(x_{i_2}, \ldots b(x_{i_{n-1}}, x_{i_n}) \ldots)))$. To take care of $(x)P$ we introduce the primitive function $0(X, P)$, and inductively define the correspondent of $(x)P$ as $0(x, \overline{P})$, where $\overline{P}$ is the correspondent of P. $\sim P$ and $P \vee Q$ may directly be introduced as primitive functions in the new system,[33] the correspondents of $\sim P$ and $P \vee Q$ of *10 being $\sim \overline{P}$ and $\overline{P} \vee \overline{Q}$, $\overline{P}$ and $\overline{Q}$ being the correspondents of P and Q. Given any enunciation

32. It would not then have occurred to the writer to define the occurrences of propositional variables etc. inductively as has since been done for free and bound variables, but such definitions clearly can be given.

33. $P \vee Q$ should rather be written $\vee(P, Q)$, as is done in an illustration of

of *10, a unique corresponding enunciation of the new system is thus determined, different enunciations of *10 always having different correspondents.

A major part of the present development is devoted to formally determining that subset of the set of all enunciations of the new system which consists of the correspondents of the enunciations of *10. For this purpose we introduce a primitive function $\alpha(P)$, and set up a canonical B basis as a result of which $\alpha(P)$ will be asserted when and only when P is the correspondent of an enunciation of *10. It will be convenient to set up this basis simultaneously with an inductive proof of its sufficiency based on the rank of an enunciation of *10.[34] This proof however will first require the setting up of a partial basis for securing the assertion of $\alpha(P)$ in the following three cases: (a) P of rank 0, (b) P of the form $0(\boldsymbol{x}, Q)$ where Q is of rank 0, (c) P of the form $Q \vee R$ where Q and R are of rank 0. By the rank of P we do not mean the rank of P as enunciation in the new system, but as translation of an enunciation of *10, i.e., we mean the rank of the latter enunciation. Likewise we shall refer to a simple functional expression P when we mean P is the correspondent of a simple functional expression.

the reduction of §4 appearing in our notes and, more extensively, in our work of 1924.

34. Considerable liberty is here taken with the order of events as compared with the development in the notes. The notes first set up the basis for $\boldsymbol{\alpha}(\boldsymbol{P})$ with $\boldsymbol{P}$ representing any matrix of *10, and carry the induction proof through to the point where it could "obviously" be completed. They then attempt to allow for the arbitrary introduction of apparent variables into $\boldsymbol{P}$; but what seemed to be a clever ending of this enterprise turns out to be quite inadequate. Perhaps because of the necessary duplication in proof caused by this order of events, the attempted proof by induction ends up with an "obviously" before this error could be discovered. Except for a relatively few changes indicated in succeeding footnotes, the actual basis as now presented is on the whole but a reshuffling of the basis developed in the notes.

Case (a) is easily covered by the following two primitive assertions and one operation.[35]

(1). $\vdash \alpha(p)$. (2). $\vdash \alpha[a(f,x)]$.

I. "$\vdash \alpha[a(F,X)]$", "$\vdash \alpha[a(X_1,F)]$" produce "$\vdash \alpha[a(F,b(X_1,X))]$".[36]

Of course, the operation of substitution of canonical form B is assumed already to have been postulated. With its help, (1) alone yields all "$\vdash \alpha(P)$"'s where P is an unmodified, consequently a propositional, variable. As for P a simple functional expression, (2) alone, under substitution, yields all "$\vdash \alpha(P)$"'s with P a simple functional expression of one argument. Note that f and x being distinct variables in (2), they remain distinct under substitution. In the inductive application of I, an assertion in the form of the first premise insures $a(F,X)$ being a simple functional expression with F a variable distinct from any variable in X, the second premise insures X_1 being a variable distinct from F so that in the conclusion $a(F,b(X_1,X))$ will be a simple functional expression satisfying the necessary condition, F distinct from any variable in $b(X_1,X)$. Whatever distinct variables F and X_1 are, that second premise can always be realized by a form of (2) under substitution. It follows that (2) and I suffice to yield all "$\vdash \alpha(P)$"'s where P is a simple

35. The notes have the added primitive assertion "$\vdash \alpha[a(f,b(x_1,x_2))]$" necessitated by their using for the second premise of I, "$\vdash \alpha[a(f,b(X_1,X_2))]$." The trick employed in the present second premise, however, is used later in the notes. Where in (2) we now think of f and x as being the variables employed, the notes rather think of them as representing those variables, and so add the condition that those variables be distinct. Perhaps because of this difference, they invariably here use small letters for those operational variables which in fact could only represent variables. Yet, in a preliminary $\alpha(P)$ development immediately preceding, they specifically point out that all operational variables must be written as capitals, a practice we have uniformly followed in the present version of the notes development.

36. While the notes invariably write productions with premises and conclusion in a vertical array, a luxury we allowed ourselves in §1, to save space we henceforth usually use the present horizontal display.

functional expression, as was desired. Our analysis, and the inapplicability of I to any other assertions, also shows that only "$\vdash \alpha(P)$"'s with P a valid enunciation of rank 0 are thus obtained.

Case (b) will be covered by further adding

(3). $\vdash \alpha\, 0(x, a(f, x))$.

II. "$\vdash \alpha 0(X, a(F, P))$", "$\vdash \alpha[a(Y, F)]$", produce "$\vdash \alpha 0(X, a(F, b(Y, P)))$".[37]

III. "$\vdash \alpha 0(X, a(F, X))$", "$\vdash \alpha[a(F, P)]$" produce "$\vdash \alpha 0(X, a(F, b(X, P)))$".

Thus, in $0(x, Q)$, with Q of rank 0, Q can only be a simple functional expression involving the individual variable x. Substitution and (3) then take care of all simple functional expressions of one argument. In II, the first premise insures X being a variable in P with $a(F, P)$ a valid simple functional expression, and hence F a variable, the second premise insures Y being a variable distinct from F, whence the validity of the conclusion. In III, the first premise makes F and X distinct variables, the second, $a(F, P)$ a valid simple functional expression, whence again the validity of the conclusion. If a simple functional expression be built up by successively inserting its arguments from right to left, (3) or III enables us to introduce the $0(x, R)$ operation the first time x is thus introduced, while II enables us to keep that operation until the desired $0(x, Q)$ is obtained.

Under case (c), three cases must be considered: (c_1) Q and R both propositional variables, (c_2) Q a propositional variable, R a simple functional expression, (c_3) Q and R simple functional expressions. The case where Q is a simple functional ex-

37. The notes incorrectly write the second premise "$\vdash \alpha\, a(f, y)$". They uniformly omit the parenthesis in $\alpha(P)$ when P is of the form $0(x, Q)$, a convention we follow here. For the sake of uniformity we have supplied their only occasional omission of the parenthesis when P is of the form $a(x, Q)$.

pression, R a propositional variable, follows from (c_2) by XII below.

For (c_1) we need merely have

(4). $\vdash\alpha(p_1 \vee p_2)$. (5). $\vdash\alpha(p \vee p)$.

for (c_2) the following suffices.

IV. "$\vdash\alpha[a(P,X)]$", "$\vdash\alpha[a(F,b(P,X))]$" produce "$\vdash\alpha[P \vee a(F,X)]$".

For the first premise insures P being a variable distinct from any variable in X, the second insures F being a variable distinct from P and any variable in X, and at the same time guarantees a form for X such that $a(F,X)$ is a valid simple functional expression. The conclusion is then valid, and may clearly be any valid "$\vdash\alpha(Q \vee R)$" with Q a propositional variable, R a simple functional expression.

Case (c_3) again subdivides into (c_3'), the functional expressions have different functional variables, (c_3'') the functional expressions have the same functional variable. (c_3') is taken care of by

V. "$\vdash\alpha[a(F_1,X)]$", "$\vdash\alpha[a(F_2,X)]$", "$\vdash\alpha[a(F_1,Y)]$", "$\vdash\alpha[a(F_2,Y)]$",

"$\vdash\alpha[a(F_1, F_2)]$" produce "$\vdash\alpha[a(F_1,X) \vee a(F_2,Y)]$";

for the premises insure $a(F_1,X)$, $a(F_2,Y)$ being simple functional expressions with the consequent variables F_1 and F_2 distinct by the last premise, and distinct from the variables in both X and Y by the first four premises. On the other hand, for case (c_3') the number of arguments must be the same in the two functional expressions. Unlike our treatment of the preceding case we do not now rely on the earlier "$\vdash\alpha(P)$"'s with P a simple functional expression, but iteratively build up the desired "$\vdash\alpha(P \vee Q)$"'s by the following.

(6). $\vdash\alpha[a(f,x_1)\vee a(f,x_2)]$. (7). $\vdash\alpha[a(f,x)\vee a(f,x)]$.

VI. "$\vdash\alpha[a(F, Y_1)\vee a(F, Y_2)]$", "$\vdash\alpha[a(F, b(X_1, Y_1))]$", "$\vdash\alpha[a(F, b(X_2, Y_2)]$"

produce "$\vdash\alpha[a(F, b(X_1, Y_1))\vee a(F, b(X_2, Y_2))]$".

We can now carry through the above suggested proof, and completion of the basis for determining all valid "$\vdash\alpha(P)$"'s. Via case (a) we have already done this for all P's of rank 0. Assume that it has been carried through for all P's of rank less than or equal to ϱ, and let P be of rank $\varrho+1$. Since P is thus of rank greater than 0, it will be in one of the three forms $\sim Q$, $0(x,Q)$, $Q\vee R$.

In the first case, since Q is of rank ϱ, we shall have "$\vdash\alpha(Q)$" by our induction. The desired "$\vdash\alpha(P)$" will then be obtained via

VII. "$\vdash\alpha(P)$" produces "$\vdash\alpha(\sim P)$".

In the second case, when Q is of rank 0, the desired assertion has already been taken care of in (b) above. Otherwise, Q will be in one of the three forms $\sim R$, $0(y, R)$, $R\vee S$. These three possibilities are covered by the following four operations.

VIII. "$\vdash\alpha 0(X,P)$" produces "$\vdash\alpha 0(X, \sim P)$"

IX. "$\vdash\alpha 0(X,P)$", "$\vdash\alpha 0(Y,P)$", "$\vdash\alpha[a(X,Y)]$" produce "$\vdash\alpha 0(X, 0(Y,P))$".

X. "$\vdash\alpha 0(X,P)$", "$\vdash\alpha(P\vee Q)$" produce "$\vdash\alpha 0(X, P\vee Q)$".

XI. "$\vdash\alpha 0(X,Q)$", "$\vdash\alpha(P\vee Q)$" produce "$\vdash\alpha 0(X, P\vee Q)$".[38]

Note that when $Q=\sim R$, $P=0(x,\sim R)$. With P of rank $\varrho+1$, R is then of rank $\varrho-1$, and hence $0(x,R)$ of rank ϱ. We will thus have "$\vdash\alpha 0(x,R)$", whence "$\vdash\alpha 0(x,\sim R)$" by VIII. Similar considerations of rank render our induction effective in the other cases. In the case $Q=0(y,R)$, with $P=0(x,Q)$, apparent variables x and y must be distinct. This is secured in IX by the

38. Where we now have X and XI, the notes insufficiently have the one operation "$\vdash\alpha 0(x, P)$", "$\vdash\alpha 0(x, Q)$", "$\vdash\alpha(P\vee Q)$" produce "$\vdash\alpha 0(x, P\vee Q)$". This is not invalid, but assumes x to be present in both P and Q.

third premise, the second premise having already insured Y being a variable. Operations X and XI together clearly take care of $Q=R\vee S$. That they are universally valid follows from our entire development, and our conditions on valid enunciations. Thus in X the second premise insures $P\vee Q$ being a valid enunciation, the first that   is a real variable in P.   is therefore a real variable in $P\vee Q$ and $0(X,P\vee Q)$ is a valid enunciation.

We come then to the third case, $P=Q\vee R$. If Q and R are of unequal ranks, the operation

XII. "$\vdash\alpha(P\vee Q)$" produces "$\vdash\alpha(Q\vee P)$"

enables us to assume that the rank of R is greater than the rank of Q. Since R is then of rank greater than 0, it will be in one of the three forms $\sim S$, $0(x,S)$, $S\vee T$. These cases will then be covered by the following three operations respectively.

XIII. "$\vdash\alpha(P\vee Q)$" produces "$\vdash\alpha(P\vee\sim Q)$".
XIV. "$\vdash\alpha(P\vee R)$", "$\vdash\alpha 0(X,R)$", "$\vdash\alpha(X\vee P)$" produce "$\vdash\alpha(P\vee 0(X,R))$".[39]
XV. "$\vdash\alpha(P\vee Q)$", "$\vdash\alpha(P\vee R)$", "$\vdash\alpha(Q\vee R)$" produce "$\vdash\alpha(P\vee(Q\vee R))$".

The applicability of XIII and XV is immediate. Thus, in the case of $P=Q\vee\sim S$, by our hypothesis on rank, $\sim S$ is of rank ϱ, Q of rank less than ϱ. S is then of rank $\varrho-1$, $Q\vee S$ therefore of rank ϱ, whence "$\vdash\alpha(Q\vee S)$" by our induction, and so "$\vdash\alpha(Q\vee\sim S)$" by XIII. Their universal validity is also readily demonstrated. Thus in XV any occurrence[40] of the same variable in P and $Q\vee R$ will be such an occurrence in P and Q, or P and R, and hence will be an occurrence of the same kind of variable by the first premise or the second premise; similarly for the other conditions.[41]

39. This operation does not occur in the notes. See footnote 42.

40. This word in its present use is foreign to our notes, though hardly the idea.

41. In fact, the notes give as the criterion for the validity of "$\vdash\alpha f(P_1, P_2, \ldots, P_n)$", where f is built up by $\sim$'s and $\vee$'s, that "$\vdash\alpha(P_i)$", "$\vdash\alpha(P_i\vee P_j)$" be valid for all i's and j's. The first set of conditions is not needed unless $n=1$

In applying XIV to the case $P=Q \vee 0(x,S)$, the premises would become "$\vdash \alpha(Q \vee S)$", "$\vdash \alpha 0(x,S)$" and "$\vdash \alpha(x \vee Q)$". Since x cannot be present in Q, the third premise as well as the first two, will be valid assertions provided the ranks agree with our induction. But since in the present case Q is at most of rank $\varrho-1$, while S is of rank $\varrho-1$, $Q \vee S$ and $0(x,S)$ will be of rank ϱ, $x \vee Q$ of rank at most ϱ, so that those premises are asserted in accordance with our induction, and "$\vdash \alpha[Q \vee 0(x,S)]$" follows. As for the validity of XIV, the second premise shows X to be an individual variable present in R. Hence, were it present in P, the first premise would make it appear in P as an individual variable, the third as a propositional variable; for the variable X explicitly thus appears in the first term of the latter disjunction. The simultaneous assertion of the three premises therefore insures X not being present in P, whence the validity of the conclusion.[42]

Finally, then, let $P=Q \vee R$ with Q and R of equal ranks. When those ranks are zero, "$\vdash \alpha(P)$" is obtained by the above special case (c). Otherwise, both Q and R are of ranks greater than zero. By XII we may therefore assume P to be in one of

and $i \neq j$, the second set was written "$\vdash \alpha(P_i, P_j)$" by an obvious slip.

42. In the notes the standard method of insuring x being a variable not present in P is to have as premise "$\vdash \alpha(0(x, a(f, x)) \vee P)$". The use of this method will be continued later. But in the present instance it would complicate the induction, and require separate treatment of additional special cases. Actually, by the addition of two added primitive assertions, and one new production, the notes correctly obtain all assertions of the above form with P a matrix. They then incorrectly conclude that with the addition of one more operation, all assertions of the above form are obtainable, and end the development with the operation having the above as premise, "$\vdash \alpha(P)$" as conclusion, claiming that all valid "$\vdash \alpha(P)$"'s are thus obtainable. Had the induction proof been carried through in detail, it would have been seen that to obtain "$\vdash \alpha(0(x, a(f, x)) \vee (Q \vee R))$ with the help of XV, "$\vdash \alpha(Q \vee R)$" must first be obtained. We may finally observe that the notes have two additional productions omitted here, one a special case of VIII, the other a production whose application is covered by XII and XV.

the following six forms. (α): $\sim S \vee \sim T$, (β): $\sim S \vee 0(x, T)$, (γ): $\sim S \vee (X \vee Y)$, (δ): $0(x, S) \vee 0(y, T)$, (ϵ): $0(x, S) \vee (X \vee Y)$, (ζ): $(S \vee T) \vee (X \vee Y)$. The first three forms may be rewritten $\sim S \vee R$ with S and R of unequal ranks. Hence $\alpha(S \vee R)$ will be asserted by the previous discussion, and so $\alpha(\sim S \vee R)$ by XII followed by XIII followed by XII. For (δ), we may first assert $\alpha[0(x, S) \vee T]$ by the unequal rank discussion, $\alpha 0(y, T)$ by our induction, $\alpha[y \vee 0(x, S)]$ by the unequal rank discussion, and then apply XIV. In (ϵ) and (ζ), which may be rewritten together as $Q \vee (X \vee Y)$, $\alpha(Q \vee X)$, $\alpha(Q \vee Y)$ may be assumed asserted by the unequal rank case, $\alpha(X \vee Y)$ by our induction, whence the results follow by XV.

Our first object is thus achieved. That is, in the above system in canonical form B, $\alpha(P)$ will be asserted when and only when P is the correspondent of an enunciation in *10 Principia Mathematica. Indeed, thus far these are the only assertions of the system.

We have observed that in a system in canonical form A, and hence in the $(\sim, \vee)$ system of *1-*5 Principia Mathematica, the operation of substitution II need merely be iteratively applied to the primitive assertions IV of the system after which only the productions III need be applied. We therefore interpret the postulational set up of *10 in the latter fashion, and assume that after the most general substitutions have been allowed for in the formal primitive propositions of *10, including those of *1-*5, all other assertions are to be obtained through the remaining rules of operation of the system.[43]

Due to our convention on apparent variables, if a propositional variable p occurs several times in an assertion, we can-

43. At least that is all that the notes allow us to say; for they do not bring up the question of an operation of substitution in *10 Principia Mathematica, but merely in fact allow for arbitrary substitutions in the primitive propositions. A quite cursory examination of the question suggests that here too this is equivalent to allowing arbitrary substitutions at any stage of a deductive process.

not merely substitute the same enunciation P of *10 for each p, but must in each instance at least rewrite P on a different set of apparent variables before making the substitution. To allow for this, we introduce a new primitive function $\beta(P,Q)$ into the system to which *10 is being reduced, and a set of postulates therefor, so that $\beta(P,Q)$ will be asserted when and only when P and Q are correspondents of enunciations of *10 obtainable from each other by a mere interchange of apparent variables, and such that, in fact, $\alpha(P\vee Q)$ is asserted. We shall also have to introduce a primitive function $\delta(x, y, P, Q)$ whose assertion is to mean that Q is obtained from P by replacing real individual variable x of P by a variable Y not present in P, or present in P as a real individual variable distinct from x, and changing apparent variables so that $\alpha(P\vee Q)$ is asserted. Considerable economy is gained by giving a simultaneous development for the resulting common basis for these two primitives.[44]

When P is of rank zero, it can have no apparent variables. We shall then have $\beta(P,Q)$ when and only when Q is identical with P. This case is then taken care of for $\beta(P,Q)$ by the following.

$\vdash\beta(p,p)$; $\quad$ $\vdash\alpha(a(F,P))$ produces $\vdash\beta(a(F,P),a(F,P))$.

In the case of $\delta(x, y, P,Q)$, with P of rank zero, P can only be a simple functional expression. However an iterative build up is now needed to allow for an arbitrary number of arguments.

$\vdash\delta(x_1,x_2,a(f,x_1),a(f,x_2))$.
$\vdash\alpha(0(X,a(F_1,X))\vee a(F,b(Y,P)))$ produces
$$\vdash\delta(X, Y,a(F,b(X,P)),a(F,b(Y,P))) .$$

44. In the notes, on the other hand, the basis for $\beta(P,Q)$ is first set up, and then, with its help, the basis for $\delta(x, y, P, Q)$ is given: This is not done with complete accuracy, and further footnotes will point out the principle corrections thus necessitated in our present combined basis. The notes here cease numbering the primitive assertions and productions, and also omit quotation marks in the latter. We have therefore done the same.

$\vdash \delta(X, Y, a(F,P), a(F,Q))$ produces

$$\vdash \delta(X, Y, a(F, b(X,P)), a(F, b(Y,Q))) .$$

$\vdash \delta(X, Y, a(F,P), a(F,Q))$, $\vdash \alpha(0(X, a(F,X)) \vee 0(Z, a(F,Z)))$

produce $\vdash \delta(X, Y, a(F, b(Z,P)), a(F, b(Z,Q))$

Note that the primitive assertion, and first production, allow for the initial introduction of the x as the simple functional expression is built up right to left, the second production allows for any later x, the third for any later variable other than x, including y.[45]

We may now assume $\beta(P,Q)$ and $\delta(x, y, P,Q)$ to have been taken care of for P of rank less than or equal to ρ, and let P be of rank $\rho+1$. The following three productions then take care of $\beta(P,Q)$ according as $P = \sim R$, $P = 0(z,R)$, $P = R \vee S$.

$\vdash \beta(P, Q)$ produces $\vdash \beta(\sim P, \sim Q)$.

$\vdash \delta(X, Y, P, Q)$, $\vdash \alpha(0(Y, a(F,Y)) \vee P)$ produce $\vdash \beta(0(X,P), 0(Y,Q))$.[46]

$\vdash \beta(P, Q)$, $\vdash \beta(R, S)$, $\vdash \alpha(P. \vee. R : \vee : Q :. \vee :. S)$ produce

$$\vdash \beta(P \vee R, Q \vee S).$$

Note that the second premise of the second production serves the purpose of excluding Y from P.

The productions for $\delta(x, y, P,Q)$ are more complicated. When $P=0(z,R)$, then $Q=0(w,S)$, so that S is obtained from R by replacing the two letters x and z by y and w respectively. This is then reduced to the δ operation by introducing a form intermediate between P and Q On the other hand, when $P = R \vee S$, we need three productions, since we must explicitly allow for x being in R and S, in R only, in S only. The concluding productions are then the following.

45. The notes do not give the first production, and in the second premise of the third production have the added term $0(Y, a(F, Z))$, here written with capitals, thus preventing Z from being Y.

46. This is new because of our fusion of the two bases; but the several productions replacing it in our notes find their counterparts in the later productions having δ in the conclusion. At this point the dot notation creeps into our notes intermingled with the parenthesis notation. Since we have retained the notation $p \vee q$, we also carry along the dots.

$\vdash \delta(X, Y, P, Q)$ produces $\vdash \delta(X, Y, \sim P, \sim Q)$.

$\vdash \delta(Z, Z', P, Q)$, $\vdash \delta(X, Y, Q, R)$, $\vdash \alpha(0(Z,P) \vee 0(Z',R))$, $\vdash \alpha(a(Z';Y))$ produce $\vdash \delta(X, Y, 0(Z,P), 0(Z',R))$.

$\vdash \delta(X, Y, P, Q)$, $\vdash \delta(X, Y, R, S)$, $\vdash \alpha(P \vee R . \vee . Q \vee S)$ produce $\vdash \delta(X, Y, P \vee R, Q \vee R)$.

$\vdash \delta(X, Y, P, Q)$, $\vdash \beta(R,S)$, $\vdash \alpha(0(X, a(F,X) \vee R)$, $\vdash \alpha(P \vee R . \vee . Q \vee S)$ produce $\vdash \delta(X, Y, P \vee R, Q \vee S)$.

$\vdash \beta(P,Q)$, $\vdash \delta(X, Y, R, S)$, $\vdash \alpha(0(X, a(F,X)) \vee P)$, $\vdash \alpha(P \vee R . \vee . Q \vee S)$ produce $\vdash \delta(X, Y, P \vee R, Q \vee S)$. [47]

The five formal primitive propositions of *1-*5 carried over to *10 under all possible substitutions of *10 are easily taken care of via the β function. Each will be replaced by a production which may be described as follows. In the given primitive proposition of *1-*5 replace all of the small letters occurring therein by different capital letters for all different positions of the small letters. For each distinct small letter, if $P_1, P_2, \ldots, P_\nu$ are the corresponding capital letters replacing it, introduce the premises $\vdash \beta(P_i, P_{i+1})$, $i = 1, \ldots, \nu - 1$ in the production. If F is the primitive proposition in terms of capital letters thus all different, add the premise $\vdash \alpha(F)$. The production is then completed by adding "produce $\vdash F$". Note that the necessary $\vdash \alpha(F)$ itself imposes the conditions of distinctness of variables as required by the β-premises, while the latter otherwise merely insure "essentially the same" enunciations of *10, i.e., apart from the particular apparent variables used, always being substituted for the same small letter of the primitive proposition for all the occurrences of that small letter in the proposition.

47. The second of these five productions is incorrectly given in the notes, the need of an intermediary between P and R being overlooked; but the corresponding situation in the independently given β basis was correctly handled. The third production has its third premise inadequately written $\alpha(P \vee Q)$. The need for the fourth and fifth productions was completely overlooked, both here, and in the corresponding situation for the β basis.

The one operation of *1-*5 other than substitution, already allowed for, is directly taken care of by adding the production

$$\vdash P,\ \vdash \sim P \vee Q \text{ produce } \vdash Q$$

to our system.[48]

The remaining formally significant primitive propositions of *10 Principia Mathematica as there given are the following.

*10.1. $\vdash :(x).\phi x.\supset.\phi y.$

*10.11. If ϕy is true whatever possible argument y may be, then $(x).\phi x$ is true.

*10.12. $\vdash :.(x).p \vee \phi x.\supset :p.\vee.(x).\phi x.$[49]

Of these, the first and third are primitive assertions of the system, the second a rule of operation. However, when subjected to all possible substitutions, the effect of all three will appear in our new system as operations. The latter are almost self-explanatory, and in order are the following.

$\vdash \delta(X, Y, P, Q)$ produces $\vdash \sim 0(X,P).\vee.Q.$

$\vdash P, \vdash \alpha 0(X,P)$ produce $\vdash 0(X,P).$

$\vdash \beta(P,R),\ \vdash \delta(X,Z,Q,S),\ \vdash \alpha[\sim 0(X,P \vee Q).\vee :R.\vee.0(Z,S)]$
produce $\vdash \sim 0(X, P \vee Q).\vee :R.\vee.0(Z,S).$

It is readily seen that in the first operation, the hypothesis insures the conclusion being a valid enunciation, so that no corresponding premise is needed.[50] In the second operation, the second premise insures X being a real individual variable of P, as desired. The complicated third operation is required by

48. Strangely overlooked in the notes. See also the next footnote. It is readily verified that thanks to the operation of substitution of canonical form B, we may thus have the same P and the same Q occurring twice, instead of using different letters connected by the β relation.

49. Primitive propositions *10.121, *10.122, and *11.07 are of a different nature, and are automatically taken care of by our use of canonical form B.

50. The notes do have a second premise $\alpha 0(x, P)$, itself changed from a first written α of the conclusion.

our convention on apparent variables. That is, $\boldsymbol{P}$ in its second occurrence must appear as an $\boldsymbol{R}$ with other apparent variables, $\boldsymbol{Q}$ as an $\boldsymbol{S}$ with not only other apparent variables than those of $\boldsymbol{Q}$, but also with $\boldsymbol{X}$ changed to a $\boldsymbol{Z}$, as $\boldsymbol{X}$ and $\boldsymbol{Z}$ are apparent variables of the conclusion. This time, unlike the situation for the first operation, the resulting two premises do not insure the conclusion being a valid enunciation, whence the third premise.

The desired reduction of *10 Principia Mathematica to a system in canonical form $\boldsymbol{B}$ is thus completed.

3. The problem of "tag."

The direction taken by the reductions of the next two sections will become clearer if we at least formulate a problem, christened "tag" by B. P. Gill, which has played a vital part in the present development. An early unpublished method of the writer completely solved the problem of determining for any two expressions in the $(\sim, \vee)$ system of Principia Mathematica, or indeed in any system in canonical form $\boldsymbol{A}$, what substitutions would make those expressions identical. Because of the form of the result, this method was termed the L. C. M. process. In passing from the $(\sim, \vee)$ system to the whole of Principia Mathematica, attention was first centered on the "matrices" thus arising, and the problem arose of determining the substitutions on the variable propositional functions occurring therein which would make two such forms identical. The general problem proving intractable, successive simplifications thereof were considered, one of the last being this problem of "tag". Again, after the finiteness problem for systems in canonical form $\boldsymbol{A}$ involving primitive functions of only one argument was solved, an attempt to solve the problem for systems going, it seemed, but a little beyond this one argument case, led once more essentially to the selfsame problem of "tag". The solution of this problem thus appeared as a vital stepping stone in any further progress to be made.

In its first form the problem may be stated as follows. Given, a positive integer ν, and μ symbols which may be taken to be $0, 1, \ldots, \mu-1$. With each of these μ symbols a finite sequence of these symbols is associated as follows.

$$\begin{array}{l} 0 \rightarrow a_{0,1} \; a_{0,2} \; \cdots \; a_{0,\nu_0} \\ 1 \rightarrow a_{1,1} \; a_{1,2} \; \cdots \; a_{1,\nu_1} \\ \cdots\cdots\cdots\cdots\cdots\cdots \\ \mu-1 \rightarrow a_{\mu-1,1} \; a_{\mu-1,2} \; \cdots \; a_{\mu-1,\nu_{\mu-1}}. \end{array}$$

It is understood that in each sequence the same symbol may occur several times, and that a particular associated sequence may be null. In terms of this basis, we set up the following operation for obtaining from any given non-null sequence

$$B = b_1 \, b_2 \ldots b_l \, .$$

on the symbols $0, 1, \ldots, \mu-1$, a unique derived sequence B' on those symbols. To the right end of B adjoin the sequence associated with the symbol b_1 in the given basis, and from the left end of this augmented sequence remove the first ν elements — all if there be less than ν elements. As long as B' is not a null sequence, this operation can then be applied to B' to yield a sequence B'', to B'', if not null, to yield B''', and so on. The problem of "tag" for the given basis is then to obtain a finite process for determining for any initial sequence B whether the resulting iterative process does or does not terminate.[51]

In the second form of the problem, the one that arose in connection with the finiteness problem, the initial sequence B may be considered part of the basis, and the problem would

51. In an early formulation of this problem, b_1 was first checked off, the corresponding associated sequence added, then the ν-th element after b_1 was checked off, corresponding associated sequence added, and so on. Whether the iterative process terminated or not then depended on whether the ever advancing check mark did or did not overtake the usually advancing right end of the sequence, whence the suggestive name proposed by Gill for the problem.

be to obtain a finite process for determining of any given sequences $c_1, c_2, \ldots, c_m$ on the μ symbols $0, 1, \ldots, \mu-1$ whether that sequence is or is not generated in the course of the above iteration of the given process, starting with $\boldsymbol{B}$. Clearly, for this second form of the problem, the problem for a given basis is immediately solvable if the process is known to terminate. Where the process does not terminate, it is readily seen that according as the lengths of the resulting sequences are bounded, or unbounded, the resulting infinite sequence of the sequences will, from some point on, become periodic, or the length of the n-th sequence will increase indefinitely with n. In the first case the second form of the problem is again immediately solvable, while in the second case the solution would follow if a method were also found for determining of any given length of sequence a point in the process beyond which all derived sequences were of length greater than that given length.[52]

The first form of the problem, emended to determine whether the iterative process was terminating, periodic, or divergent, thus seemed likely to cover both forms. In this emended form the problem of "tag" was made the major project of the writer's tenure of a Proctor fellowship in mathematics at Princeton during the academic year 1920-21. Indeed, the reduction of the last section, effected early in that academic year, sealed this determination. And the major success of that project was the complete solution of the problem for all bases in which μ and ν were both 2.[53]

While considerable effort was expended on the case $\mu=2$, $\nu>2$, but little real progress resulted, such a simple basis as $0\rightarrow00$, $1\rightarrow1101$, $\nu=3$, proving intractable.[54] For a while

52. In this analysis we may have gone somewhat further than is justified by the notes.

53. When either μ or ν is 1 the problem becomes trivial. By contrast, even this special case $\mu=\nu=2$ involved considerable labor.

54. Note of course that an arbitrary initial sequence has to be allowed for.

the case $\nu=2$, $\mu>2$, seemed to be more promising, since it seemed to offer a greater chance of a finely graded series of problems. But when this possibility was explored in the early summer of 1921, it rather led to an overwhelming confusion of classes of cases, with the solution of the corresponding problem depending more and more on problems of ordinary number theory. Since it had been our hope that the known difficulties of number theory would, as it were, be dissolved in the particularities of this more primitive form of mathematics, the solution of the general problem of "tag" appeared hopeless, and with it our entire program of the solution of finiteness problems.

This frustration, however, was largely based on the assumption that "tag" was but a very minor, if essential, stepping stone in this wider program. In the late summer of 1921, however, the reductions carried through at Princeton in proving the equivalence of canonical forms ***A*** and ***B*** suggested a further transformation of canonical form ***B***, and, indeed, led to a whole series of reductions with the final canonical form very close to the seemingly special form of "tag". As these reductions are vital in the further evolution of our thought, we turn to them in the next two sections.

4. Reduction of canonical form ***B*** to a canonical form ***C***.

In our canonical form ***B***, as well as ***A***, on the one hand, we allow primitive functions of many variables, and thus rely on the parenthesis notation; on the other hand, we assume the availability of an infinite number of variables. We show in the present section that canonical form ***B***, to be specific, can be

Numerous initial sequences actually tried led in each case to termination or periodicity, usually the latter. It might be noted that an easily derived "probability" prognostication suggested that in this case periodicity was to be expected.

reduced to a canonical form C where the boxes within a box symbolic form of the parenthesis notation is replaced merely by finite sequences of letters, and where, for a given system, the different letters so used constitute a once and for all given finite set.[55]

The basis of an arbitrary system in canonical form C is to be of the following form. There are a finite number of distinct primitive symbols $a_1, a_2, \ldots, a_\mu$. The "enunciations" of the system are simply all finite sequences of such symbols, repetitions of the same symbol being of course allowed. That is, an arbitrary enunciation of the system may be written

$$a_{i_1}\ a_{i_2} \ldots a_{i_n}$$

with n arbitrary, the a_{i_j}'s arbitrarily $a_1, a_2, \ldots, a_\mu$. A specific finite set of such enunciations is set down to constitute the "primitive assertions" of the system.[56] Furthermore, a specific finite set of "productions" of the following form is set down to yield new assertions from old:

55. This of course is a characteristic of tag; but also, essentially, of systems in canonical form A involving only primitive functions of one argument. For if $f_1(p), \ldots, f_\mu(p)$ are the primitive functions of such a system, an arbitrary enunciation thereof is in the form $f_{i_1}(f_{i_2}(\ldots f_{i_n}(q)\ldots))$, which may then as well be written $f_{i_1} f_{i_2} \ldots f_{i_n} q$. Except for the one arbitrary variable q the enunciations are then just sequences of the primitive letters $f_1, f_2, \ldots, f_\mu$. Furthermore, each production is in the form, $g_1 P_{j_1}, g_2 P_{j_2}, \ldots, g_\kappa P_{j_\kappa}$ produce $g P_j$, where the g's represent fixed sequences of the primitive f's. In the "homogeneous case", to which the more general case can be reduced, the sole operational variable of the conclusion is also the operational variable of each premise. There then turns out to be no loss of generality in assuming all enunciations to be written with the same propositional variable. This may then be deleted, leaving only sequences of f's.

56. The notes here give up using an assertion sign, a practice we feel constrained to follow in keeping our account in the spirit of the notes. In connection with the about to be described productions note that if $A, B, \ldots, E$ represent the sequences $a_1 a_2 \ldots a_l$, $b_1 b_2 \ldots b_m, \ldots, e_1 e_2 \ldots e_p$ respectively, then $A B \ldots E$ simply represents the sequence $a_1 a_2 \ldots a_l b_1 b_2 \ldots b_m \ldots e_1 e_2 \ldots e_p$.

$$g_{11}\, P_{i'_1}\; g_{12}\, P_{i'_2} \ldots g_{1m_1}\, P_{i'_{m_1}}\; g_{1(m_1+1)}$$

$$g_{21}\, P_{i''_1}\; g_{22}\, P_{i''_2} \ldots g_{2m_2}\, P_{i''_{m_2}}\; g_{2(m_2+1)}$$

$$\cdots\cdots\cdots\cdots\cdots\cdots\cdots\cdots\cdots\cdots$$

$$g_{k1}\, P_{i_1^{(k)}}\; g_{k2}\, P_{i_2^{(k)}} \ldots g_{km_k}\, P_{i_{m_k}^{(k)}}\; g_{k(m_k+1)}$$

produce

$$g_1\, P_{i_1}\, g_2\, P_{i_2} \ldots g_m\, P_{i_m}\, g_{m+1}$$

where the g's are specified sequences of the primitive a's, including the null sequence, and each P of the conclusion is present in at least one premise. In the application of these productions the P's may be identified with arbitrary sequences of the above type, it being understood however, that the conclusion may not be null.[57] The assertions of the system are then the enunciations obtainable by the repeated application of these operations to the primitive assertions, and all assertions so obtainable.

Given a system S_1 in canonical form B, we proceed to build up the basis of a system S_2 in canonical form C to which S_1 will be reducible. For precision, let the variables of S_1 be the infinite set $p_1, p_2, \ldots, p_n, \ldots$. Introduce a primitive letter a_0 in S_2. Then let the above variables in order correspond to

57. The application of these productions is not quite as automatic as in the case of the productions of canonical form B; for in the latter a given assertion can be written in the form of a given premise in one and only one way, if at all. In the present case such uniqueness is achieved only under a particular hypothesis on the ranks of the operational variables occurring in the premise. While less would suffice, actually, since the sum of the ranks of the g's and p's in a given premise must equal the rank of the corresponding given assertion, rank now being the total number of letters in a sequence, but a finite number of such hypotheses are possible, and all can uniformly be tried out. Indeed, the successive reductions of this and the next section successively analyse away most of what is nonautomatic in the present as well as in the earlier canonical forms. In this connection see footnote 78.

the enunciations, i.e., sequences, $a_0, a_0 a_0, \ldots, a_0 a_0 \ldots a_0, \ldots$ of S_2, there being n a_0's in the n-th sequence. To distinguish these enunciations of S_2 from others, we introduce a primitive letter α_0, and suitable postulates therefor, to be part of the bases of S_2, so that $\alpha_0 P$ will be asserted when and only when P is a finite sequence of a_0's. The following postulates clearly suffice.

$$\alpha_0 a_0 \; ; \quad \alpha_0 P \text{ produces } \alpha_0 a_0 P.$$

Of course, it must be seen to that further postulates do not produce other assertions of the form $\alpha_0 P$ than the above.

Let the primitive functions of S_1 be $f_i(p_1, p_2, \ldots, p_m)$, $i=1,2,\ldots,\mu$. Correspondingly introduce primitive letters a_i in S_2. We shall assume that the parenthesis notation of S_1 has been replaced by an equivalent dot notation. Corresponding to an enunciation of the form $f_i(P_1, P_2, \ldots, P_{m_i})$, the corresponding dot notation is to be $f_i \ldots P_1 \ldots P_2 \ldots \quad \ldots \ldots P_{m_i}$ where the same number of dots separate f_i and P_1 and each pair of consecutive P's, and where that number is to be one more than the largest number of dots in any P. Introduce then a primitive letter b in S_2 to correspond to the dot of S_1 so rephrases. The enunciations of S_2 corresponding to the enunciations of S_1 will be certain sequences of a_i's, $i=0,1,2,\ldots,\mu$, and b's. Before these are singled out, we introduce a primitive letter β such that the assertion of $Q\beta P$ in S_2 is to mean that P is a sequence involving no other letters than $a_0, a_1, \ldots, a_\mu, b$, that Q is a sequence of b's, and that the number of b's in Q is one more than that of the largest uninterrupted sequence of b's in P. This result is secured by adding the following postulates to S_2, each postulate involving a_i being duplicated for $i=0,1,2,\ldots,\mu$.

$$\begin{aligned} & b\beta a_i,\ bb\beta b\ ; \\ Q\beta P & \text{ produces } Q\beta a_i P, \\ Q_1\beta Q_2,\ Q_2\beta P & \text{ produce } bQ_1\beta bQ_2, \\ Q_1\beta Q_2 a_i P,\ bbQ_2\beta Q_1 & \text{ produce } bQ_1\beta bQ_2 a_i P, \\ Q_1\beta Q_2 a_i P,\ bbbQ_2 Q_3\beta Q_1 & \text{ produce } Q_1\beta bQ_2 a_i P. \end{aligned}$$

To prove this, we first show that from valid $Q\beta P$'s only valid $Q\beta P$'s result. Our first production merely annexes an a_i to the left of P in a valid $Q\beta P$, hence does not disturb the condition on the b's, and so yields a valid $Q\beta P$. In the second production, the second premise guarantees that Q_2 consists of b's only, P having the effect of an apparent variable. The two premises then show Q_1 to be a sequence of b's with one more b than Q_2. But then the same is true of bQ_1 and bQ_2. The second premise of the third production, along with the first, makes Q_2 a sequence of b's one less than Q_1, while the largest sequence of b's in P is at least one less than the number of b's in Q_1. Hence in bQ_2a_iP the largest sequence of b's is the sequence bQ_2, whence the conclusion. On the other hand, in the fourth production, the second premise, with Q_3 playing the role of apparent variable, definitely insures Q_2 being a sequence of b's at least two less than Q_1, itself a sequence of b's by the first premise. Hence, by the first premise, the largest sequence of b's in P is exactly one less than Q_1. The largest sequence of b's in bQ_2a_iP is then again exactly one less than Q, whence the conclusion. This analysis makes easy the verification that, actually, all valid $Q\beta P$'s result. Those where P consists of but one letter are given by the primitive assertions themselves. Assuming all those with P having n letters to have been found, those with $n+1$ letters can be found from them by annexing an a_i to the left of P with Q unchanged, or annexing b to the left of P, perhaps changing Q to bQ. The first effect is secured by the first operation. As for the second effect, when P has no a's, it is secured by the second op- where P has an a, and starts with a "maximal sequence" of b's, the third operation secures the desired result, while if P has an a, but does not start with a maximal number of b's, the fourth operation applies.[58]

58. The notes do not give this proof, but merely point out the two places where an operational variable acts as a substitute for an apparent variable. However, the productions speak for themselves; we have merely pointed out what they

Now introduce a primitive letter α in system S_2 so that P will be asserted when and only when P represents an enunciation in S_2 corresponding to one in S_1. The following postulates, added to our growing basis for S_2, are easily seen to produce this effect, if not later disturbed.[59]

$$\alpha_0 P \text{ produces } \alpha P, Q\beta P_1 P_2 \ldots P_{m_i}, \alpha P_1, \alpha P_2, \ldots, \alpha P_{m_i}$$
$$\text{produce } \alpha a_i Q P_1 Q P_2 \ldots QP_{m_i}, i=1,2,\ldots, \mu.$$

Note that where P is a valid translation of an enunciation in S_1, it can neither start nor end with b. Hence the largest sequence of b's in $P_1\ P_2 \ldots P_{m_i}$ is the largest such sequence occurring in the several P's. Our first production lays down the correspondents of variables as valid correspondents; and having the valid correspondents of all enunciations of rank $\leqslant \varrho$ of S_1, the μ cases of the second production yield the valid correspondents of all enunciations of S_1 of rank $\leqslant \varrho+1$.

As a result of the above, if the variables of S_1 are p_1, p_2, p_3, $\ldots$, and they are set in 1-1 correspondence with the sequences $a_0, a_0a_0, a_0a_0a_0, \ldots$, a 1-1 correspondence is set up between the enunciations of S_1, and those enunciations P of S_2 for which αP is an assertion.

We turn now to the assertions of S_1. As in the case of canonical form A it can be shown for canonical form B that every assertion in S_1 can be obtained by first obtaining a set of assertions from IV by the sole use of II$'$ and then, starting with these assertions, merely employing III$'$. The set of all assertions obtainable from IV by the repeated use of II$'$ are all assertions $\vdash h_i\,(p_{n_1}, p_{n_2}, \ldots p_{n_{l_i}})$, $i=1,2,\ldots,\lambda$, $n_1, n_2, \ldots, n_{l_i}$ arbitrary distinct positive integers. We first then show how to get correspondents in S_2 of all such assertions in S_1.[60]

say. Due to an earlier use of the letter c in what later became b, the third and fourth productions are written in the notes with c in place of b.

59. We again allow a variable of S_1 to be an enunciation thereof.

60. This part is new, but a more complicated process for achieving the same

For this purpose introduce a new primitive letter δ such that $P\delta Q$ will be asserted when and only when P and Q are both sequences of a_0's, but of unequal lengths. This result is clearly secured by the productions

$$\alpha_0 P \text{ produces } a_0\delta a_0 P, \alpha_0 P \text{ produces } a_0 P\delta a_0, \quad P\delta Q \text{ produces } a_0 P\delta a_0 Q.$$

Now our original primitive assertions in IV of S_1 are the λ particular enunciations $h_i(p_1, p_2, \ldots, p_{l_i})$, $i=1,2,\ldots,\lambda$, each with the assertion sign "⊢" prefixed to it. By the method described above, for each of these enunciations there is a corresponding enunciation in S_2. In these enunciations replace each sequence of j a_0's, i.e., those arising indeed from p_j, by P_j, and symbolize the λ corresponding expressions by $\bar{h}_i(P_1, P_2, \ldots, P_{l_i})$, $i=1,2,\ldots,\lambda$. Actually, then, $\bar{h}_i(P_1, P_2, \ldots, P_{l_i})$ is a specific sequence of letters $a_1, \ldots, a_\mu, b, P_1, P_2, \ldots, P_{l_i}$. Now introduce the symbol "⊢" as a new primitive letter in S_2 so that, ultimately, "⊢P" will be asserted in S_2 when and only when P is the correspondent of an enunciation P' of S_1 with "⊢P'" an assertion in S_1. For the above assertions in S_1 of the form "⊢$h_i(p_{n_1}, p_{n_2}, \ldots, p_{n_{l_i}})$" this result is then achieved through the productions

$$P_{j_1}\delta P_{j_2},\ j_1, j_2=1,2,\ldots l_i,\ j_1<j_2, \text{ produce}$$
$$\vdash\bar{h}_i(P_1, P_2, \ldots, P_{l_i}),\ i=1,2,\ldots,\lambda.$$ [61]

There remains then but the reproducing of the effect of III′ of S_1 in S_2.[62] We shall merely describe the κ productions thus corresponding to the κ productions of S_1, and to be added to,

effect appears in the earlier portions of the notes, referred to in connection with §1, which prove the reducibility of canonical form ***B*** to ***A***. We introduce the change since the notes show how to reduce canonical form ***A***, not ***B***, to ***C***.

61. Where $m_i=1$ the sole premise of the production would simply be $\alpha_0(P_1)$.

62. The notes are here again radically emended, but this time unnecessarily so. Except for minor slips, easily corrected, the notes' method is correct, but that correctness eluded us until the final revision of the present account. Where we now

and indeed, completing the basis of S_2. If P represents any enunciation in S_1, P' the corresponding enunciation in S_2, then the rank of P actually is the largest number of b's in a sequence of consecutive b's in P'. If then an enunciation in S_1 is of the form $f_i(P_1, P_2, \ldots, P_{m_i})$, f one of the primitive functions of S_1, and $Q_1, Q_2, \ldots, Q_{m_i}$ are sequences of b's in number equal to the rank of $P_1, P_2, \ldots, P_{m_i}$ respectively, a sequence of b's Q, in number equal to the rank of $f_i(P_1, P_2, \ldots, P_{m_i})$ will be the unique Q for which say $Q\beta Q_1 a_0 Q_2 a_0 \ldots Q_{m_i}$ is asserted in S_2. For we recall that the rank of $f_i(P_1, P_2, \ldots, P_{m_i})$ is one more than the largest of the ranks $P_1, P_2, \ldots, P_{m_i}$. It follows that if $P_1', P_2', \ldots, P_{m_i}'$ represent the correspondents in S_2 of $P_1, P_2, \ldots, P_{m_i}$ respectively, then the correspondent of $f_i(P_1, P_2, \ldots, P_{m_i})$ will be $a_i Q P_1' Q P_2' Q \ldots P_{m_i}'$. Now any expression $g(P_1, P_2, \ldots, P_n)$ such that $g(p_1, p_2, \ldots, p_n)$ is an enunciation in S_1 other than an unmodified variable can be written in one and only one way $f_i(R_1, R_2, \ldots, R_{m_i})$ with f_i a primitive function of S_1, the R_i's being P's, or similar expressions, involving some, if not all, of the P's.[63] We may then inductively define the constituents of such an expression as $g(P_1, P_2, \ldots, P_n)$ as itself, and the constituents of $R_1, R_2, \ldots, R_{m_i}$ any P being its own only constituent.[64] Consider then any production in III′ of S_1, and let its operational variables be $P_1, P_2, \ldots, P_k$. Form the distinct constituents of

explicitly allow for the rank of each constituent of the symbolic expressions occurring in an operation, the notes merely consider the ranks of the operational variables thereof, and determine the ranks of the constituents by added hypotheses on the former ranks. That the finite number of sets of hypotheses given in the notes thus suffice was not seen by us while writing the account, but now can be said to have been obviously clear when the notes were written. Since the present method is considerably simpler, we leave it in the account.

63. This is really a tacit postulate on symbolic expressions in a propositional calculus.

64. Note that we here talk of the symbolic constituents of the symbolic expression $g(P_1, P_2, \ldots P_n)$ with P's variable, whereas ranks mean the ranks of the functions of $p_1, p_2, \ldots$ thus represented.

the several premises and conclusion. They are clearly finite in number, and hence may be ordered in a finite sequence starting with $P_1, P_2, \ldots, P_k$. For the j-th constituent in this sequence introduce the operational variable Q_j which is to represent a sequence of b's equal in number to the unknown rank of that constituent. If then we introduce operational variables $P'_1, P'_2, \ldots, P'_k$ to represent the correspondents of $P_1, P_2, \ldots, P_k$ respectively, we can successively build up expressions representing the correspondents of all of the above constituents solely by means of the P's and Q's in the manner described above. The desired production in S_2 to take the place of the given one of S_1 may then be described as follows. Its conclusion will be the above built up correspondent of the given conclusion preceded by $\vdash$. Its premises will first include the correspondents of the given premises each preceded by $\vdash$. Secondly, they will include the k expressions $\alpha P'_1, \alpha P'_2, \ldots, \alpha P'_k$.[65] Finally, they will include a set of premises giving the conditions on the Q's. These will first include the k premises $bQ_j \beta P'_j$, $j = 1, 2, \ldots, k$. Furthermore, for each constituent R_j other than a P_j, and corresponding unique expression $f_i(R_{j_1}, R_{j_2}, \ldots, R_{j_{m_i}})$ therefor in terms of other constituents, we shall have the premise $Q_j \beta Q_{j_1} a_0 Q_{j_2} a_0 \ldots Q_{j_{m_i}}$. It is then readily verified that if to each premise of the given production of S_1 is arbitrarily made to correspond an enunciation of S_1, and to the $\vdash$ prefixed corresponding premise of the production of S_2 is made to correspond the corresponding enunciation of S_2, then being able to replace the P's in the first production so that the premises become the corresponding enunciations of S_1 is equivalent to being able to replace the P's and Q's in the second so that the premises beginning with a $\vdash$ become

65. These premises are probably unnecessary. That is, the succeeding conditions on the Q's, coupled with the inductive result that if $\vdash P$ is asserted, αP is asserted, probably insure their satisfaction when the remaining premises are satisfied.

the corresponding enunciations of S_2 with $\vdash$ prefixed, while the remaining premises become assertions in S_2. Furthermore, that in the favorable case the P's on the one hand, P''s and Q's on the other, are thus uniquely determined, and with them the conclusions, which are then enunciations of S_1, and of S_2 with $\vdash$ prefixed, that correspond.

With the basis of S_2, clearly in canonical form C, thus completed, it follows that an enunciation of S_1 is an assertion when and only when the corresponding enunciation of S_2 with $\vdash$ prefixed is an assertion, so that S_1 has thus been reduced to S_2. Note that we originally spoke of a system in canonical form C as having a finite number of distinct primitive symbols a_1, $a_2, \ldots, a_{\mathfrak{u}}$. For our above S_2, with different $\mathfrak{u}$, these symbols are $a_0, a_1, \ldots, a_{\mathfrak{u}}, b, \alpha_0, \alpha, \vdash, \beta, \delta$. Furthermore, each assertion of S_2 involves one and only one of the last five symbols, and but one occurrence of that one symbol.[66]

5. Successive reductions to normal form.

Starting with canonical form C, we now introduce a series of reductions such that each formulation, while being included in the preceding, eliminates some formal complexity allowed in that preceding formulation. For a given system this simplification is achieved at the expense of an increase in the number of primitive letters employed, and in the number of productions constituting its basis.

Our first reduction of an arbitrary system in canonical form C is to one in which there is but one primitive assertion, and in which each production involves but a single premise, that one premise and corresponding conclusion, however,

66. Actually, for those enunciations of S_2 not in the form "$\vdash P$", the deducibility problem is immediately solvable.

retaining all of the complexity allowed for above.[67] The general plan of the method involved is to formally allow for the logical product of arbitrary assertions in the given system, and operate within such products.

Let then S_1 be a system in canonical form C with primitive letters $a_1, a_2, \ldots, a_\mu$, S_2 the about to be described system to which S_1 is to be reduced. With $a_1, a_2, \ldots, a_\mu$ also primitive letters of S_2, introduce two new primitive letters u and a_0 in S_2.[68] When the logical product of assertions, P_1, P_2, $P_3, \ldots,$ P_n of S_1 is asserted in S_2, it will appear in the form

$$u a_0\, P_1 a_0 u u a_0\, P_2 a_0 u u u a_0\, P_3 a_0 \ldots \underbrace{u \ldots u}_{n}\, a_0\, P_n a_0\, \underbrace{u \ldots u}_{n+1},$$

each P being flanked on either side by a. The separating u sequences are thus made to increase left to right by one each to enable us by the mere form of a premise to insure that certain operational variables therein must represent assertions of S_1, if that premise is to be identified with an assertion in S_2. The final basis for S_2 will reveal the necessary source of that insurance, i.e., that the only assertions of S_2 involving u are

67. That one can do with but one primitive assertion both here and in all later reductions was not observed in the notes during the course of the corresponding development. However, in a later reference to the final formulation of this section in work corresponding to part II of this account specific reference is made to the one primitive assertion in a way that suggests that at some in between time this further simplification was noted. Likewise in the notes of 1924.

68. The introduction of a_0 is new. That by its use the notes method is made neater is not our reason for its introduction. The notes consistently and incorrectly assume that an enunciation P_1 can always be written $a_i P_2$ where a_i is the first letter of the enunciation represented by P_1. But this does not allow for P_1 being null; and since for P_1 consisting of one letter, P_2 would be null, the net effect of all the reductions of this section, if carried through exactly as in the notes, would be to impose on the P's of a canonical form C the condition that their ranks exceed a certain fairly large number, thus at least vitiating the reduction of §4. This oversight is responsible for most of the changes introduced in the present section, changes which on the whole are minor.

those of the above form. We shall call such an expression a product, the P's therein the factors of the product.

We first introduce in the basis of S_2 certain productions whereby from the assertion of a product may be obtained the assertion of all products obtainable from the given product by a mere permutation of its factors. It suffices to allow for the interchange of any two consecutive factors. For the first two factors of a product this is achieved by

$$u a_0 P_1 a_0 u u a_0 P_2 a_0 u u u a_0 S \text{ produces } u a_0 P_2 a_0 u u a_0 P_1 a_0 u u u a_0 S,$$

our system being so devised that each product appearing therein has at least three factors. This allows that last a_0 to be assumed. The u, uu, uuu of the premise are then "maximal" u sequences. As these u sequences differ by one each, P_1 and P_2 must be free from u's, and hence, by our induction, be the two initial factors of the product. The interchange then results via the production. For two consecutive factors neither starting nor ending the product,the result is achieved by

$$R a_0 u Q u a_0 P_1 a_0 u Q u u a_0 P_2 a_0 u Q u u u a_0 S$$

produces

$$R a_0 u Q u a_0 P_2 a_0 u Q u u a_0 P_1 a_0 u Q u u u a_0 S.$$

Here Q must consist of u's only. For otherwise $a_0 u Q u a_0$ and $a_0 u Q u u a_0$ would have their initial a_0's followed by identical maximal u sequences. The u sequences $uQu, uQuu$ and $uQuuu$ are then maximal, and differ in length by one each. P_1 and P_2 again then are consecutive factors of the product. Finally, for two factors ending a product the last production, rewritten with $a_0 S$ deleted, suffices.[69]

The next production to be added to the bases of S_2 allows us to pass from the assertion of a product to the assertion of the

69. We might observe that the notes go to considerable length in showing why the u method is thus effective.

first factor of a product, and hence, with the help of the previous three productions, to the assertion of an arbitrary factor of a product. The production is simply

$$u a_0 P a_0 u u a_0 R \text{ produces } P.$$

In translating the operations of S_1 into operations within products of S_2, we allow for passing from a product whose initial factors can be identified with the premises of an S_1 operation, to that product with the conclusion of the S_1 operation as additional factor. That additional factor must end the new product so as not to disturb the progression of the maximal u sequences. Let "G_1, $G_2, \ldots G_k$ produce G" represent any one of the S_1 operations. Let H represent

$$u a_0 G_1 a_0 u u a_0 G_2 a_0 \ldots \underbrace{u \ldots u}_{k} a_0 G_k a_0 \underbrace{u \ldots u}_{k+1}.$$

Then the corresponding S_2 operation may be represented by

$$H a_0 R a_0 u Q u a_0 S a_0 u Q u u \text{ produces}$$

$$H a_0 R a_0 u Q u a_0 S a_0 u Q u u a_0 G a_0 u Q u u u.$$[70]

Note that the operational variables of this production are those of the S_1 production, and Q, R, S. Since each operational variable in G occurs in at least one of the G_i's, our new production will indeed have the same operational variables in its conclusion as in its premise. The portion of the premise following H insures Q consisting of u's only. This, with the form of H, insures $G_1, G_2, \ldots, G_k$, being determined factors of the premise, G of the conclusion. Hence, the validity of our transformation of the S_1 production. The additional operational variables R and S require an assertion to which this production is applied to have at least $k+2$ factors, a requirement secured below. Of course, the basis of S_2 is to have the correspondent of each of the operations in the basis of S_1.

70. This production, as given in the notes, involves some slight errors.

With S_1 having κ productions, the above $\kappa+4$ productions constitute all of the productions in the basis of S_2. Its sole primitive assertion is then formed as follows. Let L be the largest number of premises occurring in any production of S_1. If S_1 has λ primitive assertions, let each be repeated L times to give λL sequences each involving no other letters than a_1, $\ldots, a_\mu$. If $\lambda L<L+2$, or $\lambda L<3$, again duplicate one of these sequences the one or two times needed to avoid these inequalities. If then $k_1, k_2, \ldots, k_M$ are these duplicated primitive assertions of S_1, the primitive assertion of S_1 will be their product

$$ua_0 k_1 a_0 u u a_0 k_2 a_0 \ldots \underbrace{u \ldots u}_{M} a_0 k_M a_0 \underbrace{u \ldots u}_{M+1}.$$ [71]

Now it is readily proved by induction that if at a certain point of the process for obtaining assertions in S_1 a certain finite set of assertions has been obtained, then there will be asserted in S_2 a product among whose factors are each of the above assertions repeated L times. For the primitive assertions of S_1, this is insured by the primitive assertion of S_2. Assume it to be true for the deductive process in S_1 at an arbitrary point, let P_2 be the corresponding assertion in S_2, P_1 the next assertion obtained in S_1, $P_{11}, P_{12}, \ldots P_{1k}$. the premises of the production of S_1 yielding conclusion P_1. Then each P_{1j} appears as factor of P_2 indeed L times at least. Hence from P_2, by the first three productions of S_2, an assertion P_2' can be obtained in which the first k factors are $P_{11}, P_{12}, \ldots, P_{1k}$ respectively, whatever repetitions may occur among those P's. The production of S_2 corresponding to the one of S_1 in question will then add P_1 as factor to P_2'. Mere repetition of the application of this production will then yield P_2'', which will be P_2' with L additional factors equal to P_1. The induction is thus established.

71. Instead of this single primitive assertion, the notes require the assertion of the products of $L+1$ of the h's, repeated or not, for all such choices of h's, thus, for the moment, overlooking the obvious simplification to one primitive assertion.

It follows that for each assertion P_1 in S_1 there will be an assertion P_2 in S_2 having P_1 as a factor. By the first three productions of S_2 this factor can be made the first factor of an assertion in S_2, and hence, by the fourth production of S_2, P_1 itself will be an assertion of S_2. That is, every assertion of S_1 is an assertion of S_2. Our basis for S_2 shows that the only other assertions of S_2 are products of assertions of S_1, and so not wholly written on the letters of S_1. Hence, an enunciation of S_1 is an assertion of S_1 when and only when it is an assertion of S_2, whence the reduction of S_1 to S_2.

In our second reduction of canonical form C the productions, all with single premises by the previous reduction, now take the more special form

$$g_1 P_1 g_2 P_2 \ldots g_m P_m g_{m+1}$$

produces

$$\bar{g}_1 \bar{P}_1 \bar{g}_2 \bar{P}_2 \ldots \bar{g}_m \bar{P}_m \bar{g}_{m+1}$$

where, however, m, and of course the g's, may vary from operation to operation. By contrast, in the previous productions P's could be repeated, have different arrangements in premise and conclusion, and in part be missing from the conclusion while present in the premise.

Again, let the primitive letters of the given system be symbolized $a_1, a_2, \ldots a_\mu$. Let its i-th production be

$$g_1 P_{j_1} g_2 P_{j_2} \ldots g_m P_{j_m} g_{m+1}$$

produces

$$g'_1 P_{j'_1} g'_2 P_{j'_2} \ldots g'_{m'} P_{j'_{m'}} g'_{m'+1}$$

where it is understood that each letter except P has i for additional subscript. The subscripts of the P's need not be distinct in premise or conclusion, while the different subscripts of the P's in the conclusion all appear in the premise. However, the letter P occurs exactly $m+m'$ times in the production.

We introduce a new primitive letter u, and for each such production two new primitive letters v_i, w_i. In obtaining the effect of the i-th production we will, as above, leave this subscript i understood. v_i will be used in passing from an assertion involving a's only that could be the premise of the i-th production to one which has both that premise and corresponding conclusion recognizable within it; w_i in passing from such a composite assertion to the desired conclusion only. The efficacy of our method will depend on each assertion in the new system which involves v or w having that letter only at the beginning of the assertion, and in the first case always involving exactly $2\,m+m'$ u's, in the second, m' u's. Our new productions will in every case explicitly exhibit this v and $2\,m+m'$ u's, or w and m' u's, so that we can be sure that in their application the operational variables can represent sequences of a's only. Except for a minor preliminary type, all of our "v-assertions" will be in the form

$$vug_1 P_1 uQ_1 ug_2 P_2 uQ_2 \ldots ug_m P_m uQ_m g_{m+1} ug'_1 Q_{m+1} ug'_2 Q_{m+2}$$
$$\ldots ug'_{m'} Q_{m+m'} g'_{m'+1},$$

and when so asserted will have the following properties. The sequence of a's $g_1 P_1 Q_1 g_2 P_2 Q_2 \ldots g_m P_m Q_m g_{m+1}$ is an assertion of the given, and indeed new system, while the sequences of a's $Q_1, Q_2, \ldots, Q_m, Q_{m+1}, \ldots, Q_{m+m'}$ can, in order, be identified with $P_{j_1}, P_{j_2}, \ldots, P_{j_m}, P_{j'_1}, \ldots, P_{j'_{m'}}$, that is, any two Q's corresponding to P's with identical subscripts are equal. Note that with all $2m+m'$ u's exhibited, the g's being given, the P's and Q's of such an assertion are uniquely identifiable in the assertion. Our method depends on the fact that when such an assertion is obtained in which the P's are null, then, due to the equalities forced on the Q's, $g_1 Q_1 g_2 Q_2 \ldots g_m Q_m g_{m+1}$ becomes an assertion on a's only that can be identified with the premise of the i-th production of the given system, and hence $g'_1 Q_{m+1} g'_2 Q_{m+2} \ldots g'_{m'} Q_{m+m'} g'_{m'+1}$ an expression on a's only that will be the corresponding conclusion. Of course, each production about to be described is directly seen to be in the desired newly simplified form.

Since a null assertion has been excluded from our systems, each assertion of the given system is of the form $a_j P$, $j=1,2, \ldots \mu$. The productions

$$a_j P \text{ produces } v\, a_j\, P\, u \ldots u$$

with $2m+m'$ u's in $u \ldots u$ changes each "a-assertion", i.e. assertion involving a's only, into what we shall call the intermediate v form. As all other assertions of our new system will begin with v or w, these productions will be inapplicable to them. If now an a-assertion can be the premise of the i-th production, its intermediate v form will be put into primary v form, or just v form, by the production

$$v g_1 P_1 g_2 P_2 \ldots g_m P_m g_{m+1} u \ldots u$$

produces

$$v u g_1 P_1 u u g_2 P_2 u u \ldots g_m P_m u\, g_{m+1}\, u\, g'_1\, u\, g'_2\, u \ldots u\, g'_{m'}\, g'_{m'+1}.$$

Of course this production may be applicable without the P's being identifiable with those of the premise of the i-th production. But, comparing this conclusion with our general v form, we see that it satisfies the requirement thereof with all Q's null. Now any set of a-sequences that could be identified with the $P_{j_1}, P_{j_2}, \ldots, P_{j_m}, P_{j'_1} \ldots, P_{j'_{m'}}$ of the i-th production can be built up as follows. Start with the set of null sequences. Let $Q_1, Q_2, \ldots, Q_m, Q_{m+1}, \ldots, Q_{m+m'}$ be any such derived set of a-sequences. Let $Q_{j_1}, Q_{j_2}, \ldots, Q_{j_\nu}$, j's increasing, be any subset thereof corresponding to all P's with subscripts equal to a given subscript, a_j any one of the primitive a's. Then $\ldots, a_j Q_{j_1}, \ldots, a_j Q_{j_2}, \ldots, a_j Q_{j_\nu}, \ldots,$ all other Q's unchanged, will also be such a set of a-sequences. Rewrite the subscript sequence $j_1, j_2, \ldots, j_\nu$ in the form $j_1, \ldots j_\lambda, j_{\lambda+1}, \ldots, j_\nu$ so that $j_\lambda \leq m$, $j_{\lambda+1} > m$, and let $j_{\lambda+1} - m = j'_1, \ldots, j_\nu - m = j'_{\lambda'}$. Of course we may have $\lambda = \nu$. Now for each such choice of original P subscript, and each a_j, introduce the production

$$v \ldots u g_{j_1} P_{j_1} a_j \, u Q_{j_1} \ldots u g_{\lambda} P_{j_\lambda} a_j \, u Q_{j_\lambda} \ldots u g'_{j'_1} Q_{j_{\lambda+1}} \ldots u g'_{j'_{\lambda'}} Q_{j_\nu} \ldots$$

produces

$$v \ldots u g_{j_1} P_{j_1} u a_j Q_{j_1} \ldots u g_{\lambda} P_{j_\lambda} u a_j Q_{j_\lambda} \ldots u g'_{j'_1} a_j Q_{j_{\lambda+1}} \ldots u g'_{j'_{\lambda'}} a_j \, Q_{j_\nu} \ldots$$

all of the rest of both premise and conclusion being as in the type v form above. Such a production will then change a valid v form into a valid v form, the effect being however to "drain" the P's of such a form and "swell" the Q's. If then an assertion of the given system can be put in the form of the premise of the i-th production, the corresponding intermediate v form will pass into a v form such that successive application of the above productions will completely drain the P's thereof; and, indeed, conversely. This marks the end of the first half of the passage from a-assertion to a-assertion in the new system. While the second half could be set up by means of similar productions in reverse, with interchange of emphasis on premise and conclusion of the i-th production,[72] the following method is simpler. With P's all null, the v form determines the desired a-conclusion as described above. The about to be introduced w forms each have exactly m' u's all explicitly appearing in the productions. From such a v form with P's all null the first w form is obtained via

$$v u g_1 u Q_1 u g_2 u Q_2 \ldots u g_m u Q_m g_{m+1} u g'_1 Q_{m+1} u g'_2 Q_{m+2} \ldots u g'_{m'} Q_{m+m'} g'_{m'+1}$$

produces

$$w g_1 Q_1 g_2 Q_2 \ldots g_m Q_m g_{m+1} u g'_1 Q_{m+1} u g'_2 Q_{m+2} \ldots u g'_{m'} Q_{m+m'} g'_{m'+1}$$

We can now get rid of the no longer interesting part of this w form, i.e., the part between w and the first u thereof, by the μ productions

72. This is the method of the notes. In the previous work one obvious error, discovered at the time, is corrected as directed by the notes, and a few minor changes are introduced.

$$w a_j P u g'_1 P_1 u g'_2 P_2 \ldots u g'_{m'} P_{m'} g'_{m'+1}$$

produces

$$w P u g'_1 P_1 u g'_2 P_2 \ldots u g'_{m'} P_{m'} g'_{m'+1}$$

iteratively applied till letter by letter what was the original a-assertion disappears. The desired a-conclusion then would be obtained via

$$w u g'_1 P_1 u g'_2 P_2 \ldots u g'_{m'} P_{m'} g'_{m'+1}$$

produces

$$g'_1 P_1 g'_2 P_2 \ldots g'_{m'} P_{m'} g'_{m'+1} .$$

Our final system will then be on the primitive letters $a_1, \ldots, a_\mu$, $u, v_1, w_1, v_2, w_2, \ldots, v_\kappa, w_\kappa$, κ being the number of productions of the given system. The one primitive assertion of the new system will be the one primitive assertion of the given system, the productions of the new system, all of the above productions for each of the κ productions of the given system. Our above analysis then easily shows that the assertions of the new system involving no other letters than $a_1, \ldots, a_\mu$ are exactly the assertions of the given system, and the desired reduction has been effected.

Our third and penultimate simplifying reduction of canonical form C is to one where the operations are of the form

$$g_1 P g_2$$

produces

$$\bar{g}_1 P \bar{g}_2 ,$$

i.e., involve but a single operational variable. Again let a system in the previous simplified form have primitive letters $a_1, a_2, \ldots, a_\mu$, and κ operations, the number of P's in the premise, and hence conclusion, of the i-th operation being m_i. For the i-th operation, with $i=1, 2, \ldots \kappa$, and each primitive letter

a_j we introduce 2 m_i+1 new primitive letters $a'_{ji}, a''_{ji}, \ldots,$ $a_{ji}^{(2m_i)}$, $a_{ji}^{(2m_i+1)}$.[73] We also introduce the primitive letter a_{0i} and its 2 m_i+1 primed equivalents.[74] With one such operation in mind at a time we will as above omit the extra subscript i. Apart from the use of a_0 and $a_0^{(j)}$'s, needed to take care of g's or P's that are null, the essence of our method is to pass from an a-assertion in the form $g_1 P_1 g_2 P_2 \ldots g_m P_m g_{m+1}$ to an assertion $g'_1 g'''_2 \ldots g_{m+1}^{(2m+1)} P''_1 P_2^{\overline{IV}} \ldots P_m^{(2m)}$ where the superscript k say indicates that each a_j in the corresponding expression is here written $a_j^{(k)}$. As a result our premise will now have the form gP with $g = g'_1 g'''_2 \ldots g_{m+1}^{(2m+1)}$, $P = P''_1, P_2^{\overline{IV}} \ldots P_m^{(2m)}$.

In detail, we first introduce μ productions

$$a_j P \text{ produces } a_0 a_j P, \quad j=1,2,\ldots,\mu,$$

which will be applicable in fact only to assertions on $a_1, a_2, \ldots, a_\mu$, and changes any such assertion Q into $a_0 Q$. We then introduce a finite series of finite sets of productions depending in number on m and μ. The first set has the one production

$$a_0 g_1 P \text{ produces } a'_0 g'_1 a_0 P a''_0.$$

Inductively let the conclusion of the sole production in the $(2k-1)$-st set be in the form $G_k a_0 P a_j^{(2k)}$. Then the $(2k)$-th set has the productions

$$G_k a_0 a_j P \text{ produces } G_k a_0 P a_j^{(2k)}, \quad j=1,2,\ldots,\mu,$$

the $(2k+1)$-st set the sole production

$$G_k a_0 g_{k+1} P \text{ produces } G_k a_0^{(2k+1)} g_{k+1}^{(2k+1)} a_0 P a_0^{(2k+2)}.$$

This is to hold for $1 \leq k < m$, while for $k = m$ the sole production of the $(2m+1)$-st set is to be

73. The notes observe that we could be more economical if desired.

74. This is new, and is necessary for the reason discussed in footnote 68. In fact, here the notes rely on the g's as well as P's not being null.

$$G_m a_0\, g_{m+1} a_0'' P \quad \text{produces} \quad G_m\, a_0^{(2m+1)}\, g_{m+1}^{(2m+1)}\, a_0''\, P\,.$$

We then readily see that starting with an assertion on $a_1, \ldots, \mathfrak{u}$ in the form $g_1 P_1 g_2 P_2 \ldots g_m P_m\, g_{m+1}$, one can, with the aid of these productions, obtain as an assertion

$$a_0'\, g_1' a_0''' g_2''' \ldots a_0^{(2m+1)}\, g^{(2m+1)} a_0^{(2m+1)}\, g_{m+1}^{(2m+1)}\, a_0''\, P_1'' a_0^{\overline{IV}} P_2^{\overline{IV}} \ldots a_0^{(2m)} P_m^{(2m)}.$$

Furthermore, note that starting with an assertion on $a_1, a_2, \ldots, \mathfrak{u}$, flanked on the left by a_0 as above, one can apply the above operations only in the following order, if at all. First, the sole operation of the first set; and inductively, if the operation in the $(2k-1)$-st set has last been applied, the next applicable operation can only be an operation in the $2k$-th set or the operation in the $(2k+1)$-st set, if an operation in the $(2k)$-th set has last been applied, the next applicable operation can only be an operation in the same set, or the operation in the next set. Furthermore, the last operation in its premise explicitly indicates the a_0'' , first introduced into an assertion only as a result of the first operation. It readily follows that if the last operation does enter into a possible sequence of operations, the conclusion thereof can have no letter a_j in it without a superscript. The entire given assertion has thus been translated; and it is readily seen that that last assertion, and hence given assertion, are and can be put in the forms above given.

The actual correspondent of the original i-th operation in translated form may then be written simply

$$a_0'\, g_1' a_0''' g_2''' \ldots a_0^{(2m-1)}\, g^{(2m-1)} a_0^{(2m+1)} g_{m+1}^{(2m+1)} P$$

produces

$$a_0'\, \bar{g}_1' a_0''' \bar{g}_2''' \ldots a_0^{(2m-1)}\, \bar{g}_m^{(2m-1)}\; a_0^{(2m+1)} \bar{g}_{m+1}^{(2m+1)} P\,;$$

and the passage from this translated conclusion to the actual conclusion can be effected by a set of productions the reverse of those above given. That is, in each of the above productions prior to the actual correspondent of the i-th production replace all g_j's by $\bar{g}_j$'s, and interchange hypothesis and conclusion. The resulting productions then clearly suffice to yield the conclusion yielded by the original i-th production. True, the complete set of productions thus set up to take the place of the original i-th production may now allow other paths than from assertion on $a_1, \ldots, a_\mu$, down the first group of productions, through the intermediate production, and up the second group of productions to new assertion on $a_1, \ldots, a_\mu$.[75] But it is readily seen that any departures from this progression merely constitute unravelings of parts of such a progression, or, apart from such unravelings, constitute shortcuts of valid full progressions of this type. Since, furthermore, one can change the set of productions one is working with only when an assertion on $a_1, \ldots, a_\mu$ alone is obtained, the validity of our reduction follows.

Our final reduction is to a system whose operations are in the form

$$gP$$

produces

$$Pg'$$

The present method assumes that in the productions of the previous system, all in the form

$$g_1 P g_2$$

produces

$$g'_1 P g'_2 \text{ ,}$$

g_1 and g_2 are never null. We therefore actually first need the following preliminary reduction.[76] Introduce a new primitive

75. This could have been avoided say by the v, w method used earlier.

76. This is new. The notes erroneously state that as a consequence of the previous reductions P must in fact represent sequences having at least two letters, and on that basis easily show how to replace such productions with g_1, g_2 null by an equivalent set with neither g_1 nor g_2 null.

letter a_0, and if h is the sole primitive assertion of the given system let $a_0 h a_0$ be the sole primitive assertion of the new system. Replace each of the above operations of the given system by

$$a_0 g_1 P g_2 a_0 \text{ produces } a_0 g_1' P g_2' a_0$$

and finally add the production

$$a_0 P a_0 \text{ produces } P.$$

Except for the last production the new system may be said to be simply isomorphic [77] with the old, P being an assertion in the given system when and only when $a_0 P a_0$ is an assertion in the new system. The last operation then merely recovers the assertions of the given system. Note that even that last operation is in the desired form with neither g_1 nor g_2 null.

Assume then that such is our given system with primitive letters again $a_1, a_2, \ldots, a_\mu$. We introduce new primitive letters $\bar{a}_1, \bar{a}_2, \ldots, \bar{a}_\mu$, and "translating productions"

$$a_j P \text{ produces } P\bar{a}_j, \ \bar{a}_j P \text{ produces } Pa_j, \ j = 1, 2, \ldots, \mu.$$

Starting with an assertion of the form $a_{i_1} \ldots a_{i_j} a_{i_{j+1}} \ldots a_{i_n}$ these productions will yield only assertions of the form $a_{i_{j+1}} \ldots a_{i_n} \bar{a}_{i_1} \ldots \bar{a}_{i_j}, \bar{a}_{i_1} \ldots \bar{a}_{i_j} \bar{a}_{i_{j+1}} \ldots \bar{a}_{i_n}, \bar{a}_{i_{j+1}} \ldots \bar{a}_{i_n} a_{i_1} \ldots a_{i_j}$, in addition to the original assertion. Only one of these $2n$ distinct forms consists wholly of unbarred letters, i.e., the original form, while continued application of the above operations merely keeps deriving these $2n$ "equivalent forms" cyclically, so that anyone can thus be obtained from any other.

Our reduction will then be effected if for each operation "$g_1 P g_2$ produces $g_1' P g_2'$" of the given system we introduce in the new system the operation

77. This concept was introduced in the notes in connection with the equivalence proof referred to in §1.

$$\bar{g}_2 g_1 P$$

produces

$$P g_2' \bar{g}_1',$$

where $\bar{g}_2$, for example, is g_2 with each letter replaced by the corresponding barred letter. Of course, the one primitive assertion of the given system is also the one primitive assertion of the new system. Note that if at any point as assertion without barred letters appears, then if it can be written $g_1 P g_2$, the first given translating operations derive from it $\bar{g}_2 g_1 P$, hence the above yields $P g_2' \bar{g}_1'$, and so finally $g_1' P g_2'$ is obtained as desired. That is, the new system contains all of the assertions of the given system. It further follows that the assertions of the new system consists only of the assertions of the given system and their equivalents. For, proceeding inductively, this clearly remains true under the translating operations. Now suppose it is true of an assertion in the form $\bar{g}_2 g_1 P$. Since $\bar{g}_2$ and g_1 are not null, $\bar{g}_2 g_1$ alone exhibits a change from barred to unbarred letters. P therefore must consist of unbarred letters only. $\bar{g}_2 g_1 P$ is therefore a translation of the assertion $g_1 P g_2$ of the original system, and hence the conclusion $P g_2' \bar{g}_1'$ is a translation of $g_1' P g_2'$, also an assertion of the original system. The desired reduction has thus been effected.

A system of the last given type will be said to be in normal form, any system reducible to such a system, reducible to normal form, whence the heading of the section just concluded.[78]

78. Note that in the case of systems in normal form, as also for those in the form immediately preceding, that unique identifiability of operational variables which was posessed by systems in canonical form **B**, and lost in **C**, is regained. This, of course, is due to the presence of but one operational variable in each production. Indeed, the formal mechanism for obtaining new assertions from old is far simpler for a system in normal form than for canonical forms **A** or **B**. And if it be desired to change its set of assertions from, as it were, a growing population to a uniquely determined infinite sequence of assertions, but the fol-

6. Closing the circle.

Apart from the reduction of *10 Principia Mathematica to canonical form B, our work has consisted of a series of reductions starting with canonical form A and ending with systems in normal form. Using $S_1 \rightarrow S_2$ to symbolize S_1 is reducible to S_2, our definition of reducibility clearly yields the result, if $S_1 \rightarrow S_2$ and $S_2 \rightarrow S_3$ then $S_1 \rightarrow S_3$. If then we can show that systems in normal form are reducible to canonical form A, it will follow that all of our canonical forms from first to last are equivalent, that is, for any two of them any system of either is reducible to some system of the other. Stated otherwise, the solution of the finiteness problem for all systems in one canonical form would yield the solution of the finiteness problem for all systems in the other.[79]

lowing procedure need be instituted: Order the productions of the system. Start with the primitive assertion as "given assertion", the first operation as "given operation", the primitive assertion as sole member of the "given sequence of assertions", and iterate as follows. Apply, if possible, the given operation to the given assertion as premise and add the resulting conclusion to the given sequence of assertions, thus forming the new given sequence of assertions. If the given operation is not the last, retain the given assertion as given assertion, and pass to the next operation as given operation. If the given operation is the last, pass from the given assertion to the next assertion as given assertion, and to the first operation as given operation.

79. This idea of closing the circle of reductions by reducing systems in normal form to canonical form A does not explicitly appear in the notes. It, however, was probably considered to be one of those things that could obviously be done. While, as a result, the first of the two stages in which this reduction is carried out is "apocryphal", the second of those two stages merely uses a method which in more general form was used in the notes to reduce the canonical form intermediate between A and B to A. In fact, to somewhere preserve this method is our excuse for introducing this section.

As to *10 being merely attached to this circle, the unpublished note referred to in footnote 18, categorically states that a proof of the reducibility of canonical form A to *10 "is nearly completed", and as a result even suggests that the solution of the finiteness problem for *10 would yield the solution of the finiteness problem for all of Principia Mathematica. An examination of the notes reveals the reduction to have been carried through in three stages: first the reduction to *10 of mathematical systems using *10 as logic, second the reduction to the latter

We shall perform our reduction in two stages, first to a certain type of system in canonical form B, and then just this type of system to canonical form A. Corresponding to the primitive letters $a_1, a_2, \ldots, a_\mu$ of the given system in normal form, we introduce primitive first order functions $a_1(p), a_2(p), \ldots, a_\mu(p)$. We cannot quite let an arbitrary enunciation $a_{i_1} a_{i_2} \ldots a_{i_n}$ of the given system correspond to $a_{i_1}(p) a_{i_2}(p) \ldots a_{i_n}(p)$ in the sense of a formal product of n factors. But corresponding to the way in which an ordinary product of n elements is secured by a dyadic operation and associative law thereon, we get the desired effect by introducing a primitive second order function $b(p,q)$. The correspondent of $f = a_i a_{i_2} \ldots a_{i_n}$ of the given system will then be $b(a_{i_1}(p), b(a_{i_2}(p), \ldots b(a_{i_{n-1}}(p), a_{i_n}(p)) \ldots))$, which we will symbolize by $f(p)$ for purposes of discussion. Of course, when $n = 1$ we simply have $f(p) = a_{i_1}(p)$.

Corresponding to the above mentioned associative law we introduce the operations

$$b(P, b(Q,R)) \text{ produces } b(b(P,Q),R),$$

$$b(b(P,Q),R) \text{ produces } b(P, b(Q,R)),$$

$$b(b(P,Q), b(R,S)) \text{ produces } b(P, b(b(Q,R),S)),$$

which have the effect of enabling us to pass from any mode of inserting b-parentheses in the sequence $a_{i_1}(p), a_{i_2}(p), \ldots, a_{i_n}(p)$ to any other mode. All of the resulting forms may then be considered correspondents of f, with $f(p)$ the principal correspondent.

of what were termed "algebraic systems", third the reduction of systems in canonical form A to algebraic systems. It is for the second reduction that one half of the necessary two-way proof was postponed. Not having checked the parts of the proof that are given, we cannot guarantee their correctness. In this connection see footnote 90.

If h symbolizes the one primitive assertion of the given system, our new system will have the primitive assertion $h(p)$. The operations "$g_i P$ produces Pg_i'" will be suitably taken care of if we allow for the passage from some correspondent of the hypothesis to come correspondent of the conclusion. Note that for g_i, g_i', P not null, a correspondent of $g_i P$ will be in the form $b(g_i(p), Q)$, of Pg_i', $b(Q, g_i'(p))$. We then correspondingly introduce the operation

$$b(g_i(P), Q) \text{ produces } b(Q, g_i'(P)).$$

In fact, since the primitive assertion of the new system involves but the one variable p, all assertions of the system may be considered to be written on the one variable p, since the operation of substitution of canonical form B will merely reproduce these assertions on other variables. In the application of the last given operation, P, in fact, will only be identifiable with that variable p.

To allow for P null in the operation of the given system, the separate operation

$$g_i(P) \text{ produces } g_i'(P)$$

must be added. If either g_i or g_i' is null, we may clearly assume the other not null, while P then certainly cannot be null, since the null assertion has been specifically excluded from our systems. When g_i' is null we then need but the one operation

$$b(g_i(P), Q) \text{ produces } Q.$$

When g_i is null we must explicitly insure our conclusion being written on but a single variable. For P of more than one letter, we may write $p = a_j P'$, $j = 1, 2, \ldots, \mu$ and correspondingly set up the μ operations

$$b(a_j(P), Q) \text{ produces } b(b(a_j(P), Q), g_i'(P)).$$

For P of one letter we likewise have

$$a_j(P) \text{ produces } b(a_j(P), g_i'(P)),\ j=1,2,\ldots,\nu.$$

This completely takes care of the first reduction.

For the second reduction introduce a new primitive first order function $k(p)$ and alter the preceding system as follows. Replace the primitive assertion $h(p)$ by $h(k(p))$, retain the preceding productions, and change the operation of substitution to that of canonical form A. Now it is readily proved by induction that every assertion of the resulting system will be of the form $F(k(p))$ where $F(p)$ does not involve the primitive function k. Furthermore, if $F(k(P))=F'(k(P'))$, with $F(p)$ and $F'(p)$ not involving k, we see by successively stripping away necessarily identical outmost primitive functions that $F=F'$, $P=P'$, i.e., that such a form is unique.[80] Again by induction, if $F_1(k(P_1))$, $F_2(k(P_2)), \ldots, F_n(k(P_n))$ are the successive assertions in an arbitrary proof of our system, we find that P_i is either identical with P_{i-1}, or obtainable from it by a substitution. It follows that the only assertions of our system of the form $F(k(P))$ are obtained by deductive processes in which no other substitutions are employed than that of variable for variable. But with substitution thus limited, the new system is simply isomorphic with the old under the replacement $p \leftrightarrow k(p)$, p an arbitrary variable for completeness. It follows that an enunciation $F(p)$ of the previous system is an assertion thereof when and only when $F(k(p))$ is an assertion of the new system, whence our second reduction.

For our original system we then have that $a_{i_1}a_{i_2}\ldots a_{i_n}$ is an assertion thereof when and only when

$$b(a_{i_1}(k(p)), b(a_{i_2}(k(p)), \ldots b(a_{i_{n-1}}(k(p)), a_{i_n}(k(p)))\ldots))$$

is an assertion in the system in canonical form A, as desired.

80. This is a special aspect of the L.C.M. process referred to early in §3.

We have observed in §3 how the seemingly simple problem of "tag" in fact proved intractable for $\mu=2$, $\nu>2$, of bewildering complexity for $\mu>2$, $\nu=2$. In view of our reduction of canonical form A to a form as close to that of "tag" as the normal form, the difficulty of "tag" is no longer surprising.[81] While this suggested that special cases of "tag" might well be worth consideration as major problems in themselves, the following further reduction of the normal form seemed more promising.

We merely state the result. Given any system in normal form on letters $a_1, a_2, \ldots, a_\mu$, its assertions will be the assertions, on those letters, in a system on letters $a_1, a_2, \ldots, a_\mu$, $a'_1, a'_2, \ldots a'_{\mu'}$, having a finite number of primitive assertions, and a finite number of operations of the following form

$g\,P$	$P\,g$	$g_1\,P$	$P\,g_1$
produces	produces	$g_2\,P$	$P\,g_2$
$g'P$	Pg'	produce	produce
		$g'\,P$	Pg'

While this reduction seems to undo much of the simplification that was achieved by our previous reductions, especially in that it allows productions with more than one premise, the fact that in each production the g's all occur on the same side of the operational variable makes the formulation analogous to canonical form A with primitive functions of one argument only, and the finiteness problem for that case was solved. In fact, when all the productions have the g's on the same side, the resulting system is essentially in this special canonical A form, and is actually reducible thereto by the "k method" last given. By further study, the resulting solution of the finiteness problem was extended to those of the present systems in which first

81. In fact, at one point late in the work on "tag", it seemed that the regularity induced by always removing ν elements from the beginning of a sequence was responsible for the intrusion of number theory in the development, so that it was tentatively suggested that "tag" be generalized to a form which, indeed, is exactly that of the later derived normal form.

order operations only occur, i.e., operations with but one premise, and then indeed to those systems in which only the second order operations were restricted to having the g's all on one side, e.g., to systems having only operations of the first three of the above four types.[82]

The resulting methods held out the possibility of an attack on the finiteness problem for systems having all four of the above types of operations, though certain of the difficulties of "tag" even then seemed glimmering in the distance. And just when hope was thus renewed for a solution of the general finiteness problem, a fuller realization of the significance of the previous reductions led to a reversal of our entire program.

Part II.

THE ANTICIPATION

7. Generated sets of sequences.

The power of canonical form ***B*** was demonstrated in section 2 by the reduction of *10 Principia Mathematica to a single system in that canonical form. From this experience, and the knowledge of the kind of forms and the kind of operations appearing in the whole of Principia Mathematica, or could be made to appear if a complete symbolic development thereof were given, it becomes reasonably certain that all of Principia Mathematica can in similar fashion be reduced to a system in canonical form ***B***. In the absence of the forbidding amount of work needed actually to carry out this reduction, added strength is lent to the above conclusion by the further reductions carried through in sections 4 and 5; for if the meager formal apparatus of our

82. While the solution for the first order cases is completely written up in our notes (we have not, however, checked this solution) our authority for the more general result is but a statement to that effect in the writer's diary for that date.

final normal systems can wipe out all of the additional vastly greater complexities of canonical form B, the more complicated machinery of the latter should clearly be able to handle formulations correspondingly more complicated than itself.

Granting the reducibility of the system of Principia Mathematica to a system in canonical form B, the reductions of sections 4 and 5 show that it is therefore further reducible to a system in normal form. We shall not linger, however, over the fact that the finiteness problem for the whole of Principia Mathematica would therefore be solved if the finiteness problem for the formally simple normal systems were solved.[83]

Our present interest centers in the fact that in each reduction carried through in section 5 the assertions of the given system are exactly those assertions of the new system which involve only letters of the given system. It follows that the assertions of any system in the wide canonical form C with primitive letters $a_1, a_2, \ldots, a_\mu$ are those assertions of a system in normal form with primitive letters $a_1, a_2, \ldots, a_\mu, a'_1, a'_2, \ldots, a'_\mu$ which involve only the letters $a_1, a_2, \ldots, a_\mu$. In fact, more generally, the same conclusion holds for the assertions involving only the letters $a_1, a_2, \ldots, a_\mu$ of any system in canonical form C with primitive letters $a_1, a_2, \ldots, a_\mu, a''_1, a''_2, \ldots, a''_\mu$.[84] If then we think of canonical form C as a method of generating a set of (finite) sequences on letters $a_1, a_2, \ldots, a_\mu$, i.e., the set of assertions involving only those letters, we see that the generated sets of sequences yielded by all systems in canonical form C are the same as those yielded by the formally simpler normal systems.

Now for any system in canonical form C the premises and conclusion of any production thereof could be completely de-

83. "Formally simple" refers to the bases of the normal systems.

84. This observation does not seem to have been made explicitly in the notes. In the more general discussion begun in the next paragraph we therefore assume that no other letters than $a_1, \ldots, a_\mu$ appear in canonical form C.

scribed in logical terms and the primitive relation of precedence in a sequence. This suggests the possibility of describing more complicated operations for the purpose of generating sets of sequences. Suppose each operation is of the form a certain number of premises, described in logical terms, gives rise to a certain conclusion, likewise described. Such an operation may be written

> $P_1, P_2, \ldots, P_\kappa$ produces P where $P_1, P_2, \ldots, P_\kappa, P$ have certain properties $f_1(P_1), f_2(P_2), \ldots, f_\kappa(P_\kappa), f(P)$.

Note that the P's are not the operational variables of the operation, but that the latter are allowed for in the properties mentioned.[85] Now suppose that a set of postulates is set up for sequences in letters $a_1, a_2, \ldots, a_\mu$, and Principia Mathematica is used as the logic of the resulting mathematical system. Granting the generality of Principia Mathematica, sequences $P_1, P_2, \ldots, P_\kappa, P$ will have the properties $f_1(P_1), f_2(P_2), \ldots, f_\kappa(P_\kappa), f(P)$ when and only when the latter are assertions in the above <u>sequence-Principia Mathematica</u> system. If then we add the postulates on sequences and the system of Principia Mathematica to our assumed system for generating sets of sequences on $a_1, a_2, \ldots, a_\mu$, the above type operation of the latter can be written in the form

$$f_1(P_1), f_2(P_2), \ldots, f_\kappa(P_\kappa), f(P), P_1, P_2, \ldots, P_\kappa \text{ produce } P.$$

As a result, our system for generating a set of sequences on $a_1, a_2, \ldots, a_\mu$ becomes the formal system of Principia Mathematica supplemented by certain postulates and operations of the same general type. Granting that Principia Mathematica can be reduced to a system in normal form, the same can be expected of the present system in its new form. If then, in

85. This account is taken from a notes summary of previous developments written in February 1922. Mention is made of there being a little difficulty in connecting up the variables in $P_1, \ldots, P_\kappa$ and P. We here think of these as free variables in terms of which the properties in question are expressed.

translating the enunciations of our complicated system into enunciations of the normal system the letters $a_1, a_2, \ldots, a_\mu$ are left unchanged, enunciations which are mere sequences of such letters being their own correspondents in the normal system, it will follow that the set of sequences on $a_1, a_2, \ldots, a_\mu$ generated by our given system is again but the set of assertions of the resulting normal system which involve only the letters $a_1, a_2, \ldots, a_\mu$.

In view of the generality of the system of Principia Mathematica, and its seeming inability to lead to any other generated sets of sequences on a given set of letters than those given by our normal systems, we are led to the following generalization.

> Every generated set of sequences on a given set of lotters $a_1, a_2, \ldots, a_\mu$ is a subset of the set of assertions of a system in normal form with primitive letters $a_1, a_2, \ldots, a_\mu, a'_1, a'_2, \ldots, a'_{\mu'}$, i.e., the subset consisting of those assertions of the normal system involving only the letters $a_1, a_2, \ldots, a_\mu$.

8. Unsolvability of the finiteness problem for normal systems.

In trying to test the correctness of the above generalization by our experience with sets of sequences the following counter-example seems to present itself. By that generalization, the only sets of sequences involving a single letter a would be the corresponding subsets of all normal systems on letters $a_1, a_2, \ldots, a_\mu$, $\mu = 0, 1, 2, 3.\ldots$ Now the class of all such normal systems is clearly enumerable (as will be seen in more detail in the next section). Suppose then that we define a class of a-sequences, i.e., sequences involving only the letter a, [86]

86. Not to be confused therefore with the a-assertions of Part I which could involve any or all of the letters $a_1, a_2, \ldots, a_\mu$.

as follows. Enumerate the above normal systems, and have *aa. . . a* with *m a*'s in or not in our class according as it is not in or in the set of assertions of the *m*-th normal system. The resulting class of *a*-sequences will then differ from the corresponding subset of each of our normal systems, in seeming contradiction with our generalization.

Actually that generalization is not thus contradicted. For in our example we have merely <u>defined</u> a set of *a*-sequences, whereas to yield a true counter-example we must show how to <u>generate</u> that set, i.e., set up a system of "combinatory iteration"[87] whose operations would at some time yield each and every *a*-sequence in that set, but would never yield an *a*-sequence not in the set. On the other hand, suppose that the finiteness problem were solved for the class of all normal systems. Then for each of the above normal systems that solution would in a finite number of steps tell whether *aa. . .a* with *m a*'s is or is not in that *m*-th normal system. An operation could then be set up which in order would pass down the above normal systems, for the *m*-th apply the test for *aa. . . a* with *m a*'s being or not being in the system,[88] have a production which in the latter case produces *aa. . . a* with *m a*'s for the desired system, and then in any case pass on to the next system. This operation iterated would then actually generate the

87. In the abstract referred to in footnote 19, mention is made of the method of combinatory iteration as an alternative to the truth-table method. Where the latter method involves an analysis of the logical situations a given deductive system may formalize, the former eschews all interpretation, and studies the system merely as a formal system. The operations of the system are then described as "combinatory" since they largely involve but a reshuffling of symbols; and it is through the "iteration", i.e., continued reapplication, of these combinatory operations that the entire system is obtained. We may note that the present development was entirely unaffected by the writer's published or unpublished work on the truth-table method.

88. The notes at first required this finite test to give an upper bound to the number of steps required to perform the test as a function, presumably, of *m*. This unnecessary requirement was later deleted.

above defined set of a-sequences. That is, a solution of the finiteness problem for all normal systems would yield a counter-example disproving the correctness of our proposed generalization.

Now we mentioned in §3 how an extended attempt to solve the simplified form of this finiteness problem "tag" led to ever increasing difficulties, with all the complexities of number theory in the offing. On the other hand, nothing in the above argument directly weakens the reasoning that led us to our generalization. We therefore hold on to that generalization[89] and conclude that the finiteness problem for the class of all normal systems is unsolvable, that is, that there is no finite method which would uniformly enable us to tell of an arbitrary normal system and arbitrary sequence on the letters thereof whether that sequence is or is not generated by the operations of the system from the primitive sequence of the system.[90]

The correctness of this result is clearly entirely dependent on the trustworthiness of the analysis leading to the above generalization. Apart from the details of that analysis having been

89. In thus resolving this dilemma, the writer was greatly influenced by having heard, not long before, of Brouwer's rejecting the law of the excluded middle. This revolution in the writer's thought was largely energized by the immediately prior reading of Poincare's Foundations of Science.

90. With slightly different wording, this was stated as a "Theorem" in the notes. Despite the closing of the circle of §6 not appearing explicitly in the notes, it was undoubtedly realized at the time that this result carried along with it the unsolvability of the finiteness problem for each of the canonical forms ***A*** on. Less certain, however, is our having paused at the time to realize that the completion of the proof of the reducibility of canonical form ***A*** to *10 Principia Mathematica, referred to in the second paragraph of footnote 79, would yield the unsolvability of the latter's finiteness problem. It remains uncertain, therefore, to what extent the writer anticipated Church's result on the unsolvability of the deducibility problem for the restricted functional calculus. (See Alonzo Church, "A note on the Entscheidungsproblem." Journal of Symbolic Logic, vol. 1 (1936) pp. 40-41 [this anthology, pp. 110-115]; also the reference in footnote 18.

given only in the special work of the first part of this paper, it is fundamentally weak in its reliance on the logic of Principia Mathematica, how weak will be seen in section 10. This weakness could in part be overcome by replacing Principia Mathematica as the logic to be used in connection with postulates on sequences by an operational logic based on mathematical induction in the spirit of our detailed formal reductions, though keeping the primitives of Principia Mathematica. Thus, if we have different symbols $a_1, a_2, \ldots, a_\mu$, and consider an arbitrary sequence $a_{i_1} a_{i_2} \ldots a_{i_n}$ thereof, then we can introduce the propositional function "There is an a_i in the sequence" as follows. Introduce a new letter b such that the assertion of $b\, a_i\, b\, P$ is to mean that P is a sequence on $a_1, a_2, \ldots, a_\mu$ involving the letter a_i. This will be accomplished by adding the one primitive assertion $b\, a_i b a_i$, and the two sets of productions, with $j=1, 2, \ldots, \mu$,

$$b a_i b P \text{ produces } b a_i b a_j P, \quad b a_i b P \text{ produces } b a_i b P a_j.$$

Similarly, for "There is no a_i in the sequence", etc. etc.[91]

But for full generality a complete analysis would have to be made of all the possible ways in which the human mind could set up finite processes for generating sequences. The beginning of such an attempt will be found in the appendix. In the meantime, however, assuming the correctness of our characterization of generated sets of sequences, a mathematical derivation of the unsolvability of the finiteness problem for normal systems as a consequent theorem should be feasible. This is done at least in outline in the next section, and leads us in section 10 to far reaching conclusions on the nature of logical activity, and hence of mathematics.

91. See footnote 8.

9. Outline of a minimum mathematical development.

We recall the definition of a normal system. Let $a_1, a_2, \ldots, a_\mu$ be μ distinct symbols given in order. We have given an initial sequence

$$A = a_{i_1} a_{i_2} \ldots a_{i_\lambda}$$

made up of these symbols, repeated at pleasure, and a certain finite number of operations of the form

$$g_i P \text{ produces } P g_i'$$

where g_i and g_i' are also such finite sequences. For ease in demonstrations we shall assume these operations to be given in order that they need not be distinct, that g_i and g_i' may be either or both of them null, and that P may represent a null sequence provided g_i' is not also null.

We shall call the set of sequences resulting from the iterated application of these operations starting with the initial sequence a normal system. We shall not distinguish between normal systems which differ from each other only in the specific primitive symbols employed. We may therefore imagine these symbols to be the first μ symbols in any infinite sequence of symbols, say the first μ positive integers.

With this understanding, we first show that the set of all normal systems can itself be ordered in an infinite sequence. Actually it is the set of all possible bases of normal systems that is so ordered, so that one and the same normal system, as set of sequences, may appear several times in the ordering. By the complexity of a basis of a normal system we shall mean the total number of symbols appearing (a) in the set of symbols $a_1, \ldots, a_\mu$, (b) in the initial sequence, (c) in the various operations, where we count each P as a separate symbol. We first then imagine the possible bases of normal systems divided into classes according to their complexities, and, as also below, order these classes in order of increasing complexity. Now in each class separate the bases into subclasses according to the number of primitive symbols, and correspondingly order these subclasses. Likewise in each subclass separate and order according to the number of operations. In each of the last found subclasses separate the bases according to the ranks of

$$A, g_1, g_1', g_2, g_2', \ldots, g_\kappa, g_\kappa',$$

rank of a sequence being the number of letters therein, i.e., its length, and order the resulting classes according to the rank of the first of these sequences which differ in rank for two classes. In each of the resulting classes two bases will be identical when and only when the single combined sequences $C = Ag_1 g_1' g_2 g_2' \ldots g_\kappa g_\kappa'$ are identical. As the number of primitive symbols μ is the same for all bases within a single class, if we interprete C as a number written in Arabic notation with base $\mu+1$, i.e., $a=1$, $a=2, \ldots a_\mu=\mu$,[92] the bases can finally be ordered within each class according to the number C represents. As a result the set of all bases for normal systems is ordered; and since the number of bases with given complexity is seen from the ensuing orderings to be finite, the entire ordering is that of a single infinite sequence.

We shall refer to the above ordering as the σ-<u>ordering</u> and use it in all subsequent work.

With our understanding about the primitive symbols of normal systems, all normal systems have the same primitive symbol a_1 which we shall, for simplicity, replace by a. Now consider the following set of sequences involving only the letter a: $a \ldots a$ with n a's <u>is</u> or <u>is not</u> in the set according as it <u>is not</u> or <u>is</u> in the n-th normal system in the σ-ordering. We shall refer to this set as the N-set. The mathematical (as op-

92. The notes incorrectly say "radix μ with $a_1 = 0$, $a_1 = 1, \ldots a_\mu = \mu-1$", a_2 written a_1 by a slip. Since the combined sequence may start with a_1's, we can not let $a_1 = 0$.

posed to logical) basis for the no finite method theorem lies in the following almost trivial

> Theorem. There is no normal system with the property that if its first primitive symbol be replaced by a then the set of resulting sequences involving only the letter a is the N-set.

For such a normal system would appear at some point in the σ-ordering, say it would be the m-th. But then the set of a-sequences present in the normal system, and the N-set, would differ with respect to the presence of at least the sequence $a \ldots a$ with m a's; for by the definition of the N-set, if this sequence is present in the normal system it is absent from the N-set, if absent in the normal system it is present in the N-set.

As stated this theorem would be trivial were it not for the all embraciveness of normal systems.

Stated positively the theorem amounts to the following. Given a normal system S, then there exists a normal system S' such that if S' is the m-th system in σ, then it is false that $a \ldots a$ with m a's is in S and not in S', or in S' and not in S. The proof of this existence consists in pointing out the object, to wit, S is such a system.

Our remaining "theorems" deserve that name only in the sense that a complete mathematical proof thereof clearly can be given- as contrasted with our generalization of section 7. In the absence of the details of the proof we enclose the word "theorem" in a parenthesis.[93]

93. The notes use a "*" next to "Theorem" for the same purpose. Just prior to this development, the notes had taken up the "Probably Fallacious Suggestion for a Non-provable Theorem" quoted in the appendix. This led to some misgivings concerning the no finite method theorem, and the present development was undertaken for the sole purpose of clearing up those misgivings. When, then, it

We first have the important intermediate

> (Theorem). There exists a normal system K and a correspondence C such that for each normal system and enunciation thereof there is one and only one enunciation in K by correspondence C, and such that such an enunciation in K is asserted when and only when the corresponding normal system versus enunciation is such that the enunciation is an assertion in that normal system.

The proof would have to refer back to the reduction proofs of sections 4 and 5. As these are not entirely determinate, they would have to be made so before the description of system K becomes as complete as that of the σ-ordering.[94]

We shall refer to the system K as the complete normal system because, in a way, it contains all normal systems.[95]

Now given any normal system M, we shall say that there exists a finite-normal-test for system M if there exists a normal system M' such that among the primitive letters of M' are all the primitive letters of M, and in addition, among possibly others, a primitive letter b, and such that if P is an enunciation of M we shall have P an assertion in M' when and only when it is an assertion in M, while bP is an assertion in

was obvious that a "theorem" could be proved by the same sort of methods as were used in the reductions of §§ 4, 5, the detailed proof was "postponed" so that the continuing work on rendering the basic generalization of §7 unimpeachable would not be unduly interrupted.

94. Reading between the lines, we may suggest that the proof would first set up a system K in canonical form B serving the purpose of K, this being relatively easy since the infinite number of variables used in canonical form B would allow for the infinite number of primitive symbols occurring in the totality of all normal systems. K' would then be reduced to the desired normal system by the reductions of §§4, 5. In the statement of the theorem "correspondence" must be understood as "effective correspondence", the theorem otherwise being without significance.

95. The "complete normal system" would thus correspond to Turing's "universal computing machine". See computable numbers, pp. 241-246. [This anthology, pp. 127-132].

M' when and only when P is not an assertion in M. Thus, for each P consisting wholly of primitive symbols in M one and only one of the two sequences P, bP is an assertion in M', and that according as P is or is not an assertion in M.[96] We then have the fundamental

> (Theorem). There exists no finite-normal-test for the complete normal system K.

The reductio ad absurdum proof would run as follows. Suppose there were such a system. Then out of it another normal system could be constructed which would have for its a-set the N-set. But this has been proved to be impossible. The method of proof would have to reduce the σ-ordering to a normal system operation.

We now examine the positive content of such a proof. Let L be any normal system on the letters of K and at least one additional letter b, and let L have the property that for each enunciation P of K one and only one of the two enunciations P, bP is in L. Let us write (m-th system in σ, $a \ldots a$ with m a's) for that enunciation of K which corresponds to the m-th system in σ and enunciation $a \ldots a$ thereof by the preceding theorem. Then from L, by the argument in our supposed proof, we would get a system L' such that $a \ldots a$ with m a's would be in L' when and only when b(m-th system in σ, $a \ldots a$ with m a's) is in L.[97] Now let L' be the m'-th system in σ. Then if

96. The letter b thus serves as a symbol for negation. The full generality of this definition may be seen from the following equivalence. By the negative of a set of sequences on letters $a_1, a_2, \ldots, a_\mu$ we shall mean the set of all sequences on those letters not in the given set. It is then readily proved that the existence of a finite-normal-test for a normal system M is equivalent to the negative of M being a generated set, generated set here being defined by the generalization of §7.

97. Note the positive nature of the assumed process. That is, from the presence of b (m-th system in σ, $a \ldots a$ with m a's) in L would be obtained the presence of $a \ldots a$ with m a's in L'.

a. . .*a* with *m′ a*'s in *L′*, *b*(*m′*-th system in *σ*, *a*. . .*a* with *m′ a*'s) is in *L*, if *a*. . .*a* with *m′ a*'s is not in *L′*, *b*(*m′*-th system in , *a*. . .*a* with *m a*'s) is not in *L*, whence, by our hypothesis on *L*, (*m′*-th system in *σ*, *a*. . .*a* with *m′ a*'s) is in *L*. Hence *L* as a possible finite-normal-test for *K* gives the wrong answer for *a*. . .*a* with *m′ a*'s being in, or not being in, the *m′* system in *σ*, *L′*.[98]

That is, for each normal system *L* of the above kind a normal system *L′* can be found such that if *L′* is the *m′*-th system in *σ*, then *L* considered as a finite-normal-test for *K* gives the wrong answer for *a*. . .*a* with *m′ a*'s being in *L′*. While the proof thus gives the case for which the answer is wrong, it does not of itself tell what answer is given. This will contrast strongly with our conclusion of the next section where a different hypothesis is imposed on *L*.[99]

10. Incompleteness of Symbolic Logics.

In our last discussion we forced *L* to give a unique answer to each question, as it were, put by the complete normal system *K*, and found that in at least one instance it then had to

98. That is, ***L*** as finite-normal-text for ***K*** would say that according as ***a*...*a*** with ***m'*** ***a***'s is or is not in ***L'***, (***m'***-th system in ***σ***, ***a*...*a*** with ***m'*** ***a***'s) is not or is in ***K***. But, by the construction of ***K***, the reverse is true.

99. The theorem just discussed, when combined with the generalization of §7, therefore shows not only that the finiteness problem for the class of all normal systems is unsolvable, but that the same is true for a particular one of them, namely ***K***. This, while not explicitly stated at the time, was then fully realized, and receives explicit mention in our notes for February 28, 1924. We might here also mention two theorems stated, and at least proved in outline, under the date of March 7, 1924. "There is no finite method for testing whether an arbitrary system does or does not admit of a finite test". "There is no finite method for testing whether an arbitrary system has a finite or infinite number of assertions". Some doubts were expressed, however, about one stage of the contemplated proof of the second result.

give the wrong answer. We now, rather, require L to give the right answer when it answers at all, and see what this weakening of the requirements for a finite-normal-test for K leads to.

Specifically, let L be a normal system whose primitive letters include the primitive letters of K and another primitive letter b, and which has the property that for any normal system S and enunciation P thereof (S,P) appears in L when and only when P is in S, while if $b(S,P)$ is in L, then P is not in S. This property is equivalent to the following. (S,P) is in L when and only when it is in K, while if $b(S,P)$ appears in L, (S,P) is not present in K. Any such L we shall call a normal-deductive-system adjoined to K. K itself, with merely the letter b added to its primitive letters, is such a system. While the first half of the property for L could have been made as weak as the second, there is no reason for doing so since by suitably adjoining K to such a weak L the stronger L would result.

If then L is a normal-deductive-system adjoined to K, as at the end of the last section, we can obtain from it a normal system L' such that L' has $a \ldots a$ with m a's in it when and only when $b(m$-th system in σ, $a \ldots a$ with m a's) is in L. But now it follows that if L' is the m'-th system in σ, then $a \ldots a$ with m' a's is not in L'. For $a \ldots a$ with m' a's could only appear in L' through $b(m'$-th system in σ, $a \ldots a$ with m' a's) being in L. But by our definition of a normal-deductive-system if $b(m'$-th system in σ, $a \ldots a$ with m' a's) is in L, $a \ldots a$ with m' a's is not in that m'-th system in σ, i.e., not in L'. Hence also $b(m'$-th system in σ, $a \ldots a$ with m' a's) is not in L'; for if it were, then, by the construction of L', $a \ldots a$ with m' a's would be in L'. We therefore have the very important

> (Theorem). No normal-deductive-system is complete, there always existing a normal system S and enunciation P thereof such that P is not in S while $b(S,P)$ is not in the normal-deductive-system.

We now have the still more important

> (Theorem). No normal-deductive-system is equivalent to the complete logical system (if such there be); better, given any normal-deductive-system there exists another which second proves more theorems (to put it roughly) than the first.

The proof would run as follows. Let L be the given deductive system. The proof envisaged above would be an informal proof to the effect that a system $L' = S_{m'}$, the m'-th system in σ, could be constructed such that $a \ldots a$ with m' a's is not in $S_{m'}$ while $b(S_{m'}, a \ldots a$ with m' a's) is not in L. On formalizing the proof there seems to be no doubt that it could be reduced to normal form, and on being adjoined to L would give a system where a theorem is proved which is not provable in L.[100]

When these results are expanded to the dimensions of the no finite method theorem we are led to the following.

A complete symbolic logic is impossible.[101]

100. I.e., the correspondent of $b(S_{m'}, a \ldots a$ with m' a's) in the wider system. Except for non-essentials, we have stated the theorem, and given the outline of proof, as they appear in the notes. Actually, to obtain this result, a far less drastic procedure would suffice. We need merely add $b(S_{m'}, a \ldots a$ with m' a's) as a primitive assertion to L, and reduce the resulting system to normal form - the normal form of L having been spoiled by having two primitive assertions instead of one. However, the proof first contemplated is probably far more satisfactory. For the entire development should lead away from the purely formal as the final ideal of a mathematical science, with a consequent return to postulates that are to be "self-evident" properties of the now meaningful mathematical science under consideration. Merely adding $b(S_{m'}, a \ldots a$ with m' a's) as a new postulate would then be inadmissable. And if it be said that the previous informal development constitutes a heuristic justification of such an addition, the reply would be, incorporate that development in the system itself. And so the formalization suggested in the original outlined proof; new postulates introduced in that formalization to be "self-evident" properties of the modes of symbolization etc. of the old logic.

101. This has a double content. Mere incompleteness, as in the first of the

This is an iconoclastic result from the formal logician's point of view since it means that logic must not only in some parts of its description (as in the operations), but in its very operation be informal. Better still, we may write

The Logical Process is Essentially Creative.

This conclusion, so in line with Bergson's "Creative Evolution", is not so much contrary to Russell's viewpoint (since he does not fully express himself) but to that of C. I. Lewis as given in chapter VI, section III, of his "Survey of Symbolic Logic". It makes of the mathematician much more than a kind of clever being who can do quickly what a machine could do ultimately. We see that a machine would never give a complete logic; for once the machine is made we could prove a theorem it does not prove.

---

two "Theorems" preceding, might not rule out the logic being as complete as it ever could be made. Fundamental, then, is the added effect of the second theorem, which rules out the possibility of a completed symbolic logic. That is, any symbolic logic can be made more complete. It is doubtful if the writer ever paused to note the mere incompleteness of a symbolic logic in the sense of the existence of some undecidable proposition therein, for experience with Zermelo's axiom, the axiom of infinity, and the theory of types clearly leads one to expect incompleteness in the upper reaches of a symbolic logic. Rather was the emphasis placed on the stronger concept of incompleteness with respect to a fixed subject matter, in the present instance the propositions stating whether a given sequence is or is not generated by the productions of a given normal system from its initial sequence. Likewise, Gödel would stress, for example, the incompleteness of any symbolic logic with respect to the class of arithmetical propositions. Where we say "symbolic logic" the tendency now is to say "finitary symbolic logic". However, it seems to the writer that logic should be considered essentially a human enterprise, and that when this is departed from, it is then incumbent on such a writer to add a qualifying "non-finitary".

APPENDIX

While the formal reductions of Part I should make it a relatively simple matter to supply the details of the development outlined in section 9 and the beginning of section 10, that development owes its significance entirely to the universal character of our characterization of an arbitrary generated set of sequences as given in section 7. Establishing this universality is not a matter for mathematical proof, but of psychological analysis of the mental processes involved in combinatory mathematical processes. Because these seemed to be sufficiently simple to be exhaustively described, the writer gave up a direct use of Principia Mathematica as a partial verification of the characterization in question, planning rather that the incompleteness of the logic of Principia Mathematica would be a corollary of the more general result.[102] Actually, we can present but fragments of the proposed analysis of finite processes. Being but fragments, these contributions are given by direct quotation from our notes and diary.[103]

The unsolvability of the finiteness problem for all normal systems, and the essential incompleteness of all symbolic logics, are evidences of limitations in man's mathematical powers, creative though these be. They suggest that in the realms of proof, as in the realms of process, a problem may be posed whose difficulties we can never overcome; that is that we may be able to find a definite proposition which can never be proved or disproved.[104] This theme will protrude it-

102. On the other hand, as late as September 3, 1929, among a few observations on our work of 1924, appears the following in our notes. "Plan of first proving Principia inadequate through special analysis as decided December 1925 still seems good. Will not be work wasted but help clarify above." The plan, however, included prior calisthenics at other mathematical and logical work, and did not count on the appearance of a Gödel!

103. See footnote 5.

104. See the end of footnote 1.

self every so often in our more immediate task of obtaining an analysis of finite processes.

...nature as finite intelligence.... -DIARY

...Fermat's theorem perhaps unprovable due to weakness of our logical apparatus. Have a good vision of the infinitude of integers and stateable properties of them which cannot be unravelled by our logical process of syllogism etc. -DIARY

...Think logic for finite operations.... Really only get here what would correspond to a non-growing machine operating in a discrete symbol-space. -DIARY

We begin here a derivation of the logic of finite operations and ultimately of all the logic of mathematics from first principles. These principles are supposed to be a digest of our experience of the logico-mathematical activity....

We first note that we have to do with a certain activity of the human mind as situated in the universe. As activity this logico-mathematical process has certain temporal properties; as situated or performed in the universe it has certain spatial properties.

Now the objects of this activity may be anything in the universe. The method seems to be essentially that of symbolization. It may be noted that language, the essential means of human communication, is just symbolization. In so far as our analysis of this activity is concerned, its most important feature is its ability for being self-conscious. Since our present study is this activity we shall here not consider the original objects which are symbolized in the process, but only the relations and operations upon these resulting symbols and the effect of this self-consciousness....

We are then to consider this activity of the human mind and watch it in this process of creating symbols. We may then note as before that the process itself is temporal but we are to think of the result of the process as spatial. We are then to think of these symbols as created in the flux of the universe and preserved there through time. This gives us our first principle, i.e.,

A. The Principle of the Preservation of Symbols. . . .

We are then to think of the result of logical thought as certain spatial configurations of symbols; and our study will then consist in studying the further effects brought about by the processes of symbolization and self-reflection. Now in our subject we are to regard our symbols as without properties except that of permanence, distinguishability and that of being part of certain symbol-complexes. But this latter is essential, i.e. that these symbols enter into certain spatial (not Euclidian or continuous, etc., but spatial as opposed to temporal to be described later) relations. These relations themselves can then be symbolized and the new symbols are again in space and have certain spatial relations, etc. So much for the further effect of symbolization of the spatial properties. But in addition we have this self-reflectiveness. This is a reflection of the process. This process is then itself symbolized and symbolized by a spatial symbol.

We thus have a continued activity which produces symbols which are spatial. This activity turns on itself and symbolizes its temporal character by a spatial symbol. These spatial symbols have certain spatial relations which are in turn symbolized by a spatial symbol.

Better, we have a three fold order of things.

(a) Activity in time which is creative. This is the source of the process.
(b) By reflection this activity itself is frozen into spatial properties.
(c) These spatial relations are symbolized by spatial symbols.

(d) These symbols have no further symbolizable properties internally as it were and so end the descent. (This is essentially Bergsonian.)

So much for the introductory discussion before coming to an analysis of these spatial relations etc. We may note here however, that it is in the specific bringing in of the self-consciousness and its symbolic representation that this differs from other schemes. In the old mathematics the formal and informal were confused, in symbolic logic the informal tended to be neglected. Right here we may see that the greatest ordinal etc. fallacies come just from this neglect; we may thus expect the above to give us a rational theory of types. We may note as a fundamental property of the above that everything ends up in spatial relations and these are transformed into themselves. This is at the basis of the no finite method theorem.

-NOTES

. . . but really only the productiveness and not creativeness of mind here recorded. -DIARY

A PROBABLY FALLACIOUS SUGGESTION FOR A NON-PROVABLE THEOREM.

Above we considered the question of a finite method for testing all theorems of a certain class. The no finite method theorem did not depend on how the test is shown to be a true test.

In trying to obtain a theorem which can not be proved either true or false the valid methods of proof have to be stated. Now as in finite methods the set of finite proofs probably form an illegitimate totality. In the first case however, any one could be reduced to an equivalent one which is "predicative". The state of affairs for the second is not quite as clear.

In any case consider the following enunciation: For each deductive system of the normal form and enunciation in it there exists a finite method of proving or disproving the derivability of the enunciation in the system. It will be noticed that the enunciation wants a finite method of proof for each case not a test for all. Now this may not be a mathematical theorem in that the notion of general finite method of proof is used whereas the set of them may be an illegitimate totality. If it is a definite enunciation it would seem to be neither provable nor disprovable. For if it were proved we would really have a method of proof described for all cases at once and so have a finite test.[105] On the other hand to disprove it we would have to show a case where the deducibility could neither be proved nor disproved. But if the deducibility exists then it can be proved. Hence showing it cannot be proved amounts to disproving it, and so we cannot prove that it cannot be proved nor disproved both. Hence the above theorem cannot be proven false. -NOTES

This seems to lead to a distinction between finite operations and finite methods of proof. All finite operations can be lumped into one system while all proofs can not. ... Though both the set of all finite operations and all finite proofs are illegitimate yet there is an equivalent legitimate set in the former case but not in the latter- at least that can be described. ...may not this prevent us from ever proving a theorem non-provable? The only hope seems to be a method of invariants and illegitimate totality induction. ... -NOTES

...we do not completely determine the process of proof; we simply watch it in its activity, from that note some of its properties which may enable a non-provability theorem to be possible. -NOTES

...creative totality versus Cantor's transfinite?[106] -DIARY

Think relation between the creative process[107] and transfinite ordinal numbers; see where the creative process is

105. Tarski recently pointed out to the writer that this argument is invalid since the supposed proof might be non-constructive.

106. This and succeeding entries follow all of the work corresponding to Part II. The entries are in chronological order.

107. See the end of §10.

really Principia superimposed upon types which follow each other as the transfinite ordinals. This explains the relation but brings up question of Ω_1 etc. -DIARY

Think real numbers etc. as creative totalities. . . .[108]
-DIARY

Write up this beginning of an operational theory of Mathematics.[109] -DIARY

What we must do is further analyse the process of proof. It was shown that it is creative, but the creative and non-creative parts were intermingled. What we must now do is isolate the creative germ in the thinking process.[110] -NOTES

The following suggestions came up.

(a) The conclusion that man is not a machine is invalid. All we can say is that man cannot construct a machine which can do all the thinking he can. To illustrate this point we may note that a kind of machine-man could be constructed who would prove a similar theorem for his mental acts.

(b) The creative germ seems not to be capable of being purely presented but can be stated as consisting in constructing ever higher types. These are as transfinite ordinals and the creative process consists in continually transcending them by seeing previously unseen laws which give a sequence of such numbers. Now it seems that this complete seeing is a complicated process mostly subconscious. But it is not given till it is made completely conscious. But then it ought to be constructable purely mechanically.

108. There follows a description of mental states during mathematical discovery which is probably too "cloudy" for inclusion here.

109. The surface is here but barely scratched. We may, however, note the conclusion, "The Cantor Theorem is not true in the operational theory of infinity", i. e., the Cantor-Bernstein-Schröder theorem. The notes add, "In any case this whole thing is premature at this stage".

110. The diary, at the corresponding point, takes a religious turn.

This last is an assumption which is to be fundamental in the whole discussion. It is to be the Axiom of Irreducibility[111] for Finite Operations.

(c) We prove a sequence not capable of being derived in a normal system by showing it does not satisfy a property which is shown by Mathematical Induction to be possessed by all sequences which are generated. This is old. The new thing is that this hereditary property may be of various types, and it is thus that the creative element appears in proof. (All of above before reading Brouwer) -NOTES

The following suggestions occur.

(a) The following would be the beginning of a definition of a general non-growing machine for (b)... [above]. It will have a finite number of parts a_i, i.e., $a_1, a_2 \ldots a_n$ and a finite number of relations of an arbitrary number of parts, i.e.,

$$b_j(\hat{x}_1, \hat{x}_2, \ldots \hat{x}_{m_j}).$$

At each instant certain of these parts are related and not others. Specify this. Then we would have, say, a definite rule for a certain configuration of relations producing another. This would, however, be a machine closed in itself. Instead we want it to be capable of creating, say, impressions in an outside medium and of having such impressions affect it. This I have not completely described.[112]

111. "Reducibility" probably meant.

112. Apart from the incompleteness referred to at the end of this note, this fails to be an anticipation of the concept of the Turing machine (see computable numbers) in its very attempt to allow for the structure of the machine. However, in the case of a growing machine, the number of states of the machine would no longer be finite, and reference would have to be made to the structure of the machine. The writer did not attempt to use this partial concept of a non-growing machine as an easy path towards the justification of the generalization of §7 since the clear possibility of having machines producing machines, and, more generally, of a machine directing its own growth, seemed to make any inferences drawn from the nature of a non-growing machine inconclusive.

(b) Instead of having the bald assumption of (b). . . [above] we may have the milder assumption that a new sequence may use ideas in defining it, but these ideas can only be those previously defined. This assumption looks less general. If it can be put clearly then we can prove that every such new idea will ultimately be defined mechanically since each idea used has been so defined, i.e., we use an operational-induction.

-NOTES

Think Creative process. ... See work on Theory of Deduction in three parts: (a) No finite method theorem. Almost have all of this. (b) The nature of proof. Only have certain starting points but no connected whole as yet. (c) On the question of proving theorems unprovable both ways. Have almost nothing here. ... -DIARY

... think creative process. See where a symbol for "produce" would come in in each generalization of operations; and as this symbol became "external" it would as it were changed into activity. Thus like $\supset$ but not $\sim p \vee q$ as $\sim p$ defined in terms of it.

-DIARY

... Again see a class as an operation. -DIARY

Think above. Again get picture I had before of transfinite ordinals as machines creating machines etc. (machine for operation).[113] -DIARY

113. A certain amount of detailed work occurs in the notes on the generation of ordinals considered as operations. However, as this was largely undertaken for the experience to be thus gained, we do not quote any of it. Concerning it we may quote a criticism of September 3, 1929 of the work of 1924. "In trying to see beyond this I seemed to get lost as in the former work on generating transfinite ordinals. All I could see was that every now and then with increasing difficulty a new idea comes". Also might be mentioned the appearance in the notes of a partial check of Galois theory as a finite process, and a "logical description" of the symbol forms of canonical form A.

. . . proof consists in raising the unseeing combinatory iteration which is of the same type to ever higher types where see through it all. . . . power of seeing due to things being mutually exclusive . . . difficulty of combinatory iteration is that you have destruction. . . .[114] -DIARY

. . . "transcendental definitions". Ought to be able to give such for "finite proof"— i.e. like definition of well-ordered series. . . . order of creativeness of a proof. . . . -DIARY

. . . Also get idea of constructing a theorem for each type of creativeness provable by proof of that type but not by lower type. -DIARY

. . . This leads to a new hierarchy of proofs, proofs of proofs, etc. -DIARY

Finite operations illuminated as generated by three principles (1) Symbolic "manipulation" (2) Symbolization (3) Iteration.
-DIARY

We return here to a more complete discussion and analysis of the very first part of the present research i.e. in connection with finite methods. We shall here generalize to finite methods for obtaining any results not just test for truth or falsity. . . .
We shall here first give what is at least a first approximation to a definitive solution of the difficulty of finding a natural normal form for symbolic representation.

There are three stages in the analysis we give. In the first stage we have the things symbolized. . . .

. . . .

This then gives us our second stage in our analysis, namely a system of symbolizations for corresponding mathematical states.

114. An attempt to "dig deeper" led to a vision of the "Birth of Consciousness". In our quotations we have omitted these more extravagant features of notes, and, chiefly, diary.

. . . .

First these symbolizations will be conceived of as being spatial. . . . But we shall also assume them to be finite and we might say discrete. . . . First each symbolization can be considered to consist of a finite number of unanalysable parts (unanalysable from the standpoint of the symbolization) these parts having certain properties and certain relations with each other. If we allow unary relations we need merely talk of a finite number of parts related in certain ways. The ways in which these parts can be related will be assumed to be specified for the whole system of symbolizations.

We thus have what may be called a symbol-space in which these symbolizations or symbol-complexes as we may now call them are. . . .

. . . It will be assumed that these spatial properties and spatial relations are themselves not further analysable in the given discussion, in other words that a single undivided act of judgment is involved in finding out whether a certain part has or has not a certain property or that certain parts are or are not related in a certain way. Now the system of symbolizations in question is essentially to be a human product and each symbolization a human way of describing the original mathematical state. The need for the following assumption is then readily seen: that the number of these elementary properties and relations used is finite and that there is a certain specific finite number of elements in each relation.

We are now ready to come to the third and last stage in this analysis. We can do this by assuming what seems to be the evident intent of the above analysis and that is that the symbol-complexes are completely determined by specifying all the properties and relations of its parts as above mentioned. Hence for the given system of symbol-complexes we have a certain number of relations which can be made to include properties by allowing but one variable and may be written

$$a_1(x_1, x_2, \ldots x_{\nu_1}),\ a_2(x_1, x_2, \ldots x_{\nu_2}), \ldots a_n(x_1, x_2, \ldots x_{\nu_n}).$$

Then if Π is to mean logical product of all elements, then each complex of the system can be completely described in the following form

$$\prod_i a_{n_i}(x_{j_{1i}}, x_{j_{2i}}, \dots x_{j_{v_{n_i}i}})$$

which gives all the relations enjoyed by the elements or parts $x_1, x_2, \dots x_m$ of the symbol complex.

Now these descriptions completely describe the symbol-complexes which in turn represent the original mathematical states. Hence these descriptions can be considered to represent or symbolize those mathematical states.

We thus have it that every system of symbolic representations which satisfies the assumptions we have given is equivalent to one of these systems. These latter (gotten by varying $v_1, v_2, \dots v_n$ and n) constitute then a normal set of systems of symbolic representations.[115] -NOTES

I. Finite Methods
II. Theory of Deduction
III. Theory of Conception or Concepts

-DIARY

...I study Mathematics as a product of the human mind and not as absolute. ... -DIARY

...Notion of meaning bothers me. Put it as subconscious perception of things associated with symbols. -DIARY

115. Under the date of September 4, 1929 appears the following. "The neighborhood idea applied to the general symbol space will lead to difficulties due to inter-relations. In fact, this difficulty was mentioned elsewhere. The general symbol space should be reconsidered". We rely on memory in saying that it was by then realized that even though each symbol complex was determined by the logical product of relations satisfied, the converse was complicated by the fact that the symbol space itself might force the elements to have certain relations when certain others were satisfied, so that, in fact, a finite process would have to be set up which would generate exactly those logical products of relations corresponding to symbol complexes.

... get idea of Psychic Ether, ... see "language" as a difficulty. ... -DIARY

... in all finite methods or just methods only the following principles are used: symbol-manipulation, symbolization and iteration. ... although iteration is the process, it is merely machine like, and the real thought was put into that which is iterated.

....

Corresponding to the three stages in the analysis of symbolism there were three stages in the analysis of method.[116] In the first stage were any and all methods. To allow for the most wonderful creations my image of such methods involved dark clouds pierced by flashes of lightning accompanied by rolling thunder. These methods being then regarded as existing in time were symbolized at each time (second stage).[117] Due to discreteness and finiteness we would thus have a finite sequence of symbol-complexes representing the various stages in the method. The third stage would then consist in reducing the method of passing from symbol-complex to symbol-complex to an operation of normal type on symbol-complexes of the above normal type. ...

... it was seen that even in the first stage we really have not got Mathematics but a more or less vague symbolization of it. In other words that in all cases what we are conscious of is not mathematics but a symbolization of it. That these symbols are more or less vague, that in making them more precise we really resymbolize the symbols etc. It was thus immediately evident why the process of symbolization must inevitably come in. ...

... the difficulties in connection with the notion of meaning came in. It was then recognized that a symbol has meaning because it is connected with all the things which give it meaning

116. The past tense is used because the notes here describe previous ruminations.

117. That is, were imagined to be so symbolized. There is confusion here between the method or rule, and the carrying out thereof.

though these things are in the subconscious regions. Meaning is then the result of subconscious perception. The symbol is like an island which rises above the sea of the unconscious and is itself connected with the vaster regions below.

. . . .

The question was how iteration was brought in through mere spatial symbolization. How activity at all. It was then seen that activity is symbolized by a spatial symbol. Hence in the symbol-complex representing the finite method certain parts represent temporal things and so give rise to activity. How iteration? Well, the symbol-complex which is spatial has temporal permanency, hence merely by its persisting through time, it keeps on giving directions and so keeps on stimulating the activities which are then said to be iterated.

This brought on a difficulty that all the symbol-complex can do is suggest, thus requiring a mind to interpret and direct. . . . the relation of mind and one might say matter. It was given in a more general form, i.e. not just for a definite method but in all cases. The distinction was that in a definite method we have a deterministic system whereas otherwise an indeterministic one where ideas appear without cause (i.e. cause within the system considered).

. . . .

. . . a notion which should be very fertile indeed, i.e. that of the Psychic Ether. In so far as this contains the things which give meaning to symbols it is just the unconscious. . . . it is the seat of the birth of new ideas; it is also the region where all the vaguer processes operate, especially intuition, "hunches", etc. When these become precise they crystallize from the psychic ether. Clear symbolizations are then to be regarded like atoms in this ether. In fact one may think of them as Kelvin thought of his vortex atoms in connection with the lumeniferous ether.

. . . a new difficulty came up, i.e. how to place language of the ordinary kind in all this. It loomed up as a very queer thing in all this: its fluency, etc. etc. The following suggestions came up. Its linear order can be easily associated with time. But it is rather difficult to see just what its essence is. On the one hand ... it seems to be used to call up images as when we read a novel. It then merely is the suggestion while we are conscious of the thing it suggests. On the other hand we often are conscious of only the words, but as having meaning as when we assent to the statement, "Every continuous function is integrable." Whether these two are the only aspects of language, and how they are related to each other and how they connect up with the psychic ether is yet to be determined. The relation between ordinary language and its predecessor hieroglyphics should help to bring out the connection between the two aspects above mentioned.

...on the next attack try to put everything into the psychic ether and then ... proceed to the next real difficulty. This ... has to do with just the way in which the connections between vague ideas turn into connections between precise ideas. When this is done we should not be far from the normal form for finite methods.

It is to be noted that the analysis of methods has become of much greater importance than formerly. In fact it almost looks as if analysis of proof and concepts will almost be corollaries. of the present analysis. -NOTES

...see very prettily how can avoid physical symbols etc. by noting that effect in our reasoning just as if only imagined them; i.e. Psychic Ether complete in itself.[118] Like Berg-

118. This serious error was corrected in our work of 1924 where the dualism of the physical world versus the mental world was stressed in contrast with the above monism. Fundamental is the distinction between the static outer symbol-space with its assumed capacity for bearing symbol-complexes of unbounded complexity, and the dynamic mental world with, however, its obvious limitations. This has been fully emphasized by Turing in his finite number of mental states hypothesis (referred to in footnote 9). Perhaps we should quote the following item from the diary, entered some two weeks later. "... the beauty of the self-sufficiency of the psychical system. See clearly the symbols we imagine as floating in the psychic ether etc. Raises the question may not matter similarly be the visions of God?... by saying above we can think of physical things as though just visions in our psychic ether we have a Psychic Principle of Equivalence ... ".

son's theory of memory. See my work as a complete and scientific psychical system— the first. (Mechanics is such a physical system.) -DIARY

Proceed to get the mechanism or rather mechanics of the Psychic Ether. ...we can only handle one thing at a time ...
-DIARY

...in our mental activities the following occurs repeatedly: often a given symbol has full meaning for us and we simply observe the symbol. When it has lost some of its "meaning" we replace it by the thing it symbolizes. This is the converse of symbolization... .[119]

...Really little time was spent above on the way in which operations were symbolized. What was seen was the atomic operations. ...But the way in which these combine, recombine, etc. was not noticed. ...

...mathematical operations are had when the symbols become clear cut enough to be handled as individuals and in combinatory fashion. ...

Another difficulty has not been brought out and that is the place of logic as relating to truth and falsehood. It seems that we will have to bring in some logic, that of direct verification of the existence of an elementary relation between symbols.

...recognizing the truth of a certain logical situation is a process since in general it cannot be done at a glance. It must then be reduced to elementary recognition only.

Another example of this difficulty is that of substituting a particular case for a variable. This seems to involve

(1) Recognition of the special case as a value: Elementary Recognition.

(2) Saying this in symbols— symbolization.

(3) Substitution— symbolic manipulation.

119. This is part of an "outline of complete solution" which, however, largely duplicates earlier entries, followed by a discussion of difficulties yet to be conquered.

In any case the reduction seems possible so that only elementary recognition seems to be needed for the logical aspect of the operational description of Mathematics.[120]

. . . .

A summary of the method used above is as follows. Try to give a complete description of what goes on. In this description we symbolize everything. That symbolizes away most things. But a few things cannot be symbolized away, because in the transformation produced by the symbolization they of necessity reappear. These things are meaning, symbolization, symbol manipulation, iteration, sense perception or direct verification, and perhaps a few other things. These will then constitute the elements out of which the description is built up in addition to mere symbols. -NOTES

120. While many other such evaluations of work done has been omitted in our quotations, the following should be mentioned. "... the time for complete reorganization is over. The main outline of the work is completed and we really have a case of Filling In". Actually, but the surface of the problem was thus, perhaps, barely scratched, the problem, that is, of describing "all the finite processes of the human mind", at least in so far as they might concern the generalization of §7.

Index